CBS PROBLEMS and SOLUTIONS *Series*

Problems and Solutions in ANALOG SYSTEMS

- This book is suitable for examinations of all Indian universities.
- Solutions of **280 problems** are given in most simplified manner.
- All types of problems are included and the solutions given in detail.
- To explain the basics lucidly, computerised figures have been liberally added in the solutions. This volume contains more than **510 line diagrams**.

CBS PROBLEMS and SOLUTIONS Series

Problems and Solutions in ANALOG SYSTEMS

R. GOPAL
Engineer

CBS

CBS PUBLISHERS & DISTRIBUTORS PVT. LTD.
New Delhi • Bangalore • Pune • Cochin • Chennai (India)

ISBN : 81-239-1363-X

First Edition : 2006
Reprint : 2009

Published by Satish Kumar Jain and produced by V.K. Jain for
CBS Publishers & Distributors Pvt. Ltd.,
4819/XI Prahlad Street, 24 Ansari Road, Daryaganj,
New Delhi - 110002, India.
e-mail: cbspubs@vsnl.com, cbspubs@airtelmail.in
Website: www.cbspd.com

Branches:

- ***Bangalore:*** 2975, 17th Cross, K.R. Road,
  Bansankari 2nd Stage, Bangalore - 560070
  Fax: 080-26771680 • e-mail: cbsbng@dataone.in
- ***Pune:*** Shaan Brahmha Complex, Basement, Appa Balwant Chowk,
  Budhwar Peth, Next to Ratan Talkies, Pune - 411002
  Fax: 020-24464059 • e-mail: pune@cbspd.com
- ***Cochin:*** 36/14, Kalluvilakam, Lissie Hospital Road,
  Cochin - 682018, Kerala • e-mail: cochin@cbspd.com
- ***Chennai:*** 20, West Park Road, Shenoy Nagar,
  Chennai - 600030 • e-mail: chennai@cbspd.com

Printed at :
J.S. Offset Printers, Delhi

Preface

Analog Systems is one of the essential subjects for the students of electronics, electronics and telecommunication and related fields of engineering. The author feels it a great pleasure in presenting this very complex subject through typical problems and their solutions.

This book covers all important and major topics taught at the undergraduate level in an engineering college. All the **280 problems** have been discussed at length, and detailed **510 line diagrams** given extensively to help the reader understand the *basics* and *grasp the logic*. This book is suitable for preparing for all university examinations and also to face the competitions. This feature is expected to help the reader in visualising the patterns of problems and the extent and level of difficulty in solving the questions likely to be encountered in the examinations, and thus tackling the problems with confidence and success.

Although every care has been taken to ensure accuracy, yet some errors might have crept in. The author will be grateful if these are brought to his notice so that these could be rectified in the subsequent printings and editions of the book. Readers are requested to send their suggestions for further improvement and revision to me through the publishers at the following e-mail address: cbspubs@del3.vsnl.net.in.

R. GOPAL

Contents

CHAPTER

1

Diode Circuits and Applications

Introduction

There are two types of semiconductor material:

(*i*) Intrinsic semiconductor: Pure semiconductor.

(*ii*) Extrinsic semiconductor: It is the semicon material that has been subjected to the doping process.

The extrinsic semiconductor are two types:

(*a*) *n*-type material (*b*) *p*-type material

(*a*) *n*-Type Semiconductor

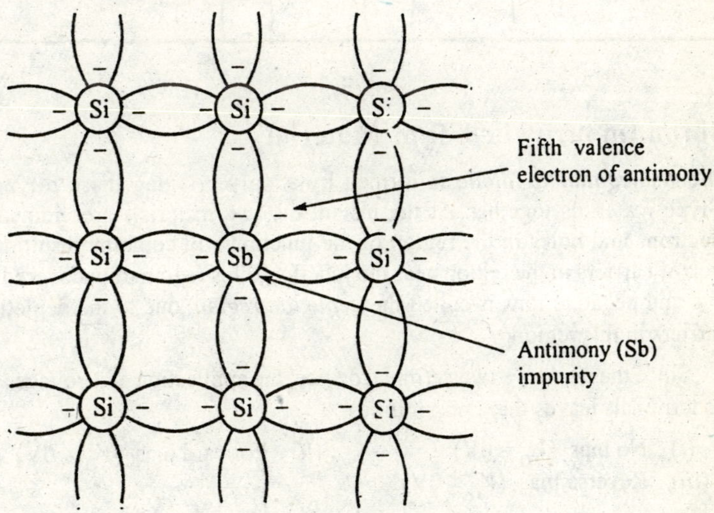

Fig. 1.1

The *n*-type is created by introducing five valence electrons impurity elements such as antimony, arsenic, and phosphorus. Diffused impuries with five electrons are called *donor atoms*. In an *n*-type material, the electron is called the *majority carrier* and the hole the *minority carrier*.

(*b*) *p*-Type Semiconductor

The *p*-type material is created by introducing three valence electrons impurity elements such as boron, gallium, and indium. The diffused impurities with the three valence electrons are called *acceptor atoms*. In a *p*-type material the hole is the *majority carrier* while the electron is the *minority carrier.*

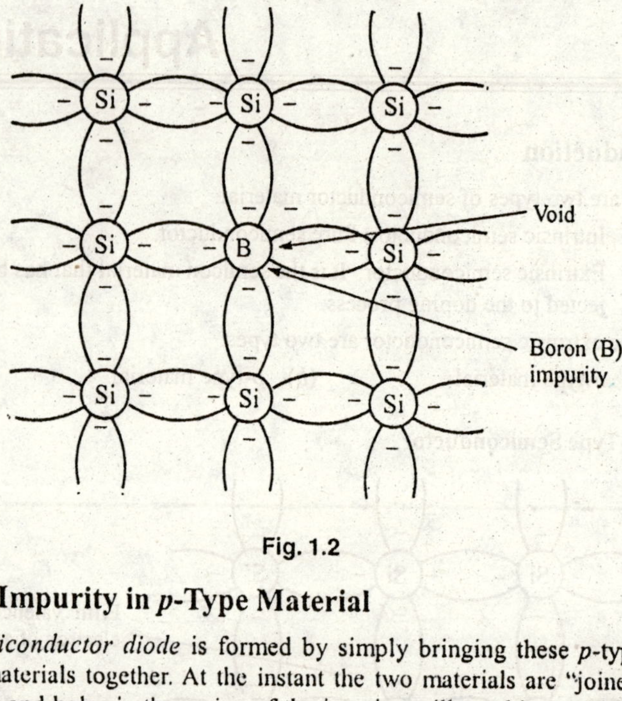

Fig. 1.2

Boron Impurity in *p*-Type Material

The *semiconductor diode* is formed by simply bringing these *p*-type and *n*-type materials together. At the instant the two materials are "joined" the electrons and holes in the region of the junction will combine resulting in a lack of carriers in the region near the junction. This region of uncovered positive and negative ions is called the *depletion region,* due to the depletion of carriers in this region.

Since the diode is a two-terminal device, the application of a voltage across its terminals leaves three possibilities.

(*i*) No bias ($V_D = 0V$) (*ii*) Forward bias ($V_D > 0V$)
(*iii*) Reverse bias ($V_D < 0V$)

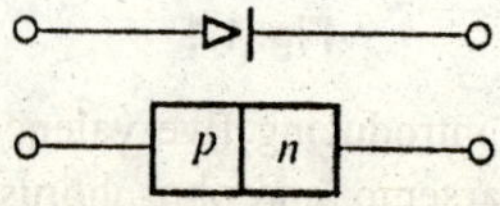

Fig. 1.3

(*i*) The general characteristics of a semiconductor diode can be defined by the following equation for the forward and reverse bias region:

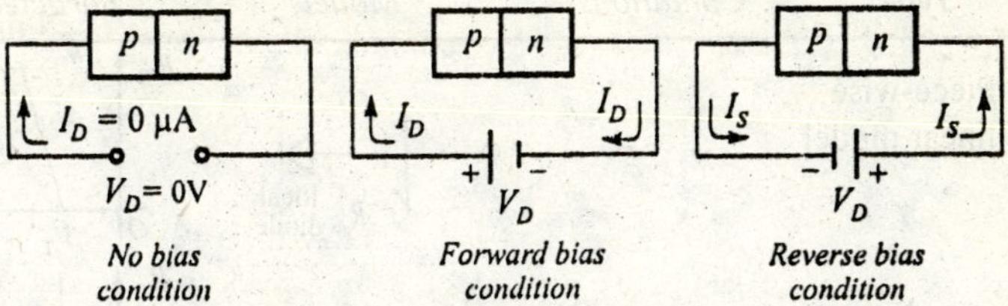

Fig. 1.4

$$I_D = I_S\left(e^{kV_D/T_k} - 1\right)$$

Where I_S = reverse saturation current

$K = 11600/\eta$ with $\eta = 1$ for Germanium (Ge)

$\eta = 2$ for Silicon (Si)

$T_K = T_C + 273°$

The diode characteristic is shown as below:

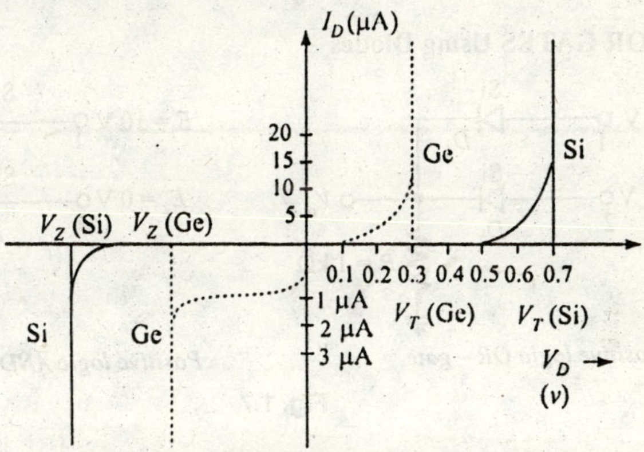

Fig. 1.5

$V_T \cong 0.7$ V for silicon diode.
$V_T \cong 0.3$ V for germanium diode.

Temperature effect: The reverse saturation current (I_S) will just double itself in magnitude for every 10° C increase in temperature.

Dynamic resistance (r_d): It is given by

$$r_d = \frac{26\ \text{mV}}{I_D} \quad \text{for Ge and Si both}$$

Diode Equivalent Circuits (models)

Type	*Conditions*	*Model*	*Characteristics*
(*i*) Piece-wise linear model		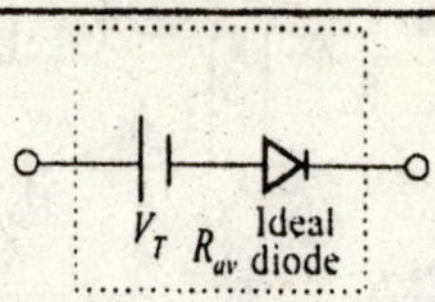	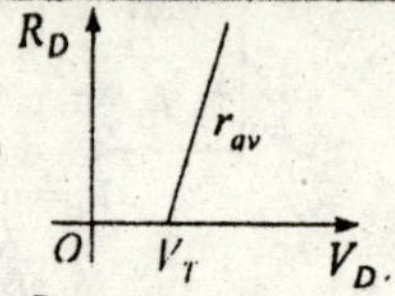
(*ii*) Simplified model	*R* network >> r_{av}	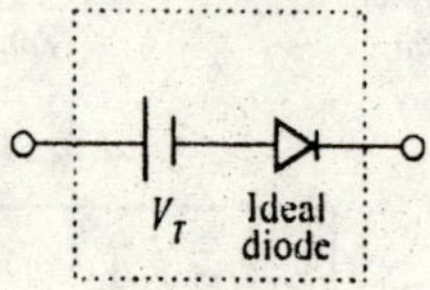	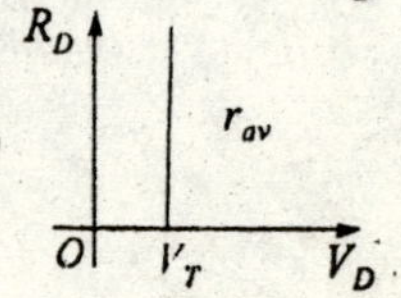
(*iii*) Ideal device	*R* network >> r_{av} *E* network >> V_T	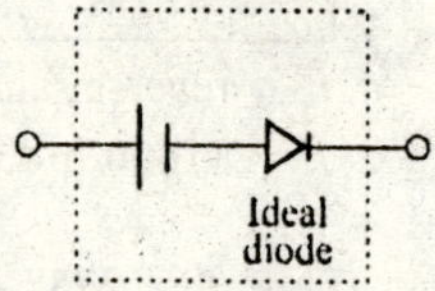	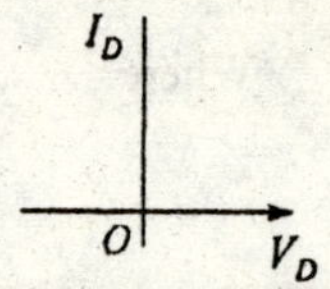

Fig. 1.6

AND / OR GATES Using Diodes

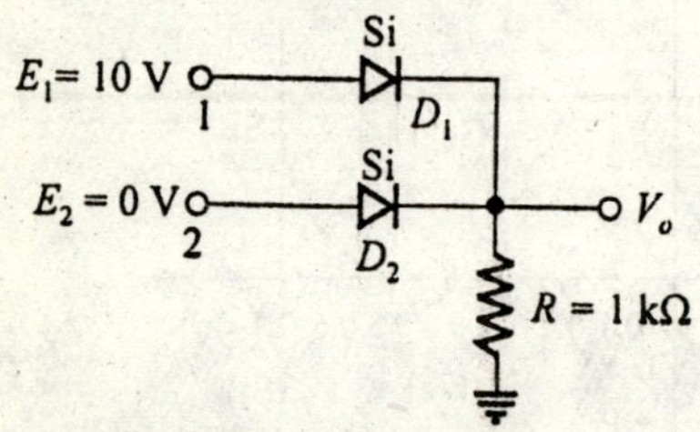

Positive logic OR – gate

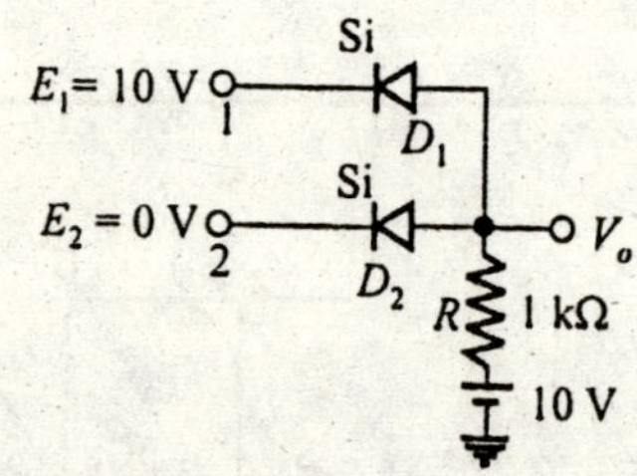

Positive logic AND – gate

Fig. 1.7

Sinusoidal Inputs — Half Wave Rectification

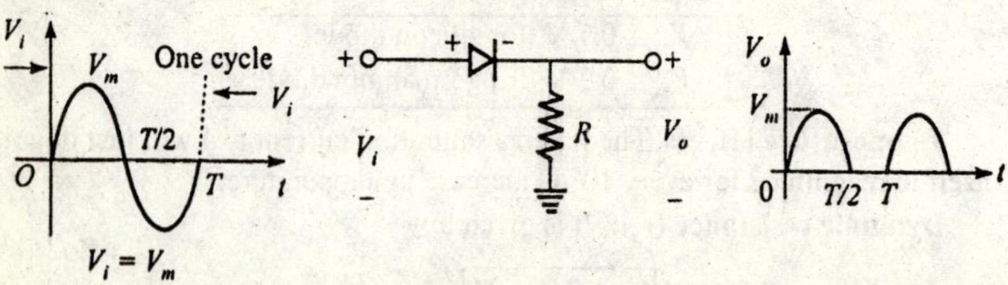

Fig. 1.8

The output of half wave rectifier is

Thus, the process of removing one-half of the input signal to establish a d.c. level is apply called a *half wave rectification.*

The average voltage of a half wave rectifier is

$$\boxed{V_{\text{d.c.}} = 0.318\ V_m}$$

Effect of V_T (threshold voltage of diode) on one voltage is

$$\boxed{V_{\text{d.c.}} = 0.318\ (V_m - V_T)}$$

The peak inverse voltage (PIV) rating of the diode in a half wave rectification is

$$\boxed{\text{PIV rating} \geqq V_m}$$

Full-wave Rectification : The d.c. level obtained from a sinusoidal input can be improved 100%, using a process called *full wave rectification.*

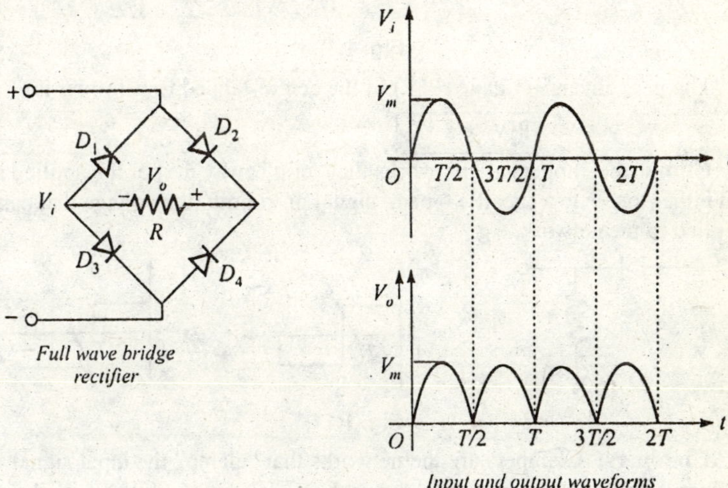

Fig. 1.9

The average voltage of full wave rectifier is

$$\boxed{V_{\text{d.c.}} = 0.636\ V_m}$$

The effect of V_T in the average voltage is

$$\boxed{V_{\text{d.c.}} = 0.636\ (V_m - 2\ V_T)}$$

The peak inverse voltage of each diode is

$$\boxed{\text{PIV} \geqq V_m}$$

Centre-tapped Transformer: It is a popular full wave rectifier with only two diodes, but requiring a centre-tapped transformer to establish the input signal across each section of the secondary of the transformer.

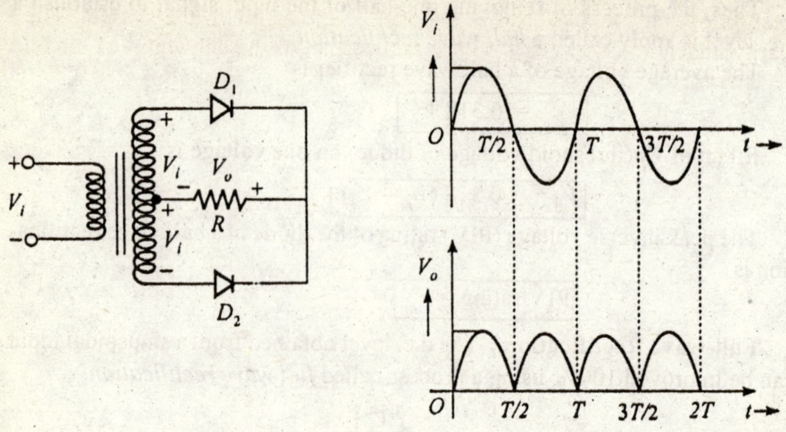

Fig. 1.10

The peak inverse voltage (PIV) of the centre-tapped transformer is

$$\boxed{\text{PIV} \geq 2\,V_m}$$

Clippers: Clippers are networks that "clip" away part of the applied signal either to create a specific type of signal or to limit, the voltage that can be applied to the network, e.g.

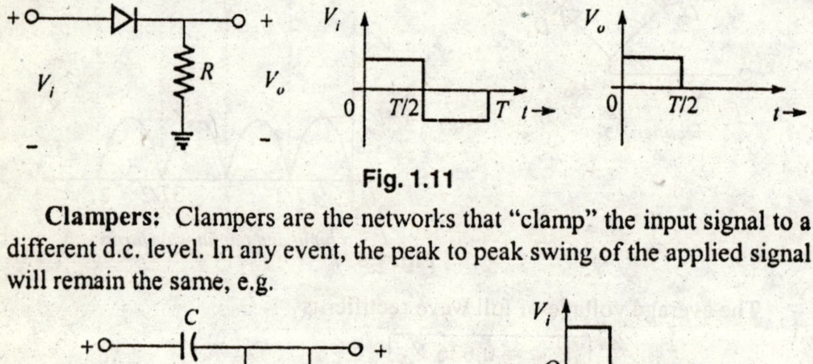

Fig. 1.11

Clampers: Clampers are the networks that "clamp" the input signal to a different d.c. level. In any event, the peak to peak swing of the applied signal will remain the same, e.g.

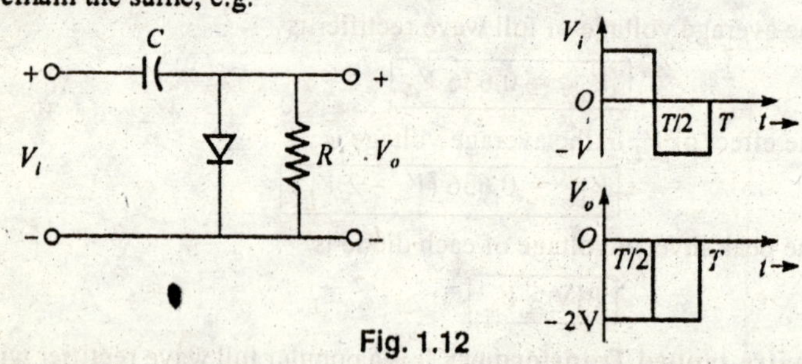

Fig. 1.12

Zener Diode: These are the diodes that make effective use of the zener breakdown potential of an ordinary *p-n* junction characteristic to provide a device of wide importance and application. For zener diode conduction, the direction of conventional flow is opposite to the arrow in the symbol. The polarity under conduction is also opposite to that of the conventional diode. The zener diode is used as voltage stabilizer or voltage regulator.

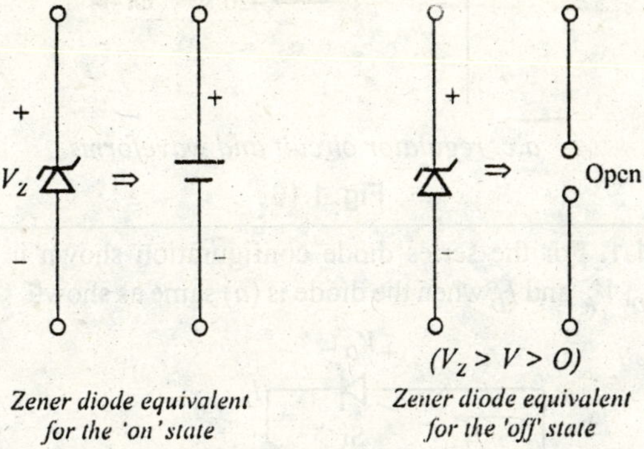

Fig. 1.13

Square Wave Generator and a.c. Regulator : The zener diodes with two in numbers when connected back-to-back can be used as an a.c. regulator and a square wave generator. The circuit diagram is shown below:

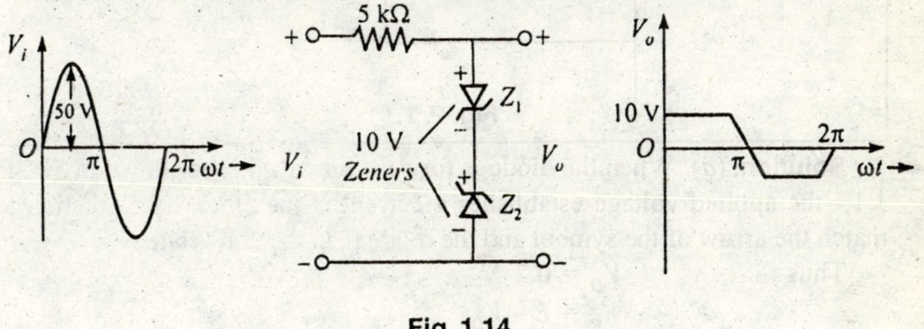

Fig. 1.14

Waveforms and circuit of square wave generator

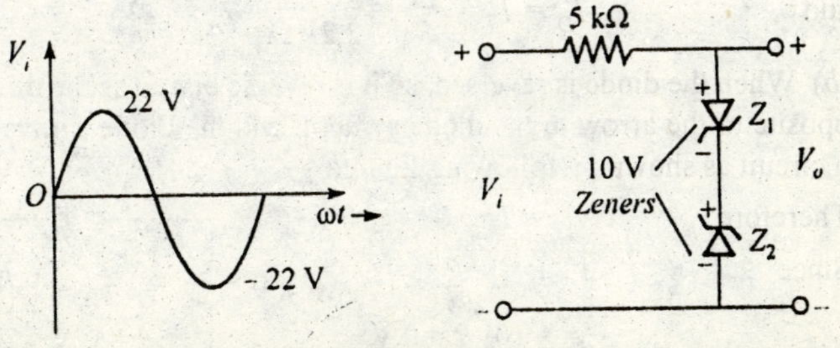

Fig. 1.15

The output waveform will be as such

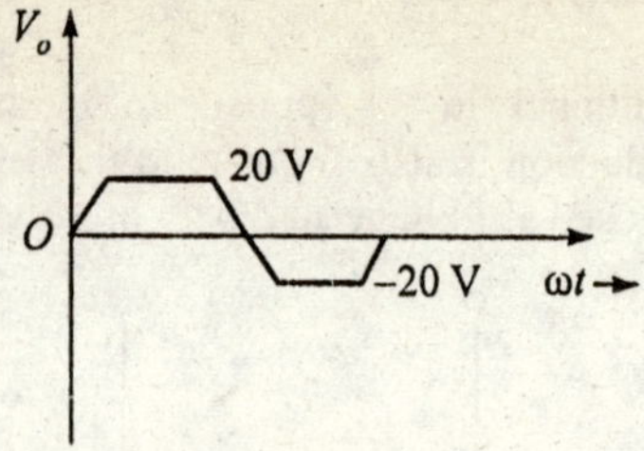

a.c. regulator circuit and waveforms

Fig. 1.16

Problem 1.1. For the series diode configuration shown in Fig. P. 1.1 determine V_D, V_R and I_D when the diode is (*a*) same as shown, (*b*) reversed.

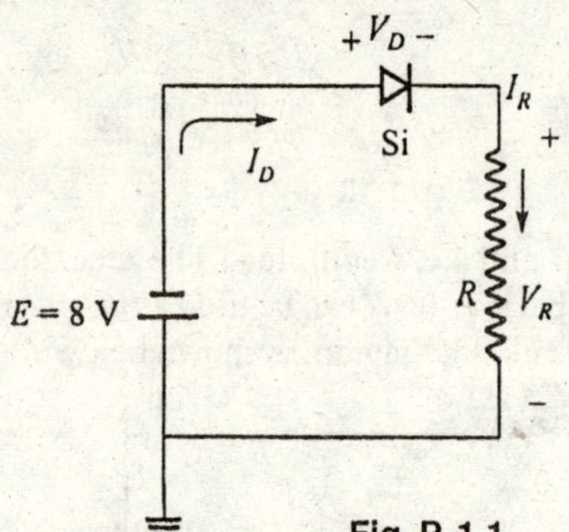

Given, $R = 2.2$ kΩ

Fig. P. 1.1

Solution. (*a*) When the diode is forward biased, i.e. as shown in Fig. P. 1.1, the applied voltage establishes a current is the clockwise direction to match the arrow of the symbol and the diode is in the "on" state.

Thus
$$V_D = 0.7 \text{ V}$$
$$V_R = E - V_D$$
$$= 8 \text{ V} - 0.7 \text{ V} = 7.3 \text{ V}$$

and
$$I_D = I_R = \frac{V_R}{R} = \frac{7.3\text{V}}{2.2\text{k}\Omega} = 3.32 \text{ mA}$$

(*b*) When the diode is reversed, so it is reverse biased as the direction of *I* is opposite to the arrow in the diode symbol, and the diode equivalent is the open circuit as shown in following figure:

Therefore $I_D = I_R = 0$

Since $V_R = I_R \cdot R$

$= (0)\, R = 0$ V.

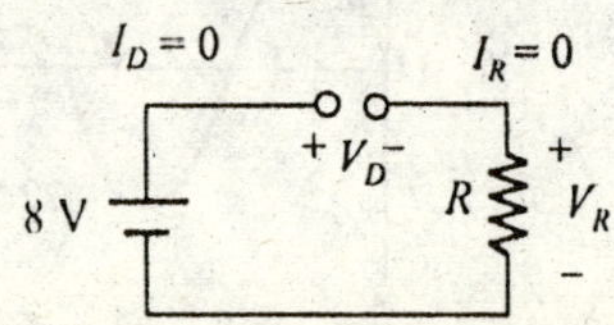

Fig. 1.1 (a)

Applying KVL to the loop

$$8 - V_D - V_R = 0$$

$$\Rightarrow \quad 8 - V_D - 0 = 0 \quad \Rightarrow \quad V_D = 8 \text{ V}$$

Problem 1.2. For the series diode configuration shown in Fig. P. 1.2, determine V_D, V_R and I_D.

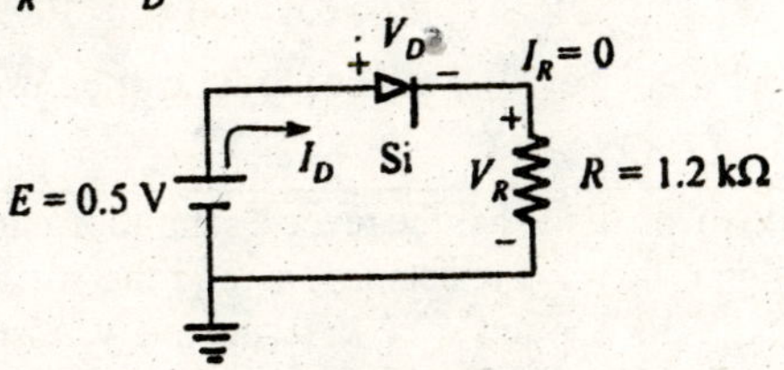

Fig. P. 1.2

Solution. Although the pressure establishes a current with the same direction as the arrow symbol, the level of applied voltage is insufficient to turn the silicon diode "on". Establishing the open circuit equivalent as the appropriate approximation.

Resulting voltage and current levels are therefore

$$I_D = 0 \text{ A}$$

$$V_R = I_R\, R = I_D \cdot R = (0)\, 1.2 \text{ k}\Omega = 0 \text{ V}$$

and $$V_D = E = 0.5 \text{ V}$$

The pt. of operation on the characteristic is shown below

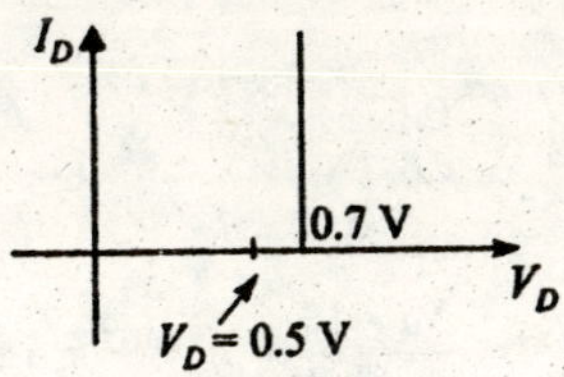

Fig. 1.2 (a)

Problem 1.3. A *pn*-junction silicon diode is forward biased through a resistance $R = 5\text{ k}\Omega$ in series with a d.c. source $V = 5$ V. Find V_{AK}, the voltage drop across the diode and the diode current I_A by using (*i*) the constant voltage drop $V_D = 0.7$ V model, and (*ii*) the large signal model with $V_r = 0.6$ V and $R_F = 100\ \Omega$.

Solution. (*i*) Assuming that $V_{AK} = V_D = 0.7$ V

$$I_A = \frac{V - V_D}{R} = \frac{5 - 0.07}{5\text{K}}$$

$$= 0.86 \text{ mA}$$

(*ii*) Assuming that $V_r - R_F$ model is valid, we have

$$I_A = \frac{V - V_r}{R + R_F} = \frac{5 - 0.06}{5K + 100}$$

$$= \frac{4.4}{5100} = 0.86 \text{ mA}$$

Then $\quad V_{AK} = V_r + I_A R_F = 0.6 + 0.86 \times 10^{-3} \times 100 = 0.686 \text{ V}$

Thus, because $V >> V_r$ or V_D, either model can be used in d.c. circuit analysis.

Problem 1.4. A silicon diode has the reverse saturation current $I_R = 50$ nA. Find the small signal forward resistance, r_f and the small signal reverse resistance r_r at $V_{AK} = 10.7$ V and – 0.7 V, respectively. Assume the thermal voltage $V_T = 26$ mV at $T = 300$ K.

Solution. For a silicon diode, we have

$$I_A = I_R\left(e^{V_{AK}/2V_T} - 1\right)$$

The dynamic conductance will therefore be

$$g_f = \frac{1}{r_f} = \frac{dI_A}{dV_{AK}} = \frac{I_R}{2V_T} e^{\frac{V_{AK}}{2V_T}}$$

For $\quad V_{AK} = +0.7$ V

$$g_f = \frac{50 \times 10^{-9}}{2 \times 0.026} e^{\frac{0.7}{0.052}} = 0.675 \text{ mho}$$

or

$$r_f = \frac{1}{0.675} = 1.48\ \Omega$$

For $\quad V_{AK} = -0.7$ V

We have the reverse conductance as

$$g_r = \frac{50 \times 10^{-9}}{2 \times 0.026} e^{-\frac{0.7}{0.052}} = 1.37 \times 10^{-12} \text{ mho}$$

$$\Rightarrow \quad r_r = \frac{1}{1.37 \times 10^{-12}} = 73 \times 10^{10}\ \Omega$$

In practice, r_r is much smaller ($\approx 10^8$) due to surface leakage currents.

Problem 1.5. In the full wave 2-diode rectifier circuit shown in Fig. P. 1.5, we have $V_r = 10$ V (rms), the sum of the d.c. resistance of the transformers winding and the forward d.c. resistance of the diode is R_S = 0.2 Ω the load resistance $R_L = 10\ \Omega$, the forward diode drop $V_D = 0.7$ V. Find (*i*) the peak load current I_{lm}, (*ii*) the d.c. load current I_{LDC}, (*iii*) the individual diode d.c. current I_{ADC}, (*iv*) the peak inverse voltage (PIV) across the diodes, and (*v*) the d.c. output voltage V_{LDC}.

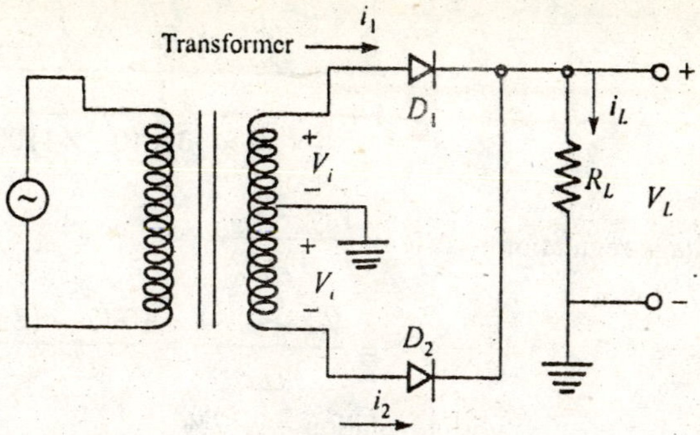

Fig. P. 1.5

Solution.

$$(i) \quad I_{lm} = \frac{V_{im} - V_D}{R_L + R_S} = \frac{10\sqrt{2} - 0.7}{10 + 0.2} = 1.32 \text{ A}$$

$$(ii) \quad I_{LDC} = \frac{2 I_{lm}}{\pi} = \frac{2 \times 1.32}{\pi} = 0.839 \text{ A}$$

$$(iii) \quad I_{ADC} = \frac{1}{2} I_{LDC} = \frac{0.839}{2} = 0.419 \text{ A}$$

$$(iv) \quad \text{PIV} = 2 V_{im} = 2 \times 10\sqrt{2} = 28.28 \text{ V}$$

$$(v) \quad V_{LDC} = I_{LDC} \times R_L = 0.839 \times 10 = 8.39 \text{ V.}$$

Problem 1.6. A full wave rectifier produces rms voltage of 10 V from a 50 Hz line source, and feeds a resistive load of 1100 Ω. If the filter uses a capacitance C = 50 μF, find the resulting (*i*) output ripple factor, (*ii*) d.c. output voltage V_{LDC}, (*iii*) voltage regulation, and (*iv*) the ripple output voltage V_r.

Solution. (*i*) We know that

$$\text{Ripple factor, RF} = \frac{1}{4\sqrt{3}\, f_1 R_L C}$$

$$\Rightarrow \quad \text{RF} = \frac{1}{4\sqrt{3} \times 50 \times 1100 \times 50 \times 10^{-6}} = 0.053$$

(*ii*) d.c. output voltage $V_{LDC} = \dfrac{V_{im}}{1 + \dfrac{1}{4 f_1 C R_L}}$

$$= \frac{10\sqrt{2}}{1 + \dfrac{1}{4 \times 50 \times 50 \times 10^{-6} \times 1100}} = 12.96 \text{ V}$$

(*iii*) Voltage regulation $\text{VR} = \dfrac{V_{NL} - V_{FL}}{V_{FL}}$

$$\Rightarrow \quad \text{VR} = \frac{V_{im} - V_{LDC}}{V_{LDC}} = \frac{10\sqrt{2} - 12.96}{12.96} = 0.0912$$

or volt. regulation = 9.12%

(*iv*) Ripple output voltage

$$V_x = \frac{V_{LDC}}{4\sqrt{3}\, f_1 R_L C_1} = V_{LDC} \times \text{RF}$$

$$\Rightarrow \quad V_r = 12.96 \times 0.053 = 0.687 \text{ V.}$$

Problem 1.7. A silicon diode circuit is shown in Fig. P. 1.7. Calculate the load current I_L as well as the output voltage V_o for:

(*i*) $V_i = +0.2$ V and (*ii*) $V_i = +5$ V. Assume the voltage drop V_D across a conducting diode to be 0.7 V. $R_L = 5$ kW and $V_{cc} = 5$ V.

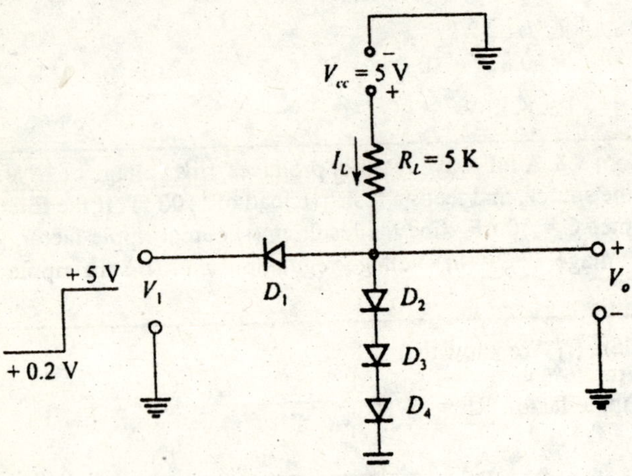

Fig. P. 1.7

Solution. (*i*) For $V_i = +0.2$ V, the diode D_1 is forward biased, and the output voltage can be calculated as

$$V_o = VD_1 + V_i = 0.7 + 0.2 = 0.9 \text{ V}$$

The diodes D_2, D_3 and D_4 can turn on only when

$$V_o = 3 \times V_D = 3 \times 0.7 = 2.1 \text{ V}$$

and thus for our case these diodes will not conduct.

Load current $$I_L = \frac{V_{cc} - V_o}{R_L}$$

$$\Rightarrow \quad I_L = \frac{5 - 0.9}{5\text{K}} = 0.82 \text{ mA}$$

(*ii*) For $V_i = +5\text{V}$, D_1 will be reverse-biased and D_2, D_3 and D_4 will be forward biased.

The voltage output, $V_o = 0.7 + 0.7 + 0.7 = 3 \times 0.7 = 2.1 \text{ V}$

and the resulting load current

$$I_L = \frac{V_{cc} - V_0}{R_L} = \frac{5 - 2.1}{5\text{K}} = 0.592 \text{ mA}$$

The voltage across the diode D_1 is

$$V_{AK} = V_o - V_i$$
$$= 2.1 - 5 = -2.9 \text{ V}$$

Thus, D_1 is reverse biased as assumed.

Problem 1.8. Determine the current I for each of the following configurations using the approximate equivalent model for the diode.

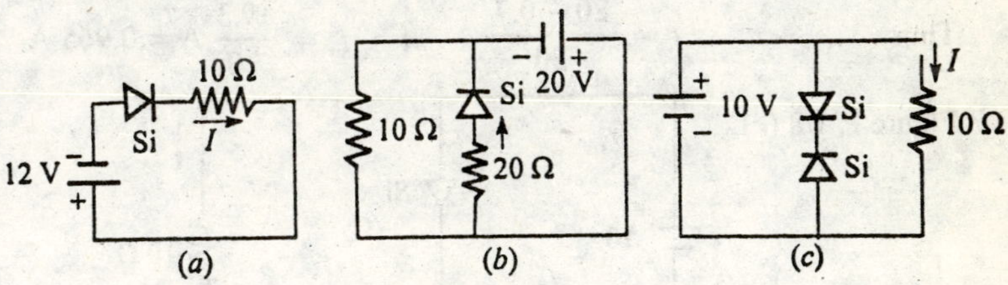

Fig. P. 1.8

Solution. Figure P. 1.8 (a) since the diode is reverse biased as anode is connected to –ve terminal of the d.c. source, the equivalent of diode is an open circuit

Thus $$I = 0$$

$$V_D = -10 \text{ V}$$

and $$V_R = (I_R)(R)$$
$$= (0)(10) = 0 \text{ V}$$

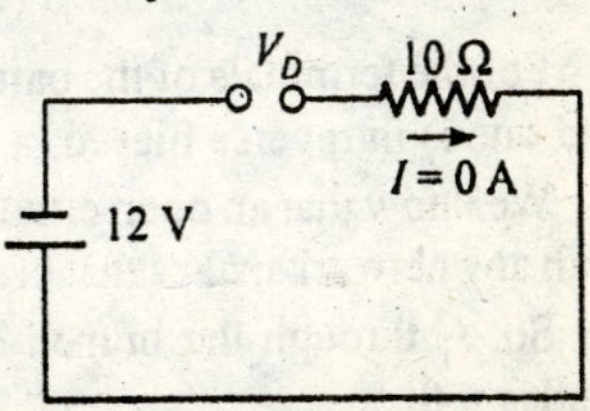

Fig. 1.8 (a)

Figure P. 1.8 (b) the circuit is redrawn as below.

Since the cathode of the diode is connected to –ve terminal of the battery and anode of the diode is connected to + ve terminal of the battery.

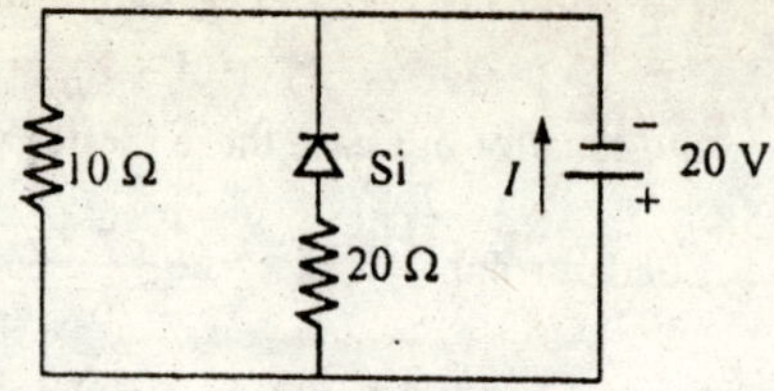

Fig. 1.8 (b)

The diode is in forward biased and its equivalent will be a voltage source of 0.7 V (for Si) as shown below:

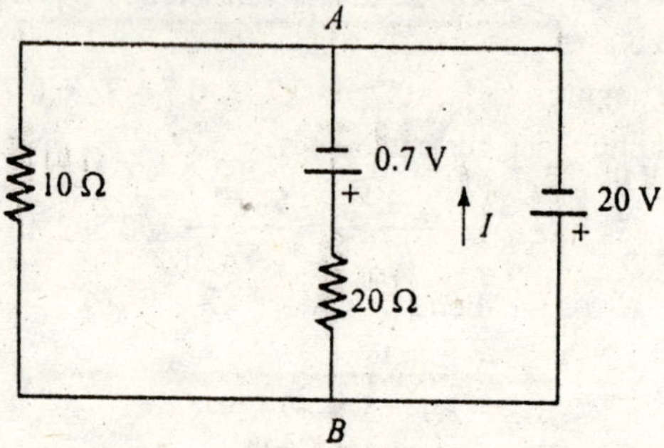

Fig. 1.8 (c)

Applying kVL, we get that

$$V_{AB} = -20 \text{ V} \quad \Rightarrow V_{BA} = +20 \text{ V}$$

Thus $$I = \frac{20 - 0.7}{20} \quad \Rightarrow \quad I = \frac{19.3}{20} \text{ A} = 0.965 \text{ A}$$

Figure P. 1.8 (*c*)

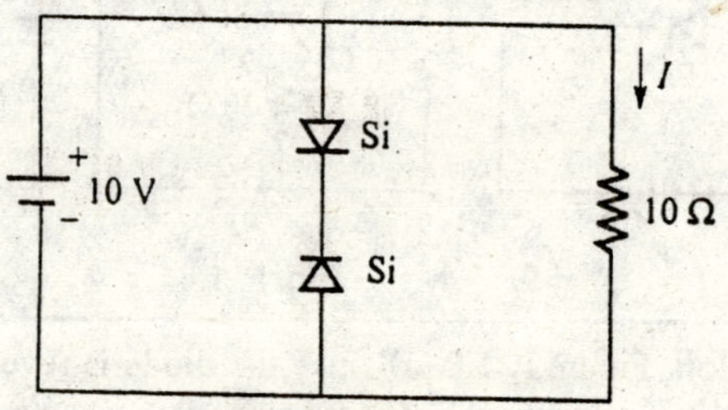

Fig. 1.8 (d)

For the terminals of the battery set as shown one diode is in forward biased and one is in reverse biased.

We know that an open circuit in series with any network makes that branch open.

So, I_1 through the branch containing diodes = 0

Thus $$I = \frac{10\text{V}}{10\Omega} = 1\text{A}$$

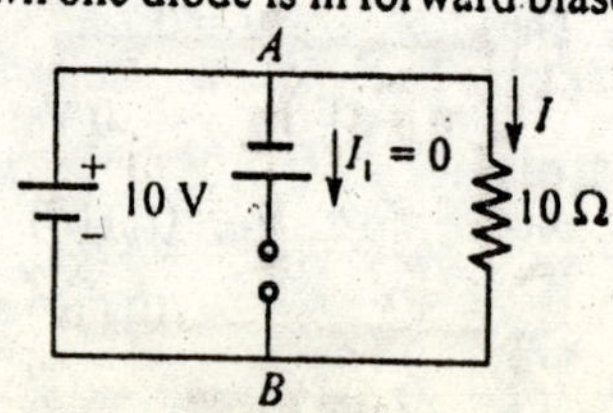

Fig. 1.8 (e)

Problem 1.9. Determine V_o and I_D for the networks shown in Fig. P. 1.9.

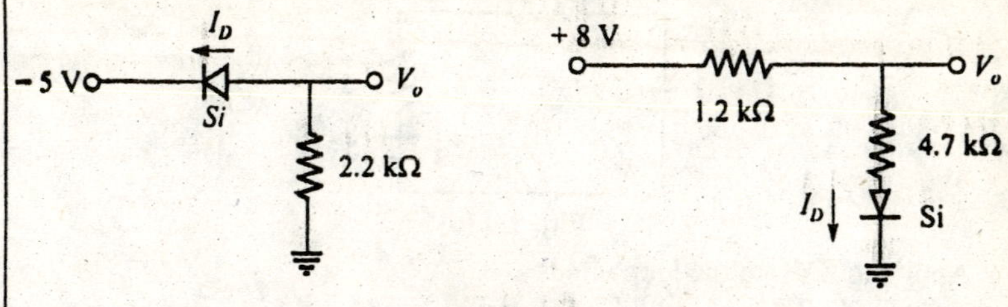

Fig. P. 1.9

Solution. (*a*) The given network is shown below:

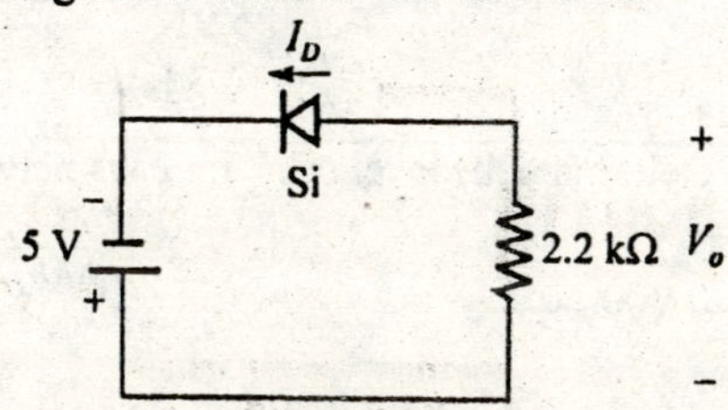

Fig. 1.9 (a)

Since the diode is in forward biased it should be replaced by its equivalent circuit as "ON" state.

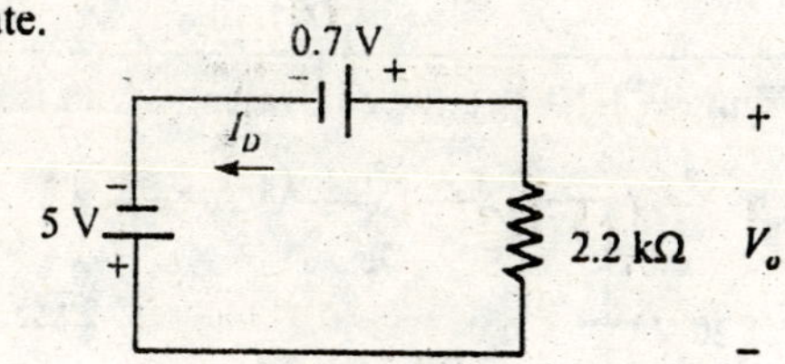

Fig. 1.9 (b)

Applying KVL to the given loop

$$5 + V_o - 0.7 = 0 \qquad \Rightarrow V_o = -4.3 \text{ V}$$

Thus, V_o = 4.3 V with the polarity opposite to the as shown.

$$I_D = I_R = \frac{V_o}{2.2 \text{ K}} = \frac{4.3}{2.2 \text{ K}} = 1.954 \text{ mA}$$

in the same direction as shown.

(*b*) Redrawing the given network

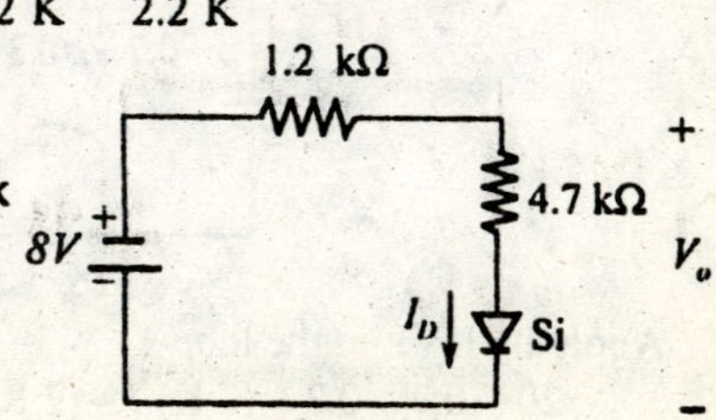

Fig. 1.9 (c)

Replacing the diode by its equivalent voltage source since it is forward biased.

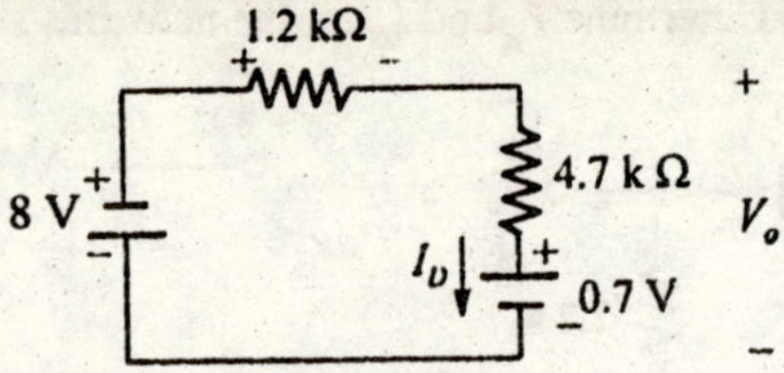

Fig. 1.9 (d)

Applying KVL to the loop yields

$$8 - (1.2\text{ K})\, I_D - (4.7\text{ K})\, I_D - 0.7 = 0 \Rightarrow (5.9\text{ K})\, I_D = 7.3$$

$$\Rightarrow \qquad I_D = \frac{7.3}{5.9\text{K}} = 1.237\text{ mA}$$

$$V_o = (4.7\text{ K})I_D + 0.7 = 6.154\text{ V}$$

Problem 1.10. Determine the level of V_o for each network shown below:

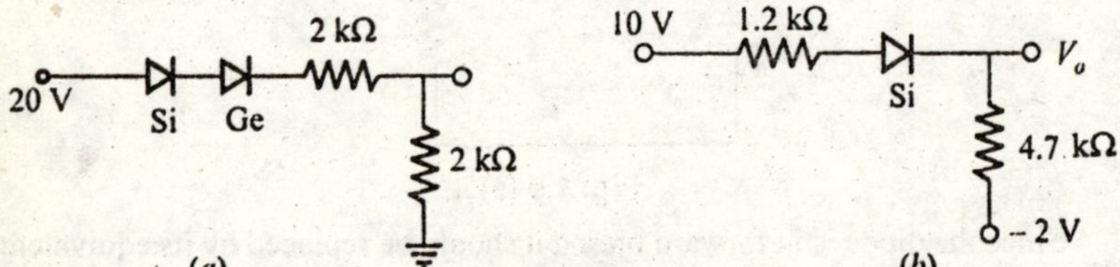

Fig. P. 1.10

Solution. Figure P. 1.10 (*a*), the equivalent circuit is redrawn as below:

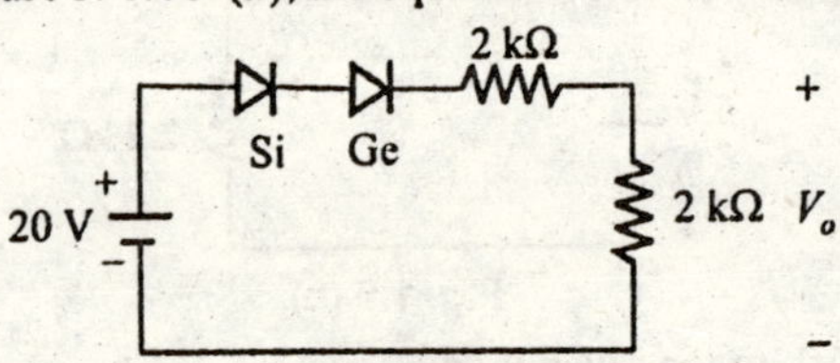

Fig. 1.10 (c)

Since both the diodes are forward biased so they should be replaced by their equivalent voltage sources.

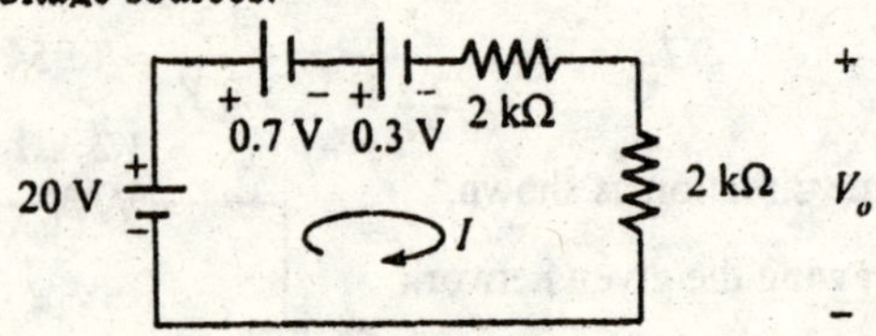

Fig. 1.10 (d)

Applying KVL to the loop

$$20 - 0.7 - 0.3 - (2\text{ K})\, I - (2\text{ K})\, I = 0$$

$$\Rightarrow \qquad (4\text{ K})\, I = 19$$

$$\Rightarrow \quad I = \frac{19}{4\text{ K}} = 4.75\text{ mA} \Rightarrow V_o = (2\text{ K})\, I = 9.5\text{ V}$$

Figure P. 1.10 (b), since the diode is in forward bias so replace it by a voltage source in the equivalent circuit.

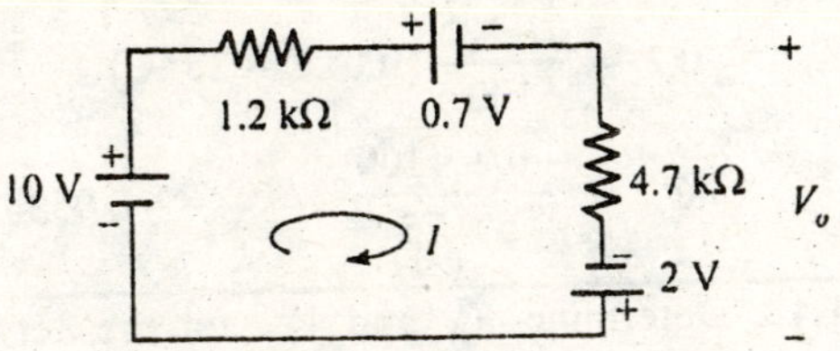

Fig. 1.10 (e)

Applying KVL to the loop

$$10 - (1.2\text{ K})\, I - 0.7 - (4.7\text{ K})\, I + 2 = 0$$

$$\Rightarrow \qquad (5.9\text{ K})\, I = 11.3 \Rightarrow I = \frac{11.3}{5.9\text{K}} = 1.915\text{ mA}$$

$$V_o = (4.7\text{ K})\, I - 2 = 9 - 2 = 7\text{ V}$$

Problem 1.11. At what temperature wil the 10 V zener diode have a nominal voltage of 10.75 V? Given that the temperature coefficients $T_C = 0.072\%/°\text{C}$, $T_0 = 2.5°$ C.

Solution. Given $V_Z = 10$ V, $T_C = 0.072\ \%/°\text{C}$

$V_Z = 10.75$ V, $T_1 = ?$, $T_0 = 25°$ C

We know that

Nominal voltage of a zener diode is given by

$$\Delta V_Z = \frac{T_C V_Z}{100}(T_1 - T_0)$$

Substituting the given values, we get

$$10.75 - 10 = \frac{0.072 \times 10}{100}(T_1 - 25°\text{ C})$$

$$\Rightarrow \qquad T_1 - 25°\text{ C} = 104.167 \Rightarrow T_1 = 104.167 + 25 = 129.167\ °\text{C}.$$

Because of the positive temperature coefficient, the new zener potential is given by

$$V'_Z = V_Z + \Delta V_Z = 10.75$$

$$\Rightarrow \qquad \Delta V_Z = 0.75\text{ V}$$

Problem 1.12. Determine the temperature coefficient of a 5 V zener diode (rated 25° C value), if the nominal voltage drops to 4.8 V at a temperature of 100° C.

Solution. Given nominal voltage = 4.8 V, and

$$V_Z = 5\text{ V}$$

$$\therefore \qquad \Delta V_Z = (4.8 - 5)\text{ V} = -0.2\text{ V},\ T_0 = 25°\text{C},\ T_1 = 100\ °\text{C}$$

Now, from the relation

$$\Delta V_Z = \frac{T_C V_Z}{100}(T_1 - T_0)$$

We get

$$-0.2 = \frac{T_C \times 5}{100}(100°\text{ C} - 25°\text{ C})$$

$$\Rightarrow \qquad T_C = \frac{-0.2 \times 100}{5 \times 75}\%/°\text{C} = -0.0534\%/°\text{C}$$

Problem 1.13. Determine V_{o1} and V_{o2} for the network shown in Fig. P. 1.13.

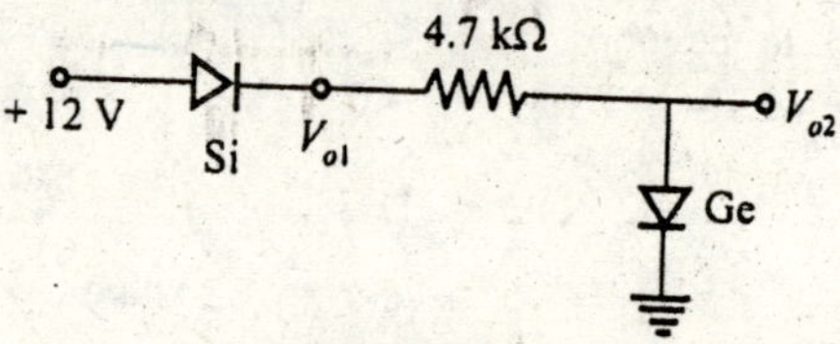

Fig. P. 1.13

Solution. Since both the diodes are forward biased, they should be replaced by their equivalents in the equivalent circuit, taking $V_{T-\text{Si}} = 0.7$ V and $V_{T-\text{Ge}} = 0.3$ V.

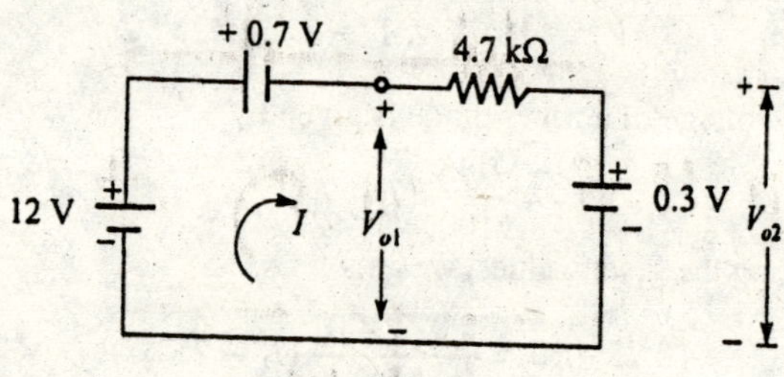

Fig. 1.13 (a)

Applying KVL to the loop

$$12 - 0.7 - (4.7\text{ K})\,I - 0.3 = 0 \Rightarrow (4.7\text{ K})\,I = 11$$

$$\Rightarrow \qquad I = \frac{11}{4.7\text{ K}} = 2.34\text{ mA}$$

Now, from the circuit, it is easily seen that

$$V_{o1} = (12 - 0.7)\text{ V} = 11.3\text{ V}$$

and

$$V_{o2} = V_{o1} - (4.7\text{ K})\,I$$

$$= 11.3 - 4.7\text{ K} \times \frac{11}{4.7\text{ K}} = 0.3\text{ V}$$

V_{o2} is directly seen to be the voltage across the Ge diode, which is = 0.3.

Problem 1.14. Determine V_o and I_D for the networks of shown in Fig. P. 1.14.

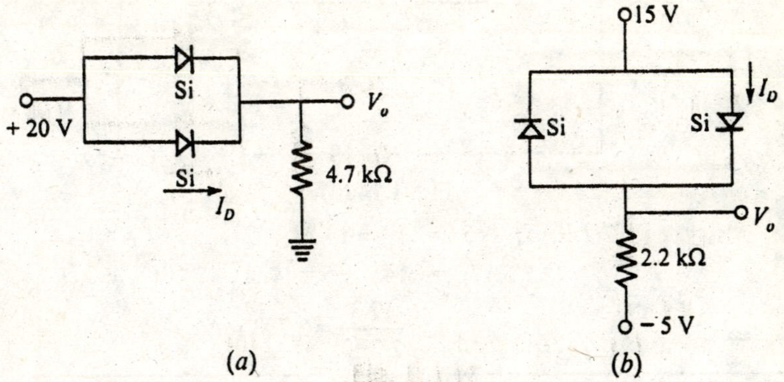

Fig. P. 1.14

Solution. Figure P. 1.14 (*a*), we see that the direction of arrows of both the diodes are same as that of the resultant current, so both the diodes are forward biased.

Redrawing the equivalent, we get

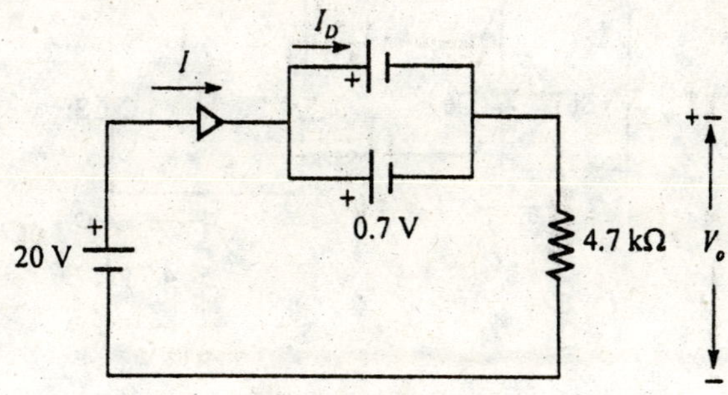

Fig. 1.14 (c)

Thus $$V_o = 20 - 0.7 = 19.3 \text{ V}$$

and $$I = \frac{V_o}{4.7 \text{ K}} = \frac{19.3}{4.7 \text{ K}} = 4.106 \text{ mA}$$

Since both the diodes are identical

$$I_{D_1} = I_{D_2} = \frac{I}{2} = 2.053 \text{ mA}$$

Figure P. 1.14 (b), direction of arrow of one diode is opposite to the resultant current, so that diode would act as an open circuit.

From the equivalent circuit, we can directly see that

$$V_o = 15 - 0.7 = 14.3 \text{ V}$$

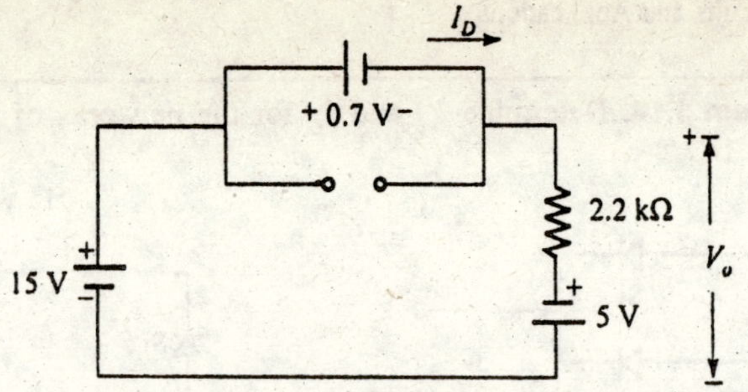

Fig. 1.14 (d)

Also $$(2.2\text{ K})\,I_D - 5 = V_o = 14.3$$

$$(2.2\text{ K})\,I_D = 14.3 + 5 = 19.3$$

$$\Rightarrow \quad I_D = \frac{19.3}{2.2\text{ K}} = 8.773\text{ mA}$$

Problem 1.15. Determine V_o and I for the networks shown in Fig. P. 1.15:

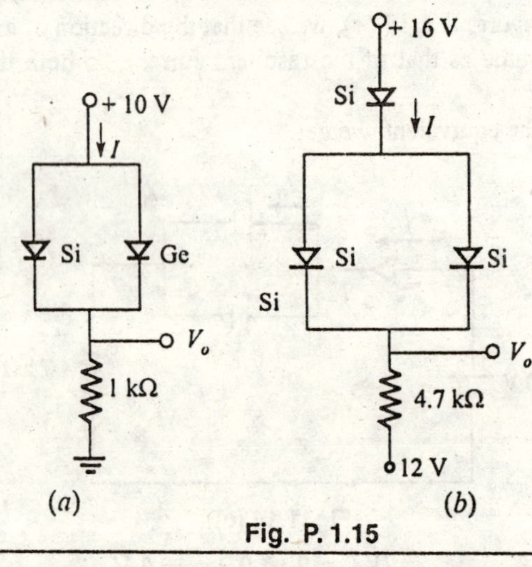

Fig. P. 1.15

Solution. Figure 1.15 (a), initially, it would appear that the applied voltage will turn both diodes "ON". But, if both were "ON", the 0.7 V drop across the silicon diode would not match the 0.3 V across the Ge diode as required by the parallel elements. The resulting action can be explained simply by realizing, that when the supply is turned on it will increase from 0 V to 12 V over a period of time — although probably measurable in milliseconds. At the instant during the rise that 0.3 V is established across the Ge diode, it will turn "ON" and maintain a level of 0.3 V. The Si diode will never have the opportunity to capture its required 0.7 V, and therefore remains in its open-circuit state.

Thus, the circuit may be drawn as:

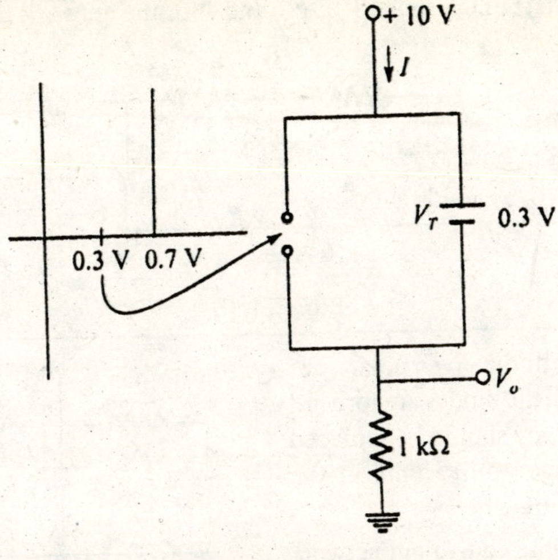

Fig. 1.15 (c)

Thus

$$V_o = (10 - 0.3)\text{ V} = 9.7\text{ V}$$

i.e. $$V_R = V_o = 9.7\text{ V}$$

$\therefore$ $$I = \frac{V_R}{R} = \frac{9.7\text{V}}{1\text{k}\Omega} = 9.7\text{ mA}$$

Figure 1.15 (*b*) since all the diodes are forward biased they must be replaced by their equivalent voltage sources. Redrawing the circuit

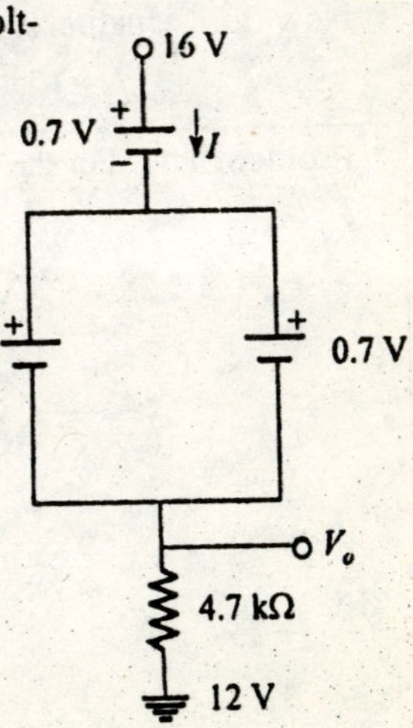

Fig. 1.9 (a)

Thus

$$V_o = 16\text{ V} - 0.7\text{ V} - 0.7\text{ V}$$

$$= 14.6\text{ V}$$

$$I = I_R = ?$$

$$V_R = V_o - 12\text{ V}$$

$$= (14.6 - 12)\text{ V} = 2.6\text{ V}$$

So, $$I_R = \frac{V_R}{R} = \frac{2.6\text{V}}{4.7\text{k}\Omega} = 0.553\text{ mA}$$

or, required $I = 0.555$ mA

Problem 1.16. Determine V_{o1}, V_{o2} and I for the network shown below:

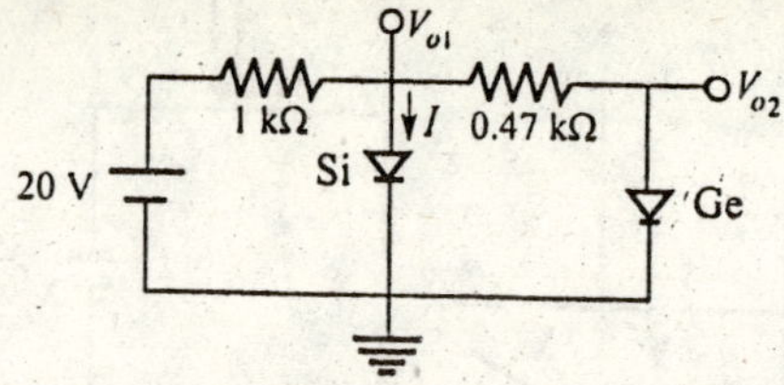

Fig. P. 1.16

Solution. For the given polarity of the battery, both the diodes are forward biased, hence they should be replaced by their voltage equivalents in the equivalent circuit.

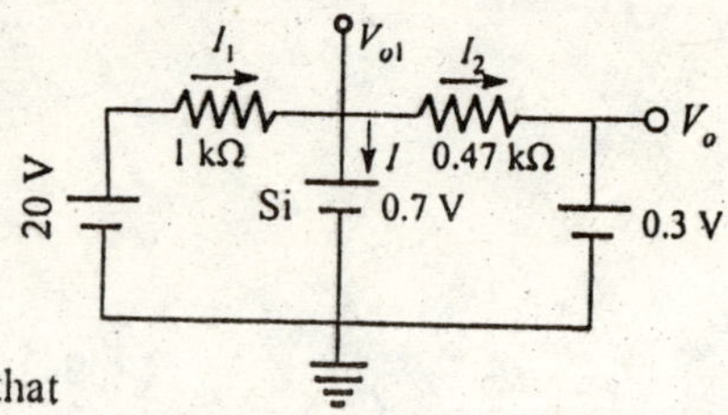

Fig. 1.16 (a)

Redrawing the equivalent network

From the circuit, it is directly seen that

$$V_{o1} = 0.7 \text{ V and } V_{o2} = 0.3 \text{ V}$$

Now $\quad I_1 = I_2 + I \Rightarrow I = I_1 - I_2 \quad (1)$

$$I_1 = \frac{20 - V_{o1}}{1\text{K}} = \frac{20 - 0.7}{1\text{K}} = 19.3 \text{ mA}$$

and $\quad I_2 = \dfrac{V_{o1} - V_{o2}}{0.47\text{K}} = \dfrac{0.7 - 0.3}{0.47 \text{ K}} = 0.851 \text{ mA}$

Now from equation (1), we get

$$I = I_1 - I_2 = 19.3 \text{ mA} - 0.851 \text{ mA} = 18.45 \text{ mA}$$

Problem 1.17. For the figure shown below find the values of V_o and I_D.

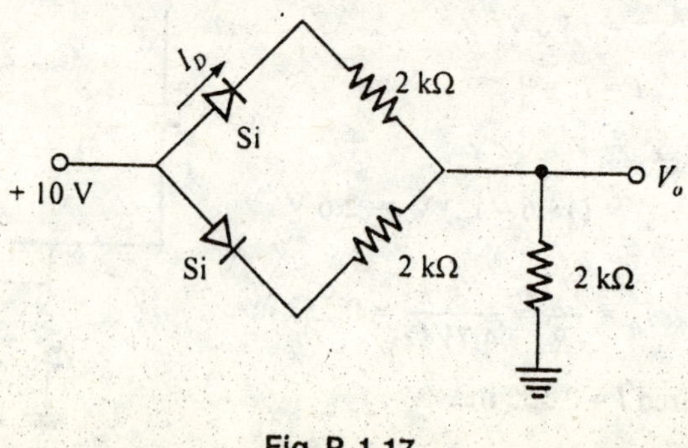

Fig. P. 1.17

Solution. Since, the direction of the resultant current is the same as that of arrows of both diodes so, both the diodes will be forward biased.

Replacing the diodes by their voltages, we get the circuit as

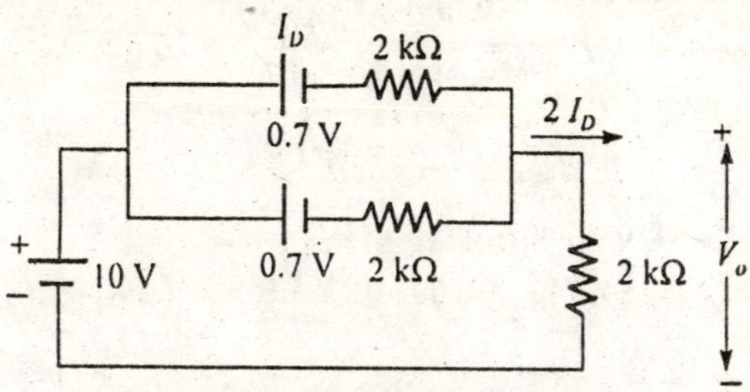

Fig. 1.17 (a)

Voltage of each of the parallel branches = 0.7 V + (2 K) I_D

Now, applying KVL to the loop containing any of the parallel branches, we find

$$10 - 0.7 - (2\text{ K})\ I_D - (2\text{ K})\ 2\ I_D = 0$$

$$\Rightarrow \quad (6\text{ K})I_D = 9.3$$

$$\Rightarrow \quad I_D = \frac{9.3}{6\text{ K}} = 1.55\text{ mA}$$

Now $\quad V_o = 2\ I_D \times 2\text{ K} = 2 \times 1.55 \times 10^{-3} \times 2 \times 10^3 = 6.2\text{ V}$

Problem 1.18. (*a*) Given P_{max} = 14 mW for each diode of the following figure determine the maximum current rating of each diode.

(*b*) Determine I_{max} for V_{imax} = 160 V.

(*c*) Determine the current through each diode at V_{imax} using the results of part (*b*).

(*d*) If only one diode were present, determine the diode current and compare it to the maximum rating.

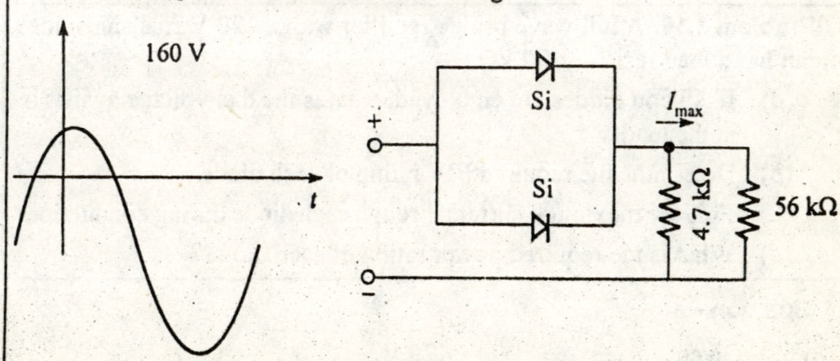

Fig. P. 1.18

Solution. (*a*) Given P_{max} = 14 mW for each diode so, the maximum current rating of each diode

$$I_{Dmax} = \frac{P_{max}}{V_T} = \frac{14\,\text{mW}}{0.7} = 20\text{ mA}$$

(*b*) For $V_{imax} = 160$ V

The voltage across the two resistors

$$= 160 - 0.7 = 159.3\text{ V}$$

So $I_{max} = I_1 + I_2$

Where $I_1 = \frac{159.3\,\text{V}}{56\text{ k}\Omega} = 2.84$ mA

and $I_2 = \frac{159.3\,\text{V}}{4.7\,\text{k}\Omega} = 33.89$ mA

Thus, $I_{max} = 2.84 + 33.89 = 36.73$ mA

(*c*) For $V_{imax} = 160$

Since both the diodes are identical to each other

$$\therefore \quad I_{D_1} = I_{D_2} = \frac{I_{max}}{2} = \frac{36.73}{2}$$

$\Rightarrow$ Diode current = 18.36 mA

(*d*) If only one diode is present

$$I_D = I_{max} = 36.73\text{ mA}$$

i.e. $I_D = 36.73\text{ mA} > I_{Dmax} = 20$ mA

Problem 1.19. A full wave bridge rectifier with a 120 V rms sinusoidal input has a load resistor of 1 kΩ.

(*a*) If silicon diodes are employed, what is the d.c. voltage available at the load?

(*b*) Determine the required PIV rating of each diode.

(*c*) Find the maximum current through each diode during conduction.

(*d*) What is the required power rating of each diode?

Solution.

(*a*) Given $V_{irms} = 120$ V

$$\therefore \quad V_{imax} = 120 \times \sqrt{2}\text{ V}$$

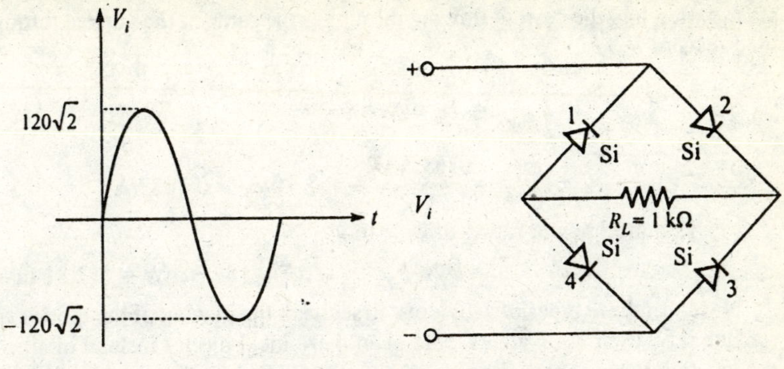

Fig. 1.19 (a)

For +ve value of voltage supply, diodes 2 and 4 will be forward biased, 1 and 3 will be reverse biased. Thus, our equivalent circuit will be as

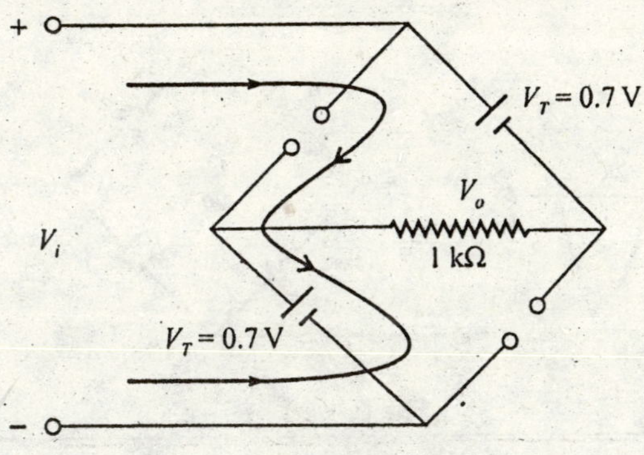

Fig. 1.19 (b)

applying KVL to the loop as shown by the arrow path direction.

$$V_i - 0.7 - V_o - 0.7 = 0$$

$$\Rightarrow \quad V_o = V_i - 1.4$$

$$V_{o\ max} = (V_{i\ max} - 1.4)\ \text{V} = (120\sqrt{2} - 1.4)\ \text{V} = 168.3\ \text{V}$$

For full wave rectification, $V_{d.c.} = \dfrac{2V_m}{\pi}$

So, here $\quad V_{d.c.} = \dfrac{2}{\pi}V_{o\max} = 0.636 \times 168.3\ \text{V} = 107.04\ \text{V}$

(*b*) The required peak inverse voltage (PIV) rating of each diode in full wave bridge rectifier is given by

$$\text{PIV} \geq V_{i\max}$$

i.e. PIV rating of each diode $= V_{i\ max} = 120 \times \sqrt{2}\ \text{V} = 169.7\ \text{V}$

(*c*) During conduction, the maximum current through each diode $= I_{D\ max}$.

During conduction as is shown by the previous diagram, two diodes operate in series, and the current flowing through is the same as the current through load resistor R_L.

$$\text{So} \qquad I_{D\,max} = I_R\,\text{max} = \frac{V_{o\,max}}{R_L}$$

$$\Rightarrow \qquad I_{D\,max} = \frac{168.3\text{ V}}{1\text{ K}} = 168.3\text{ mA} = 0.1683\text{ A}$$

(*d*) Required power rating of each diode

$$= V_T \times I_{D-max} = 0.7 \times 168.3\text{ mW} = 117.81\text{ mW}$$

Note. In these types of problems, first check the biasing of the diodes and replace it by their equivalents. Had been there ideal diodes instead of silicon diodes, simply put a short-circuit in place of the diodes, which are forward biased.

Problem 1.20. Determine V_o and the required PIV rating of each diode for the configuration shown in Fig. P. 1.20:

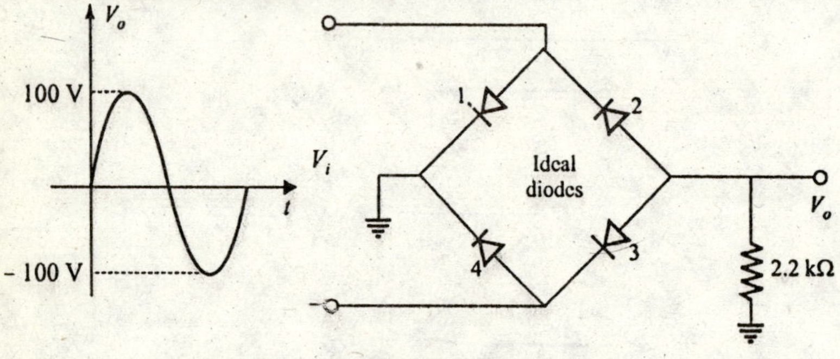

Fig. P. 1.20

Solution. For the positive half cycle of the supply voltage diodes 1 and 3 will be forward biased and diodes 2 and 4 will be reverse biased. Opposite is the case for negative half cycle. Thus, it is full wave rectification.

When the supply has +ve half cycle, the equivalent diagram is

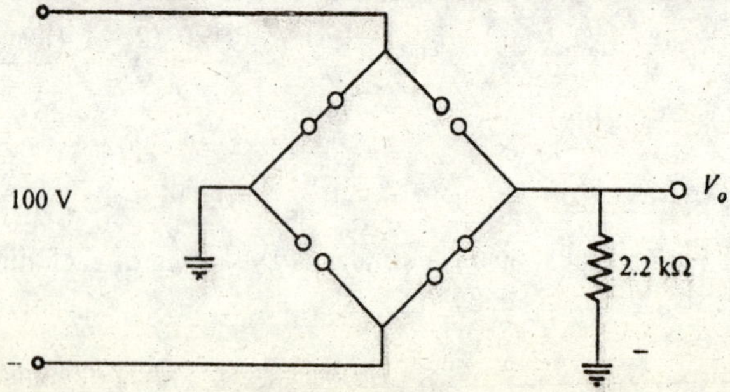

Fig. 1.20 (a)

Thus $\qquad V_o = -100$ V.

For the –ve half cycle of the supply, the equivalent circuit can be drawn and V_o is once again – 100 V.

i.e. $\qquad V_o = -100$ V

required PIV rating of each diode $\geq V_m$

i.e. $\qquad \text{PIV} = V_m = 100$ V

Problem 1.21. Determine the output waveform and determine the d.c. output level and the required PIV rating of each diode (assuming the diodes to be ideal).

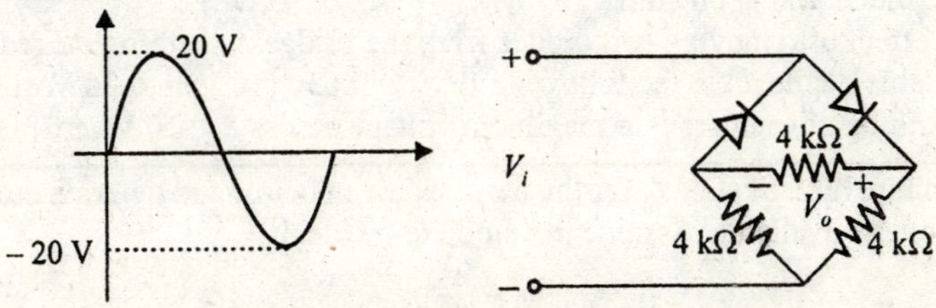

Fig. P. 1.21

Solution. For the positive half cycle of the input voltage the network will appear as

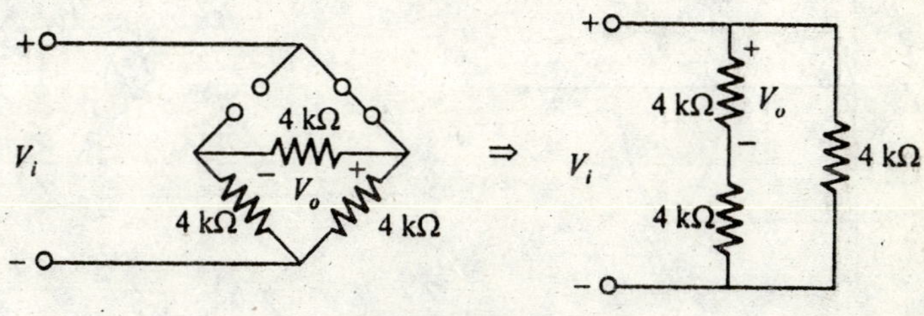

Fig. 1.21 (a)

From the equivalent diagram, it is quite evident that

$$V_o = \frac{1}{2} V_i$$

or $$V_{o\max} = \frac{1}{2} V_{i\max} = \frac{1}{2} \times (20\,\text{V}) = 10\,\text{V}$$

Thus, waveform of V_o for +ve V_i will be as

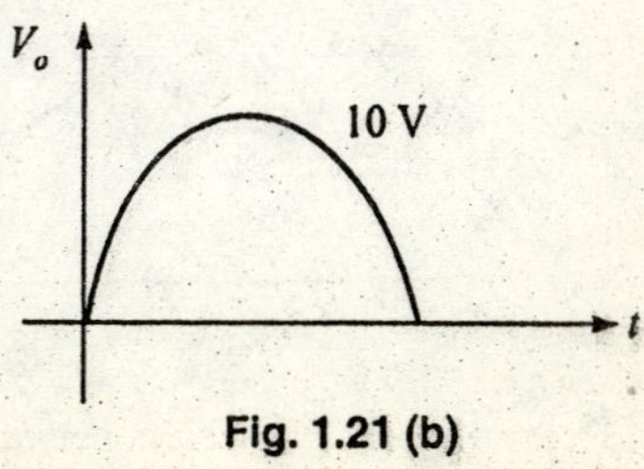

Fig. 1.21 (b)

For the negative half cycle of the input voltage the roles of the diodes will be interchanged and once again $V_{o\max} = 10$ V.

So, the resultant waveform of V_o for the complete cycle will be as below:

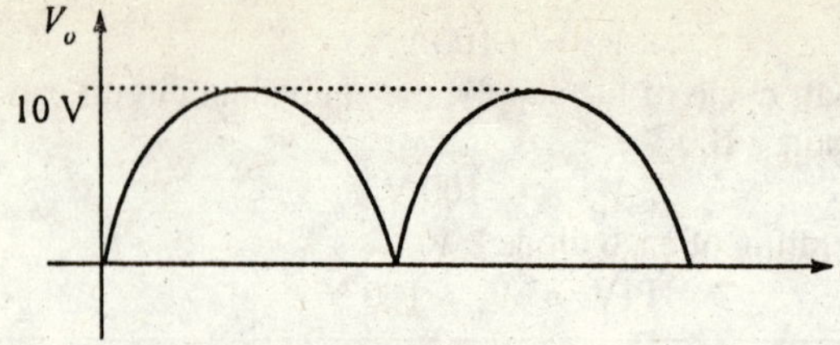

Fig. 1.21 (c)

The waveform clearly indicates that it is full wave rectification even though only one diode acts at one time.

The effect of removing two diodes from the bridge is therefore to reduce the available d.c. level to the following $V_{d.c.} = 0.636$ (10 V) = 6.36 V.

PIV rating of each diode = maximum voltage across R = 10 V.

Problem 1.22. Sketch V_o for the network given below and determine the d.c. voltage available. Assume the diode to be ideal.

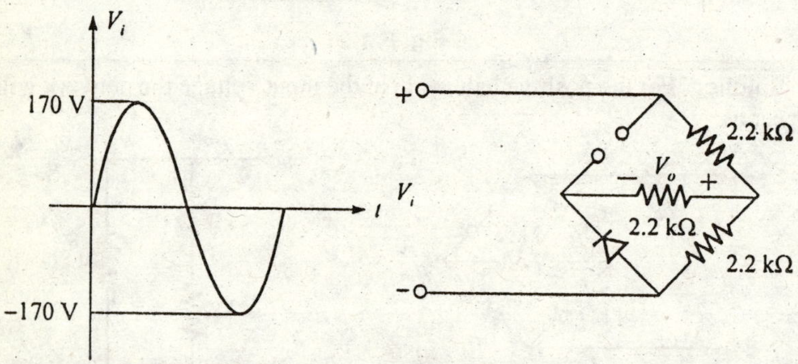

Fig. P. 1.22

Solution. For the positive region of the input voltage the network will appear as:

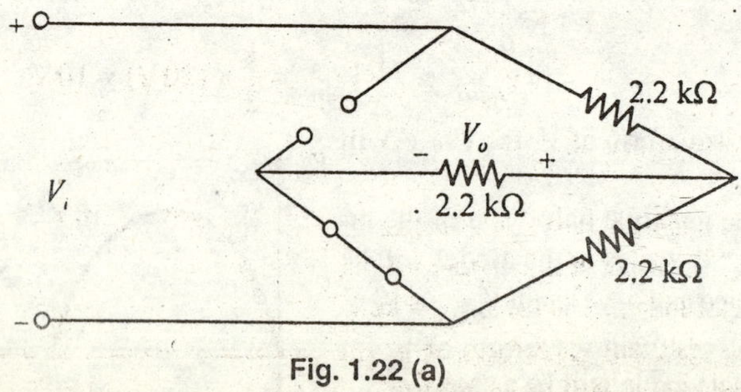

Fig. 1.22 (a)

Redrawing the equivalent network, we get the following diagram:

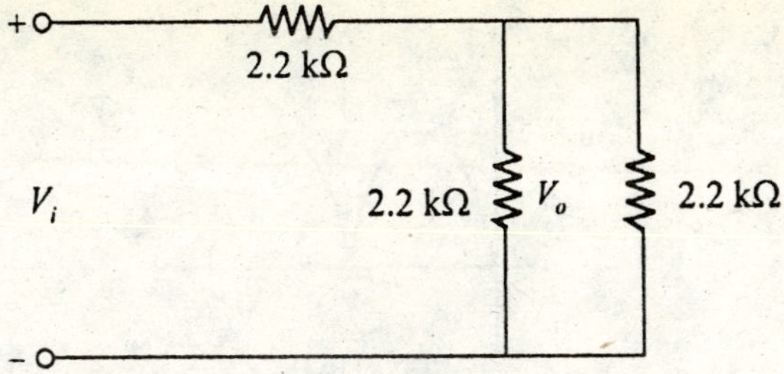

Fig. 1.22 (b)

To get V_o solve the above network as below:

$$R_{eq} = (2.2\text{ k}\Omega) + (2.2\text{ k}\Omega \parallel 2.2\text{ k}\Omega)$$
$$= 2.2\text{ k}\Omega + 1.1\text{ k}\Omega = 3.3\text{ k}\Omega.$$

Thus, total I_{max} will be $I_{max} = \frac{V_{max}}{R_{eq}} = \frac{170\text{ V}}{3.3\text{ k}\Omega} = 51.51\text{ mA}$

From KVL, we get $V_o = V_i - (2.2\text{ K})\, I$

$\Rightarrow$ $V_{omax} = 170 - (2.2\text{ K}) \times (51.59\text{ m}) = 56.67\text{ V}$

and the V_o will have waveform for first half cycle as

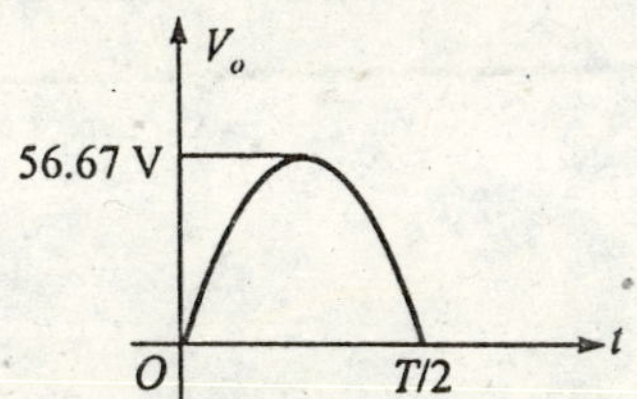

Fig. 1.22 (c)

For the negative region of the input voltage, the role of the diodes are reversed and applying KVL, we get once again

$$V_o = V_i - (2.2\text{ K})\, I$$

$\Rightarrow$ $V_{o\,max} = 56.67\text{ V}.$

having waveform as

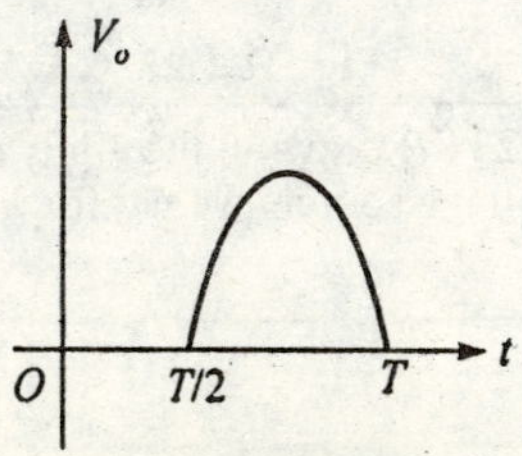

Fig. 1.22 (d)

Thus, for the complete cycle of the input voltage the waveform of the output voltage V_o will be as

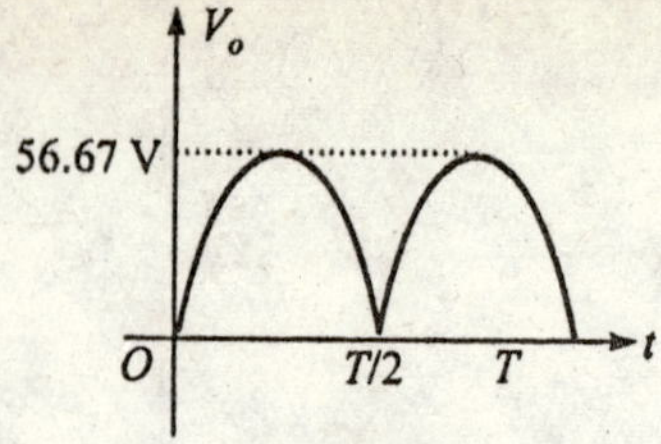

Fig. 1.22 (e)

and peak value of V_o = 56.67 V

From the sketch of V_o, we confirm that it is full wave rectification, where

$$V_{d.c.} = \frac{2V_m}{\pi} = 0.636V_m$$

Thus, required $V_{d.c.}$ = 0.636 × 56.67 V = 36.04 V

Problem 1.23. Determine V_o for each network shown in Fig. P. 1.23 for the input shown:

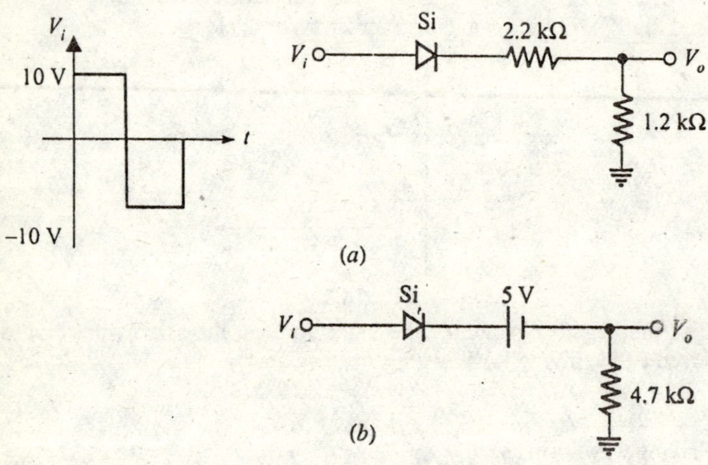

Fig. P. 1.23

Solution. Figure P. 1.23 (*a*) drawing the equivalent diagram replacing Si diode by its equivalent voltage source, we get for +ve V_i as

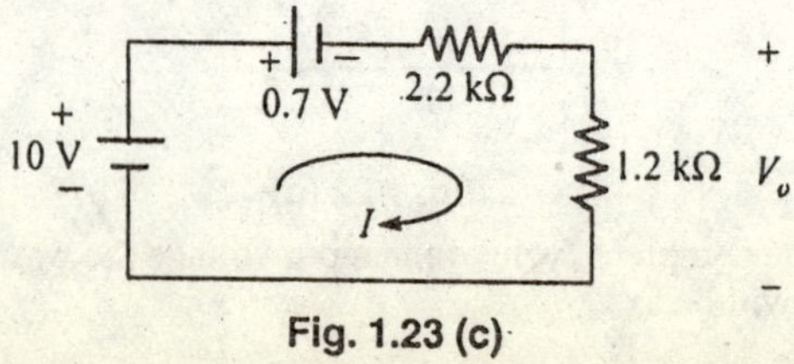

Fig. 1.23 (c)

$$R_{eq} = 2.2\ k\Omega + 1.2\ k\Omega = 3.4\ k\Omega$$

Applying KVL to the loop

$$10 - 0.7 - (3.4\ K)\ I = 0$$

$$\Rightarrow \quad (3.4\ K)\ I = 9.3$$

or
$$I = \frac{9.3}{3.4\ K} = 2.73\ mA$$

Hence
$$V_o = 10 - 0.7 - 2.73 \times 10^{-3} \times 2.2 \times 10^{3} = 3.28\ V.$$

For –ve region of the input signal, we find the open circuit due to reverse biasing of the diode, and hence

$$i_R = 0 \Rightarrow V_o = 0$$

∴ net V_o for complete input cycle will be a positive pulse of value

$$V_o = 3.28\ V$$

Figure P. 1.23 (*b*) when the input voltage V_i is in the positive region, i.e. $V_i = +10$ V, the diode will be forward biased and must be replaced by its equivalent voltage source of magnitude equal to its threshold voltage.

Drawing the equivalent network, we get

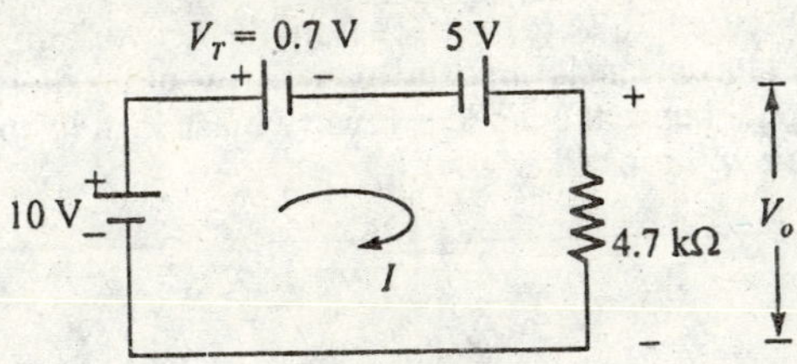

Fig. 1.23 (d)

Applying KVL to the loop, we get the equation

$$10\ V - 0.7\ V + 5\ V - V_o = 0$$

Which yields
$$V_o = 14.3\ V$$

For the negative region of the input pulse, the diode is reverse biased. Thus, there results an open circuit, and hence

$$I_R = 0$$

$$\therefore \quad V_o = I_R \times R = (0)\ (4.7\ K) = 0$$

So, for complete cycle of the input voltage, V_o will have a positive pulse of value = 14.3 V.

Problem 1.24. Determine V_o for each network given below for the input shown in Fig. P. 1.24.

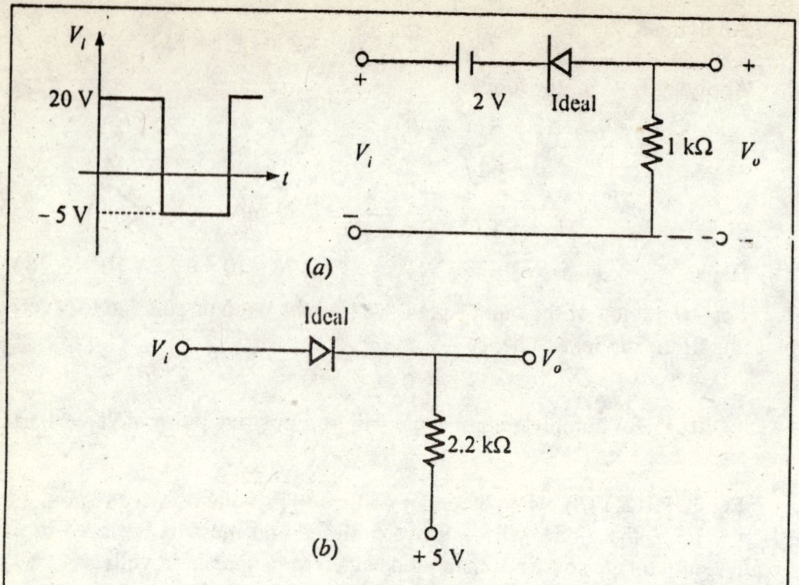

Fig. P. 1.24

Solution. Figure P. 1.24 (a) we see from the network given that the direction of the resultant current is opposite to the arrow of the diode for the positive pulse of the input, hence the diode is reverse biased, and hence open circuit.

The network will look like

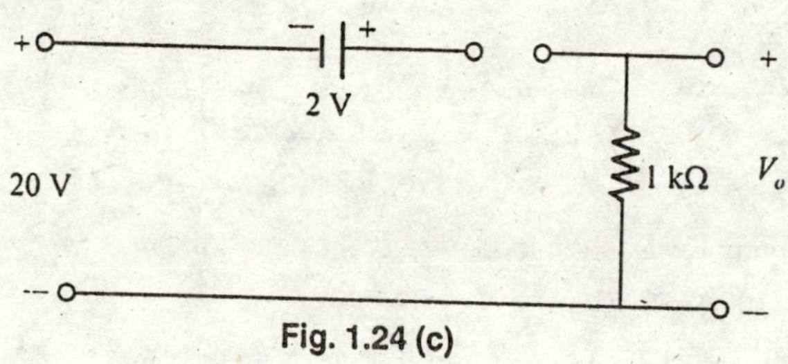

Fig. 1.24 (c)

Thus
$$V_o = I_R \times (1\text{ K}) = 0 \times (1\text{ K}) = 0\text{ V}$$

For negative region of the input supply, the diode is forward biased and the network will look like

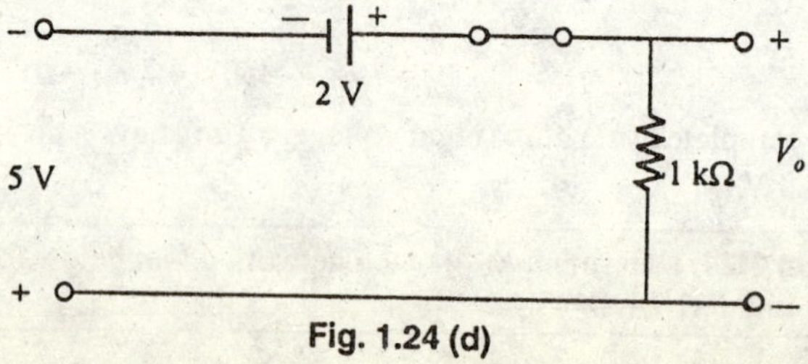

Fig. 1.24 (d)

Applying KVL to the loop, we get

$$-5+2-V_o = 0 \Rightarrow V_o = -3\text{ V}$$

So, for complete one cycle of the input, V_o will have a negative pulse of -3 V.

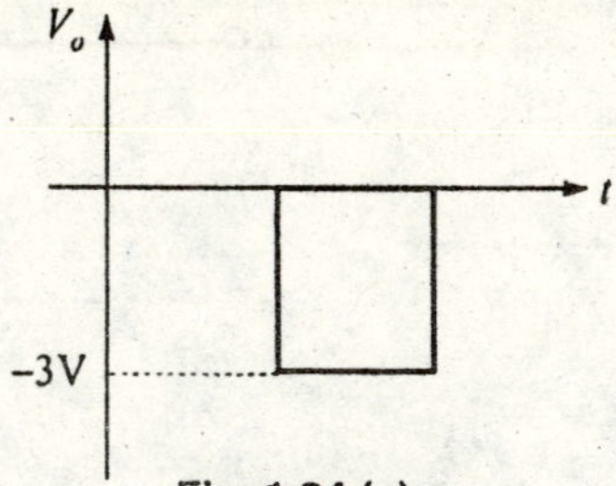

Fig. 1.24 (e)

Figure P. 1.24 (*b*) for the positive pulse of the input V_i, the diode is forward biased, and hence it is short-circuit as the diode is ideal.

Drawing the network as

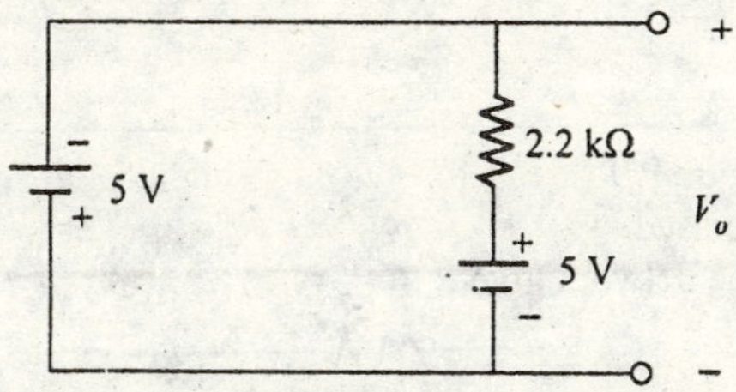

Fig. 1.24 (f)

Here, the direction of resultant current matches that of the arrow of the diode.

Hence $\quad 20 - V_o = 0$

$\Rightarrow \quad V_o = 20$ V

For the negative pulse of the input V_i, the network will be as

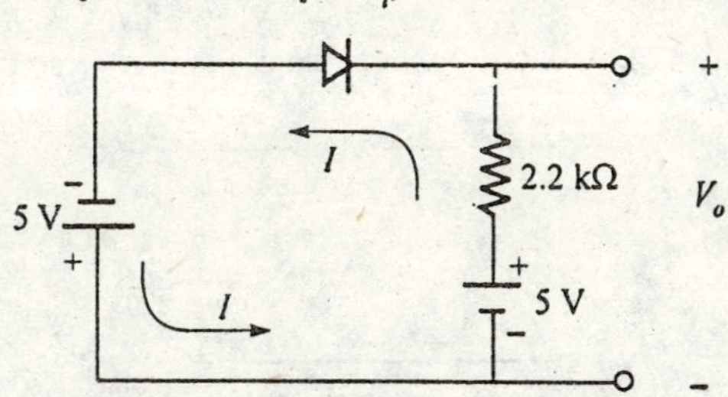

Fig. 1.24 (g)

Since, both the voltage sources are additive and the direction of the resultant current is opposite to that of arrow of the diode, and hence open circuit.

$$\Rightarrow \quad V_o = -I_R \times (2.2\text{ K}) + 5\text{ V} = 0 + 5\text{ V} = 5\text{ V}$$

Problem 1.25. Determine V_o for each network of figures given below for the input shown in Fig. P. 1.25.

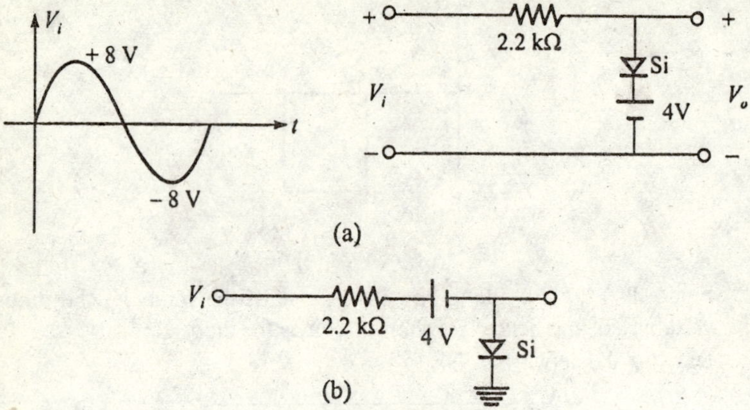

Fig. P. 1.25

Solution. Figure P. 1.25 (*a*) for +ve pulse of the input voltage, the Si diode will be forward biased.

Redrawing the network given

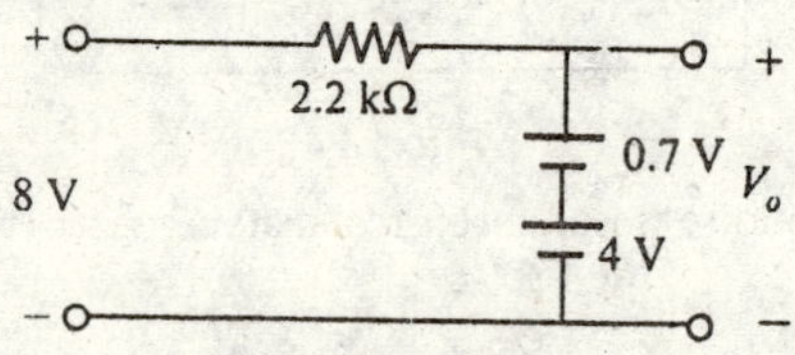

Fig. 1.25 (c)

Hence $\quad V_o = (0.7 + 4)\text{ V} = 4.7\text{ V}$

For −ve pulse of the input voltage, the diode is reverse biased, and hence open.

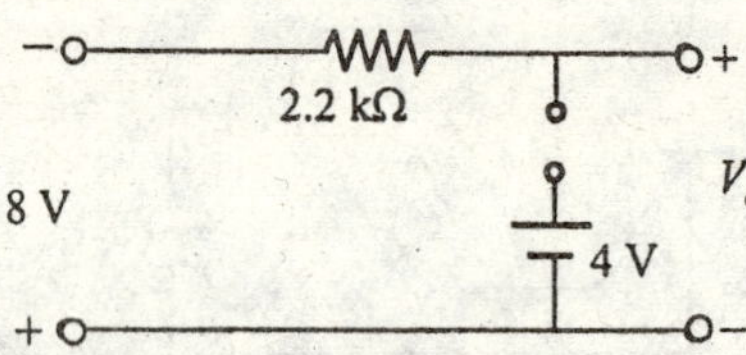

Fig. 1.25 (d)

$\Rightarrow \quad V_o = 0\text{ V}$

Thus, the resultant waveform of V_o will be a positive pulse of magnitude 4.7 V and is clipped at 4.7 V.

Figure P. 1.25 (*b*) for +ve pulse of the input signal V_i the network will look like

Which implies $V_o = 0.7$ V

For negative pulse of the input V_i, the diode is reverse biased, and hence open.

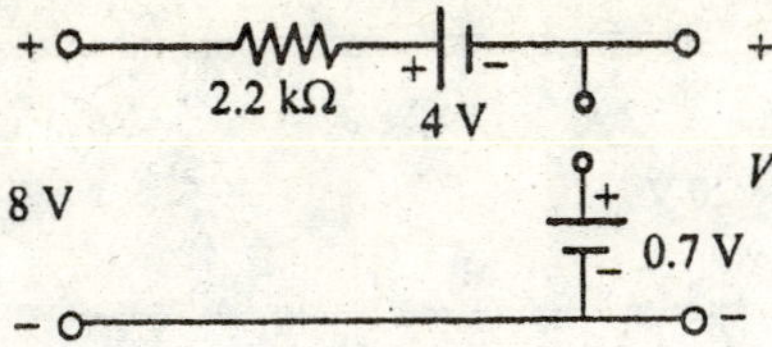

Fig. 1.25 (e)

The resulting network will be as

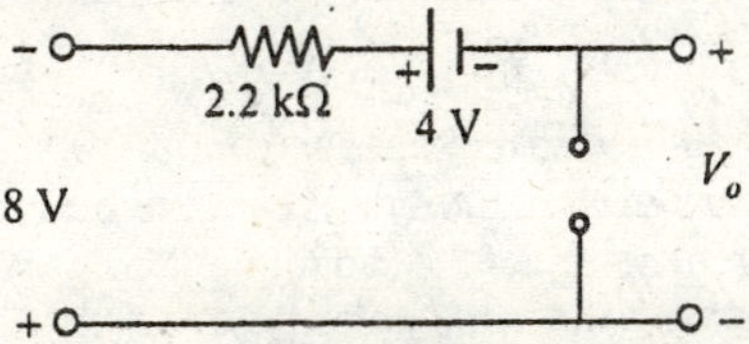

Fig. 1.25 (f)

Since, there is not any closed circuit, hence the current in any branch is zero.

and $V_o = (-8-4)$ V $= -12$ V

i.e. for complete cycle of the input signal, the V_o waveform will have a +ve pulse of magnitude 0.7 V (peak) and a –ve pulse of – 12 V peak.

Problem 1.26. Sketch V_o for the network shown below for the given input.

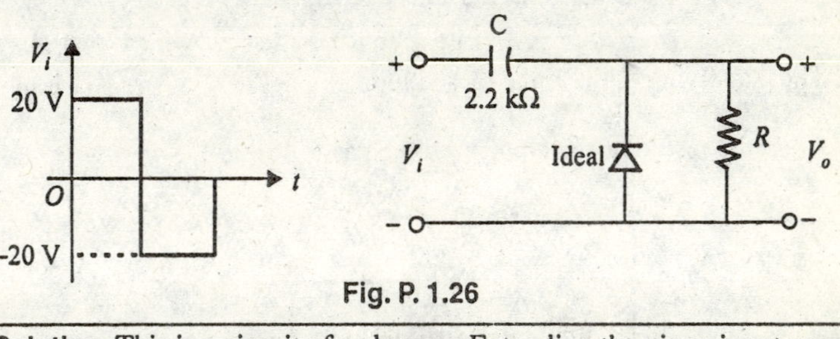

Fig. P. 1.26

Solution. This is a circuit of a clamper. Extending the given input, we get

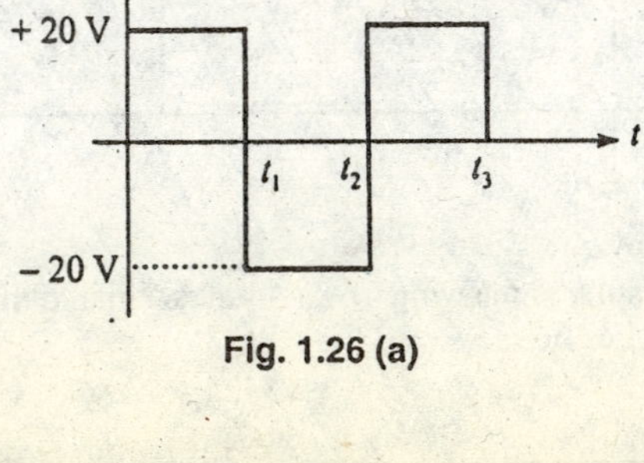

Fig. 1.26 (a)

First, we consider the region of the input, which forward bias the diode which in our case is –ve pulse.

So, for $t_1 \to t_2$, diode is forward biased and the network appears as

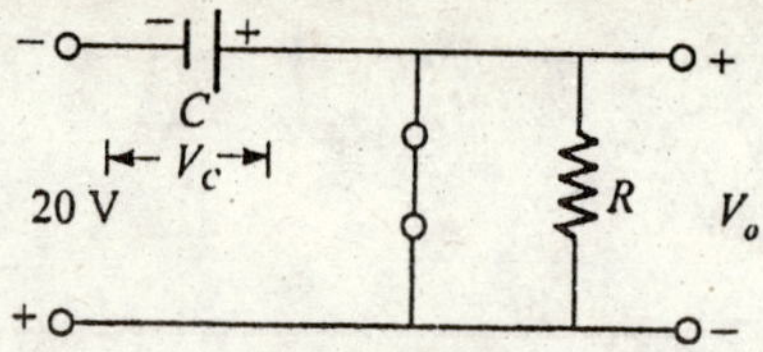

Fig. 1.26 (b)

For this region, R is shorted out by the diode, hence $V_o = 0$.

Applying KVL to the loop, we get

$$-20 + V_C = 0$$

$$\Rightarrow \quad V_C = 20 \text{ V}$$

i.e. capacitor is charged instantaneously up to 20 V.

Now, for region $t_2 \to t_3$, the diode is reverse biased, and hence is open. But, due to the inherent nature of the capacitor, it holds its charge or voltage up to considerable time with the same polarity.

Hence, the network will appear as

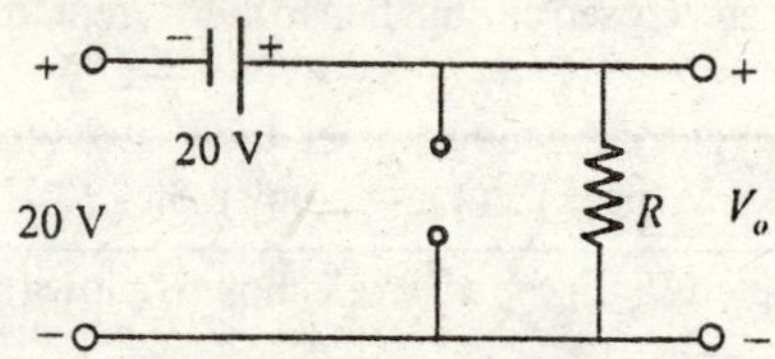

Fig. 1.26 (c)

Thus, applying KVL to the closed loop, we get

$$20 \text{ V} + 20 \text{ V} - V_o = 0 \quad \Rightarrow \quad V_o = 40 \text{ V}$$

And, the resultant sketch of V_o will be as follows:

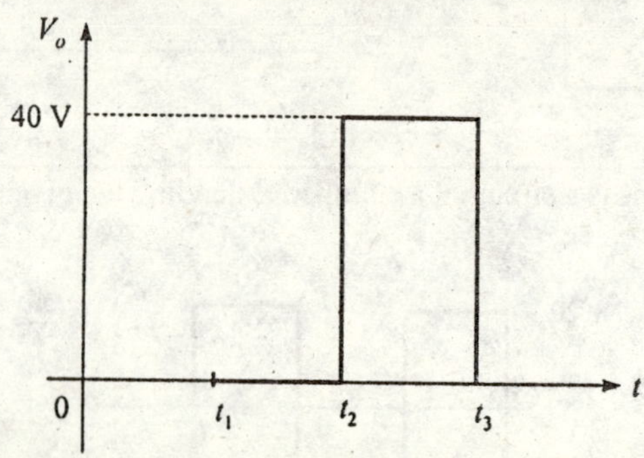

Fig. 1.26 (d)

In clamper circuits, total swing of the input is equal to the total swing of the output which matches here

Because, total swing of V_i = 40 V (from −20 to +20)
And, total swing of V_o = 40 V (from 0 to 40 V)
The time constant of the discharging network is given by $C = RC$.

Problem 1.27. For the network given below for the input shown in Fig. P. 1.27.

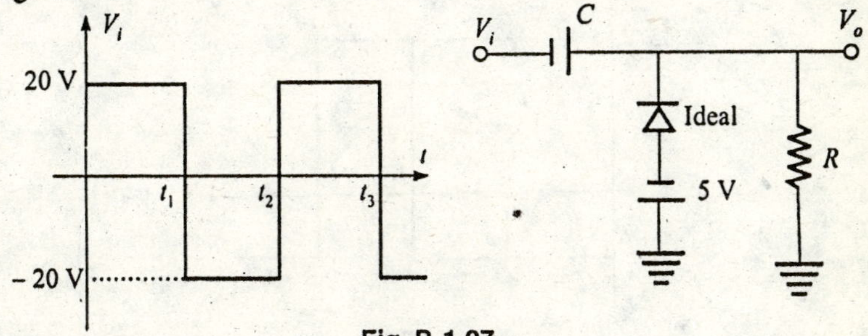

Fig. P. 1.27

Solution. Since, the diode is forward biased for the negative pulse of the input.

So consider the portion of the input waveform for $t_1 \to t_2$ first.

The network will appear as

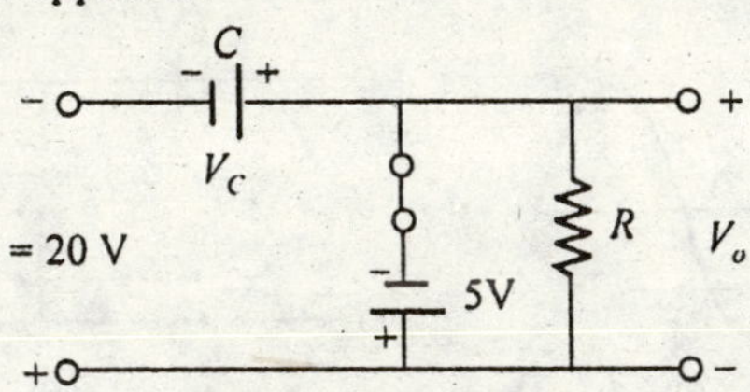

Fig. 1.27 (a)

Since, resistor R and 5 V battery are in parallel and for given polarity of V_o.

$$V_o = -5 \text{ V} \quad \text{for} \quad t_1 \to t_2$$

Also, Applying KVL to the first loop, we get

$$20 \text{ V} - 5 \text{ V} - V_C = 0$$

Which yields $\qquad V_C = 15 \text{ V}$

For the region $t_2 \to t_3$, the diode is reverse biased and hence is open, but the capacitor holds the voltage 15 V till the considerable time.

For $t_2 \to t_3$ region of the input signal, the network will appear as

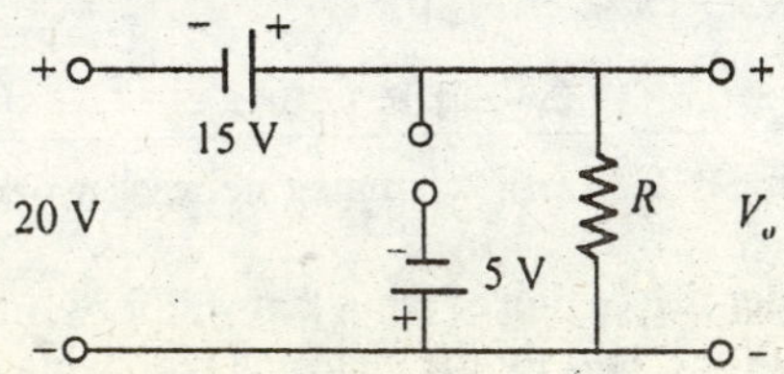

Fig. 1.27 (b)

Due to open circuit, 5 V battery will not have any effect on V_o, and hence applying KVL to the loop including R, we find

$$20\text{ V} + 15\text{ V} - V_o = 0 \quad \Rightarrow \quad V_o = 35\text{ V}$$

The waveform of V_o will be as shown below:

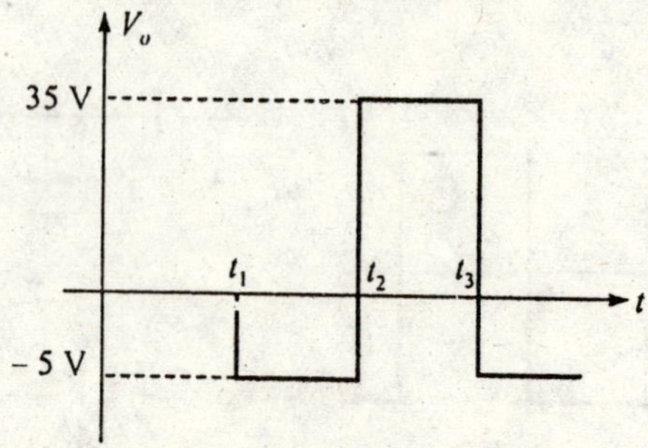

Fig. 1.27 (c)

Here also, it can be checked that the total swing of the input signal is equal to the total swing of the output V_o.

Problem 1.28. Sketch V_o for each network shown in Fig. P. 1.28:

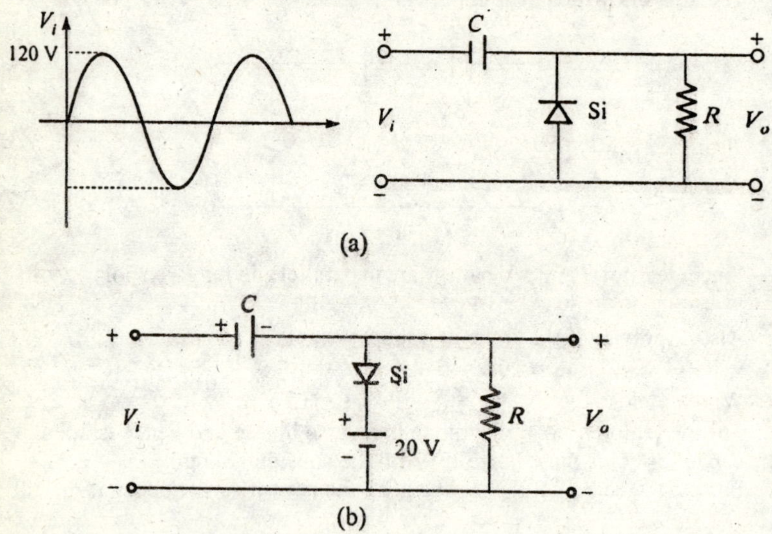

Fig. P. 1.28

Solution. Figure P. 1.28 (a), clamping network works equally well for sinusoidal inputs.

First, we consider that region of input that will forward bias the diode.

So, considering for t_1 to t_2 for the input and redrawing the network as

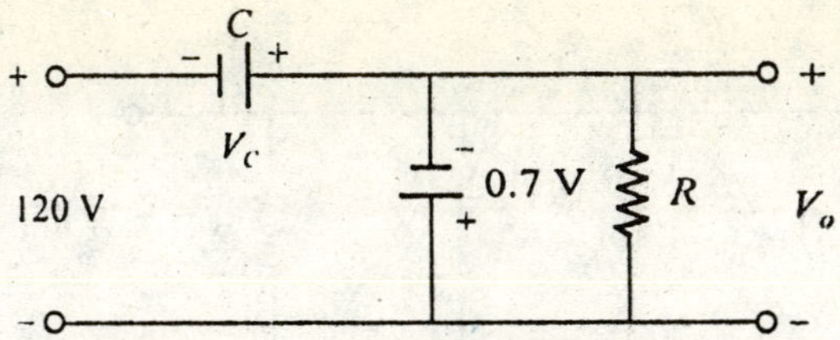

Fig. 1.28 (c)

Thus $V_o = -0.7$ V

Applying KVL to the first loop, we get

$$120 \text{ V} - 0.7 \text{ V} - V_C = 0$$

Which yields $V_C = 119.3$ V

Now, for t_2 to t_3 region of the input signal, the network will appear as

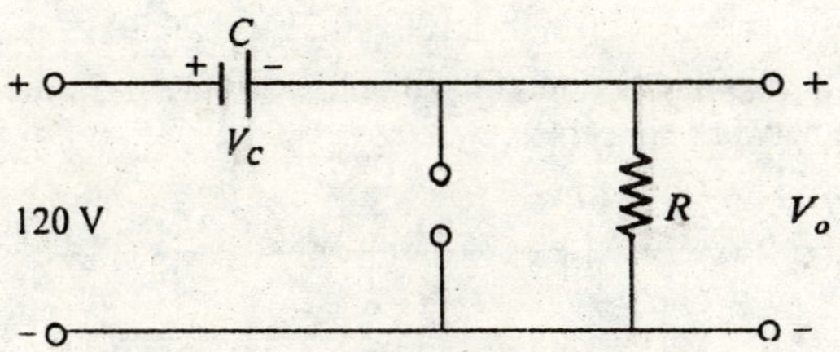

Fig. 1.28 (d)

From KVL, we get

$$120 \text{ V} + V_C - V_o = 0$$

$$\Rightarrow \quad V_o = (120 + 119.3) \text{ V} = 239.3 \text{ V}$$

Thus, the resultant sketch of V_o will be as

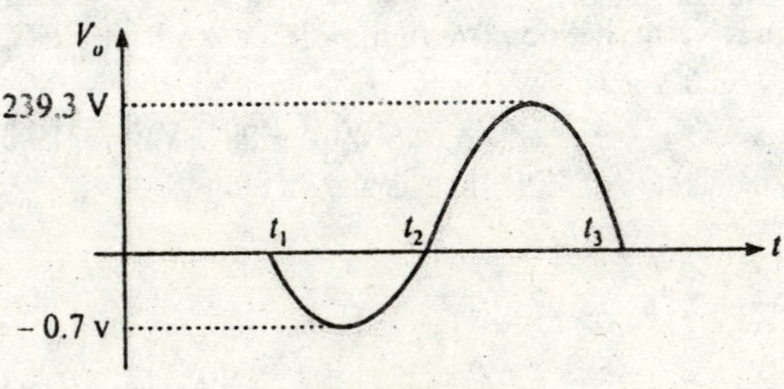

Fig. 1.28 (e)

Figure P. 1.28 (*b*), for the region $0 \rightarrow t_1$, the diode is forward biased, and hence we start with that region.

In this region, the network will appear as

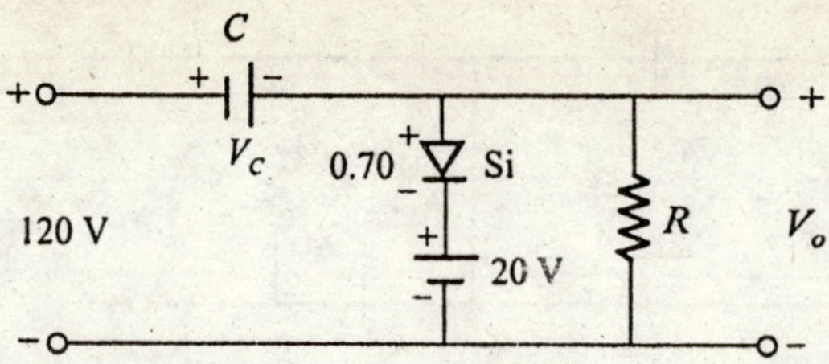

Fig. 1.28 (f)

Since, the branch containing the diode, and the branch containing R are parallel to each other.

Hence, voltage across them will be same.

$$\therefore \qquad V_o = +0.7\text{ V} + 20\text{ V} = 20.7\text{ V}$$

Applying KVL to the loop including capacitor

$$120\text{ V} - V_C - 0.7\text{ V} - 20\text{ V} = 0$$

$$\Rightarrow \qquad V_C = (100 - 0.7)\text{ V} = 99.3\text{ V}.$$

Now, for the negative region of the input, i.e. for $t_1 \rightarrow t_2$, diode is reverse biased and the network appears as

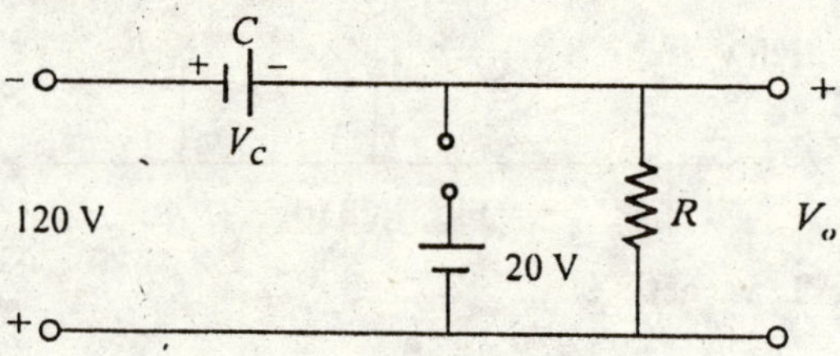

Fig. 1.28 (g)

Due to open circuit, 20 V battery has no effect on V_o

Also, capacitor maintains its voltage for the time considence.

Now, from KVL to the loop, we find

$$-120\text{ V} - V_C - V_o = 0$$

$$\Rightarrow \qquad V_o = -(120\text{ V} + 99.3\text{ V}) = -219.3\text{ V}.$$

Thus, the net sketch of V_o will be as shown below:

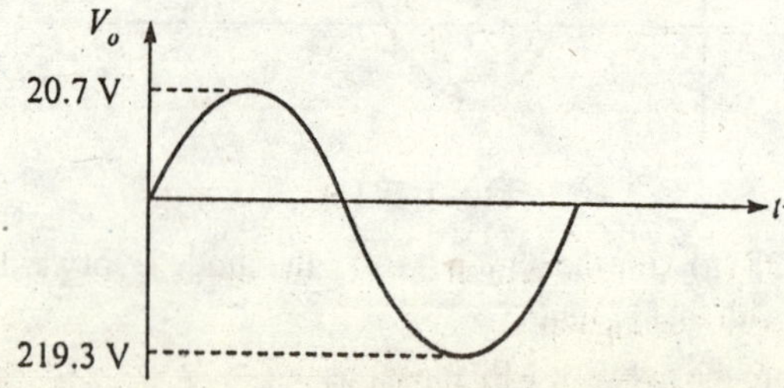

Fig. 1.28 (h)

Since, the total swing of V_o = $20.7 + |-219.3| = 240$ V.

Which matches that of input so, our analysis is correct.

Problem 1.29. For the network given below:

(*a*) Calculate 5.

(*b*) Compare 5 to half the period of the applied signal.

(*c*) Sketch V_o.

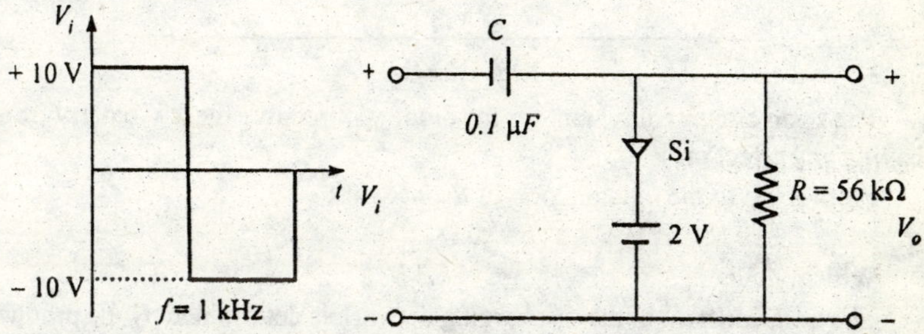

Fig. P. 1.29

Solution. Since, the diode is forward biased for the positive half cycle of the applied signal, so we shall start with that region.

The network will appear as

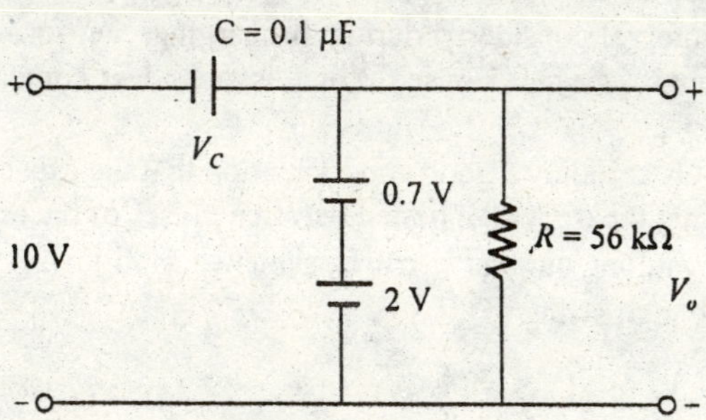

Fig. 1.29 (a)

Since, *R* is parallel to the branch containing diode

Hence $V_o = (0.7 - 2)\text{ V} = -1.3\text{ V}$

Also, from KVL to the first loop

$$10\text{ V} - V_C - 0.7\text{ V} + 2\text{ V} = 0$$

$$\Rightarrow \quad V_C = 11.3\text{ V}$$

For the negative half cycle of the input signal the network will appear as

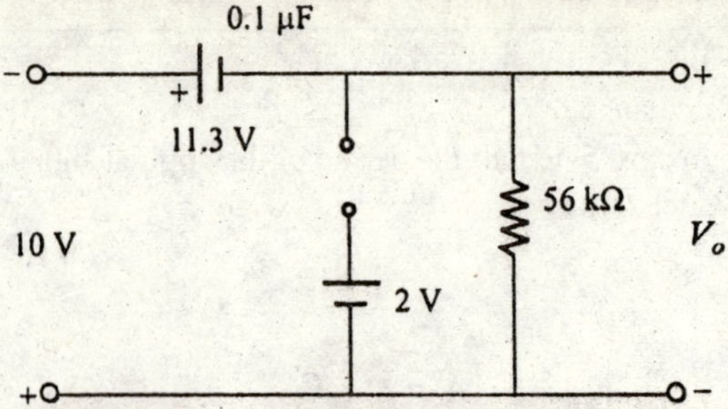

Fig. 1.29 (b)

The open circuit equivalent for the diode will remove the 2 V battery from having any effect on V_o.

From KVL to the outside loop of the network

$$-10\text{ V} - 11.3\text{ V} - V_o = 0$$

Which yields $V_o = -21.3$ V

The time constant of the discharging network is determined by the product RC and has the magnitude

$$C = RC = (56\text{ k}\Omega)(0.1\ \mu\text{F}) = 5.6 \text{ m sec}$$

(*a*) The total discharge time is, therefore

$$5C = 5(5.6\text{ m sec}) = 28\text{ m sec}$$

(*b*) Given frequency $f = 1000$ Hz resulting in a period of 1 m sec and a half period, i.e. an interval of 0.5 m sec between levels.

For the interval considered during discharging, i.e. for –ve half cycle of the input signal, time = 0.5 m sec, which is much less compared to total discharge time 5 *C*.

Hence, it is certainly a good approximation that the capacitor will hold its voltage, during the discharge period between pulses of the input signal.

(*c*) The resulting outputs V_o can be sketched as follow:

Here, the total swing of

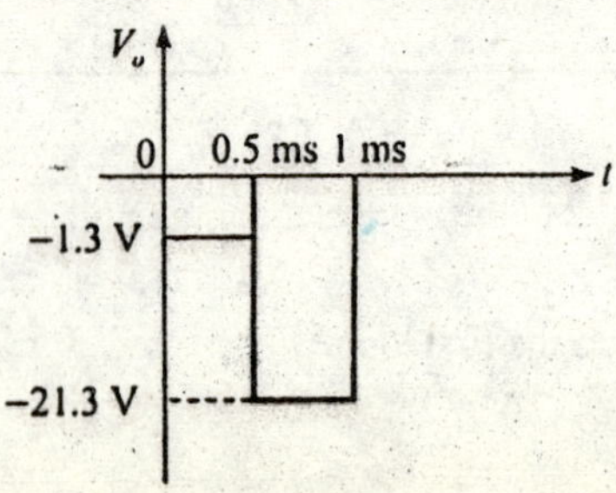

Fig. 1.29 (c)

$$V_o = |-21.3| - |-1.3| = 20 \text{ V}$$

which matches that of input.

Problem 1.30. Design the network of the Fig. P. 1.30, to maintain V_L at 12 V for a load variation (I_L) from 0 to 200 mA. That is determine R_S and V_Z. (*b*) Determine $P_{Z\max}$ for the zener diode of part (*a*).

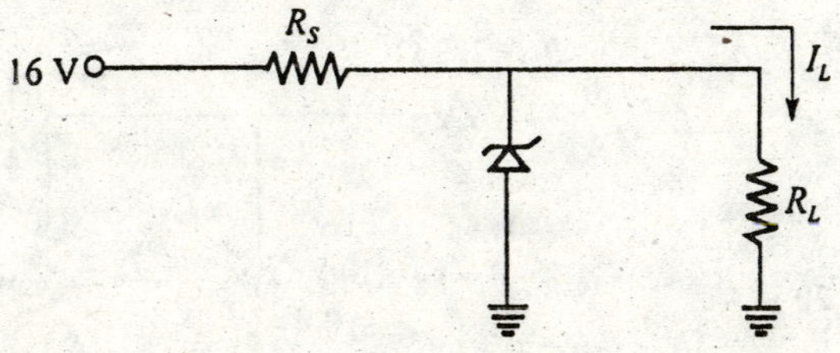

Fig. P. 1.30

Solution. Given, load voltage V_L = 12 V

Redrawing the network as below:

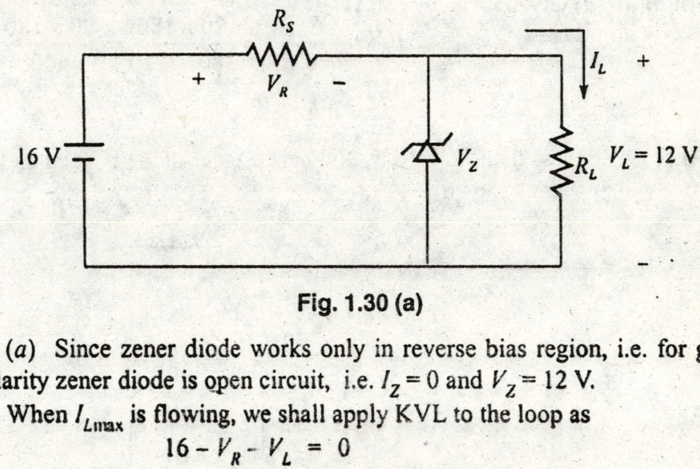

Fig. 1.30 (a)

(*a*) Since zener diode works only in reverse bias region, i.e. for given polarity zener diode is open circuit, i.e. $I_Z = 0$ and $V_Z = 12$ V.

When $I_{L\max}$ is flowing, we shall apply KVL to the loop as

$$16 - V_R - V_L = 0$$

$$\Rightarrow \quad V_R = 16 - 12 = 4 \text{ V}$$

or, $$I_{L\max} \times R_S = 4 \text{ V}$$

$$\Rightarrow \quad R_S = \frac{4 \text{ V}}{200 \text{ mA}} = \frac{4}{0.2}\Omega = 20\Omega$$

(*b*) $$P_{Z\max} = V_Z \times I_{Z\max}$$

when zener diode is "ON", $I_{Z\max}$ flows through it which is 200 mA here.

So $$P_{Z\max} = 12 \times 200 \times 10^{-3}$$

$$= 12 \times 0.2 \text{ W} = 2.4 \text{ W}$$

Problem 1.31. (*a*) Determine V_L, I_L, I_Z and I_R for the network shown below, if $R_L = 180\ \Omega$.

(*b*) Repeat past (*a*) if $R_L = 470\ \Omega$.

(*c*) Determine the value of R_L that will establish maximum power conditions for the zener diode.

(*d*) Determine the minimum value of R_L to ensure that the zener diode is in the "ON" state.

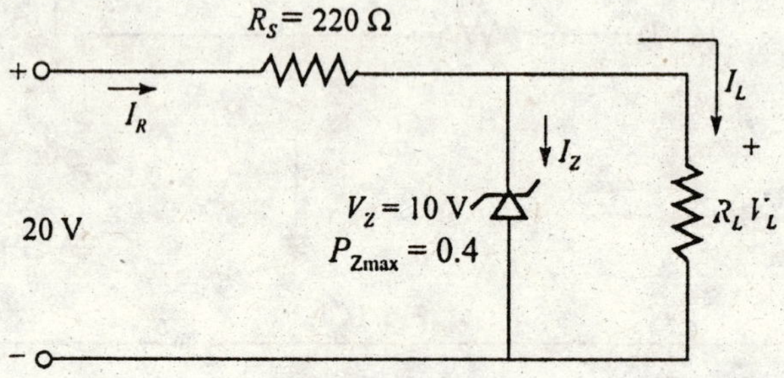

Fig. P. 1.31

Solution. (*a*) Given $R_L = 180\ \Omega$

$$V_L = V_i \times \frac{R_L}{R_S + R_L} = \frac{20 \times 180}{180 + 220} = \frac{20 \times 180}{400}$$

$$= \frac{180}{20} = 9\text{ V}$$

Since, 9 V is less than $V_Z = 10$ V, and hence zener diode is open and the network will appear as shown

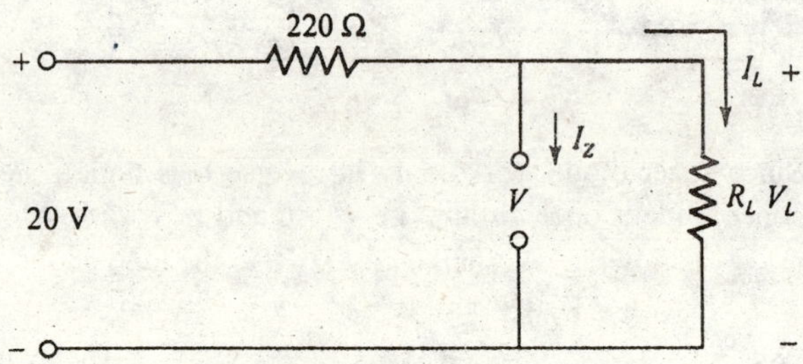

Fig. 1.31 (a)

i.e. $V = V_L = 9\text{ V}$

$$I_L = \frac{V_L}{R_L} = \frac{9\text{ V}}{180\ \Omega} = 50\text{ mA}$$

$$I_Z = 0\text{ mA}$$

$$V_R = V_i - V = 20 - 9 = 11\text{ V}$$

$$\therefore \qquad I_R = \frac{V_R}{R_S} = \frac{11\text{V}}{220\Omega} = 50\text{mA}$$

(*b*) Now, given $\quad R_L = 470\ \Omega$

$$V = V_L = V_i \times \frac{R_L}{R_L + R_S} = 20 \times \frac{470}{470 + 220} = 13.6\text{ V}$$

Since, 13.6 V is greater than $V_Z = 10$ V, and hence the zener diode will get "ON" and the network will appear as

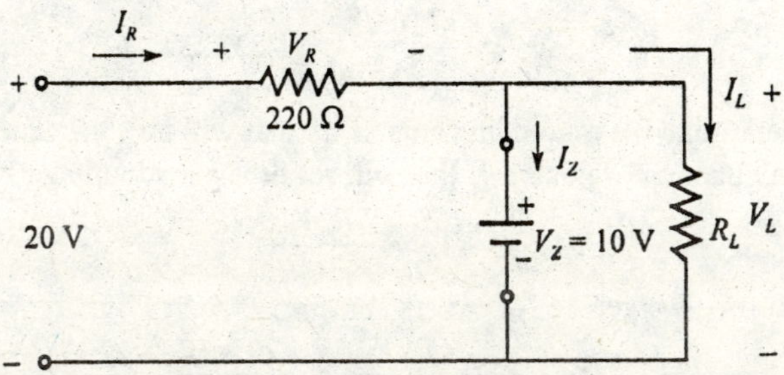

Fig. 1.31 (b)

$\therefore\ V = V_L = 10$ V as at $V_Z = 10$ V, breakdown occurs according to the graph shown

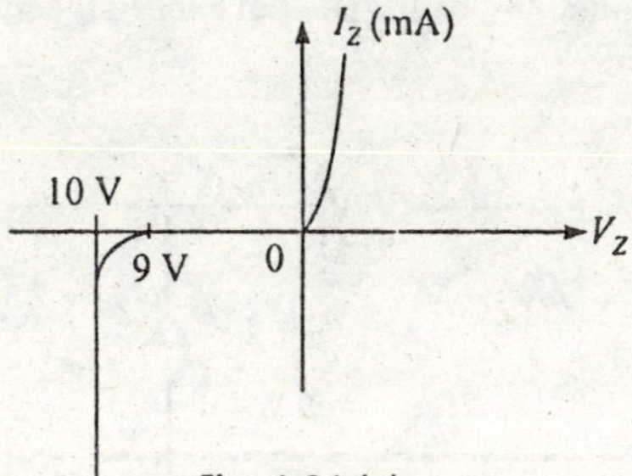

Fig. 1.31 (c)

Now $\qquad I_L = \frac{V_L}{R_L} = \frac{10\text{V}}{470\Omega} = 21.28$ mA

Also $\qquad V_R = V_i - V_L$

$\qquad\qquad = 20\text{ V} - 10\text{ V} = 10\text{ V}$

Hence $\qquad I_R = \frac{10\text{V}}{2.20\Omega} = 45.45$ mA

Since $\qquad I_Z = I_R - I_L$

$\qquad\qquad = (45.45 - 21.28)\text{ mA} = 24.17\text{ mA}$

(c) Given maximum power for the zener diode

$$P_{Zmax} = 0.4 \text{ W} = 400 \text{ mW}$$

Also $$V_Z = 10 \text{ V}$$

$$\therefore \quad I_{Zmax} = \frac{P_{Z\,max}}{V_Z} = \frac{400 \text{ mW}}{10 \text{ V}} = 40 \text{ mA}$$

Now $$I_L = I_R - I_{Zmax} = (45.45 - 40) \text{ mA}$$

$$= 5.45 \text{ mA}$$

$$V_L = V_Z = 10 \text{ V}$$

$$\therefore \quad R_L = \frac{10 \text{ V}}{5.45 \text{ mA}} = 1.833 \text{ k}\Omega$$

(d) To determine the minimum value of R_L that will turn the zener diode on, simply calculate the value of R_L that will result is a load voltage $V_L = V_Z$.

i.e. $$V_L = V_Z = \frac{R_L V_i}{R_L + R_s}$$

Solving for R_L, we get

$$R_{L\,min} = \frac{R_S V_Z}{V_i - V_Z} = \frac{220 \times 10}{20 - 10} = \frac{2200}{10} = 220 \Omega$$

Problem 1.32. For the network shown in Fig. P. 1.32, determine the range of V_i that will maintain V_L at 8 V and not exceed the maximum power rating of the zener diode.

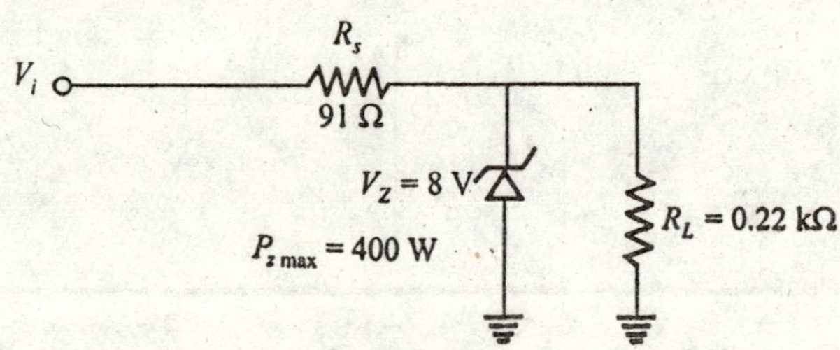

Fig. P. 1.32

Solution. For fixed values of R_L, the voltage V_i must be sufficiently large to turn the zener diode on. The minimum turn-on voltage $V_i = V_{imin}$ is determined by

$$V_L = V_Z = V_i \cdot \frac{R_L}{R_L + R_S}$$

or $$V_{imin} = \frac{(R_L + R_S)V_Z}{R_L}$$

Putting the respective values in the above equation

$$V_{i\min} = \frac{(220 + 91)}{220} \times 8\,\text{V} = 11.3\,\text{V}$$

The maximum value of V_i is limited by the maximum zener current I_{ZM}.

$$I_{ZM} = \frac{P_{Z\max}}{V_Z} = \frac{400\text{ mW}}{8\text{ V}} = 50\text{ mA}$$

Now $\quad I_R = I_Z + I_L$

or $\quad I_{R\max} = I_{ZM} + I_L = 50 + \frac{V_L}{R_L} = \left(50 + \frac{8 \times 1000}{220}\right)\text{mA}$

$= 86.36$ mA

Now $\quad V_{i\max} = V_{R\max} + V_Z$

or $\quad V_{i\max} = R_S \times I_{R\max} + V_Z$

$$= \left(\frac{91 \times 86.36}{1000} + 8\right)\text{V} = (7.86 + 8)\text{ V} = 15.86\text{ V}$$

Problem 1.33. Design a voltage regulator that will maintain an output voltage of 20 V across a 1 kW load with an input that will vary between 30 and 50 V. That is determine the proper value of R_S and the maximum current I_{Zm}.

Solution. Given, $V_{i\min} = 30$ V and $V_{i\max} = 50$ V.
Now, we know that

$$V_{i\min} = \frac{(R_L + R_S) \times V_L}{R_L} V = \frac{(1000 + R_S)}{1000} \times 20$$

$\Rightarrow \quad 30 = \frac{(1000 + R_S)}{1000} \times 20$

Which yields $\quad R_S + 1000 = \frac{30 \times 1000}{20}$

$\Rightarrow \quad R_S = 1500 - 1000 = 500\ \Omega$

Now $\quad V_{i\max} = R_S \times I_{R\max} + V_Z$

$\Rightarrow \quad 50 = 500 \times I_{R\max} + 20$

$\Rightarrow \quad I_{R\max} = \frac{30}{500}\text{A} = 60\text{ mA}$

Now $\quad I_L = \frac{V_L}{R_L} = \frac{20\text{ V}}{1000} = 20\text{ mA}$

Since $\quad I_{R\max} = +I_{ZM} + I_L$

$\Rightarrow \quad I_{ZM} = I_{R\max} - I_L = 60 - 20 = 40$ mA

Problem 1.34. The circuit shown in Fig. P. 1.34 represents a d.c. voltmeter which reads 25 V full scale. The meter resistance is 560 Ω and full scale current is 0.2 mA. The zener diode is used to prevent over loading of the sensitive meter movements without affecting meter linearity. If the diode is a 20 V zener, find R_1 and R_2 so that when $V_i > 25$ V, the zener diode conducts and overload current is shunted away from the meter.

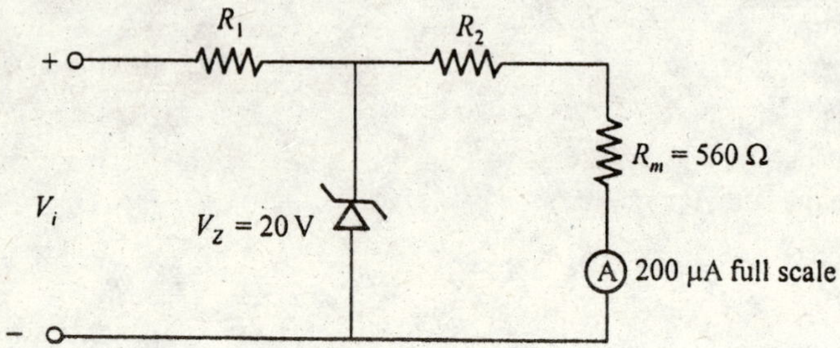

Fig. P. 1.34

Solution. $V_Z = V_o = 20$ V, $\dfrac{V_o}{R_2 + R_m} = 0.2 \times 10^{-3}$

$$\frac{20}{R_2 + 560} = 0.2 \times 10^{-3}, \; R_2 = \frac{20 - 0.112}{0.2 \times 10^{-3}} = 99.44\,\text{k}\Omega$$

$$R_1 = \frac{25 - 20}{0.2 \times 10^{-3}} = \frac{5}{0.2 \times 10^{-3}} = 25\,\text{k}\Omega$$

Therefore $R_1 = 25$ kΩ and $R_2 = 99.44$ kΩ

CHAPTER

2

BJT and Its Modelling

Transistor Construction: The transistor is a three-layer semiconductor device consisting of either two *n* and one *p*-type layers of material or two *p*-and one *n*-type layers of material. The former is called an *npn* transistor while the latter is called a *pnp* transistor. The diagrams of both the types are as follows:

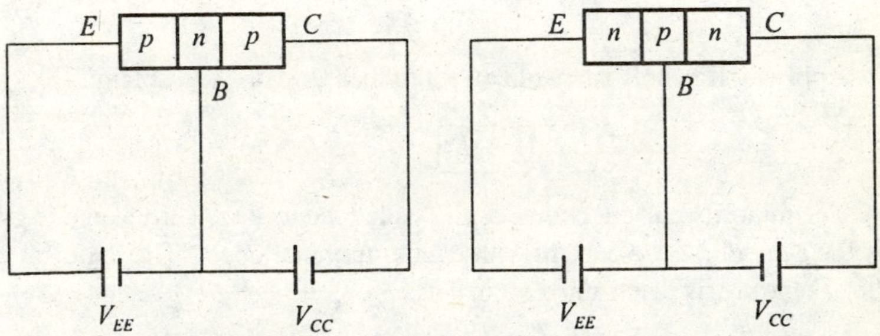

(*a*) *pnp* transistor (*b*) *npn* transistor

Fig. 2.1

The emitter layer is heavily doped, the base is lightly doped and the collector only lightly doped. For the transistors the ratio of the total width to that of the centre layer is 0.15/0.001 = 150 : 1. The doping of the sandwiched layer is also considerally less than that of the outer layer (typically, 10 : 1 or less). This lower doping level decreases the conductivity (increases the resistance) of this material by limiting the number of "free" carriers.

The BJT, which is abbreviation of bipolar junction transistor, is a three terminal device. The term bipolar reflects the fact that holes and electrons participate in the injection process into the oppositely polarized material. If only one carrier is employed (electron or hole), it is considered a unipolar device.

Transistor operation: One *pn*-junction of a transistor is forward biased while the other is reversed biased. The majority carriers flows, when the *pn*-junction is forward biased and minority carriers flows when the *pn*-junction is reverse biased.

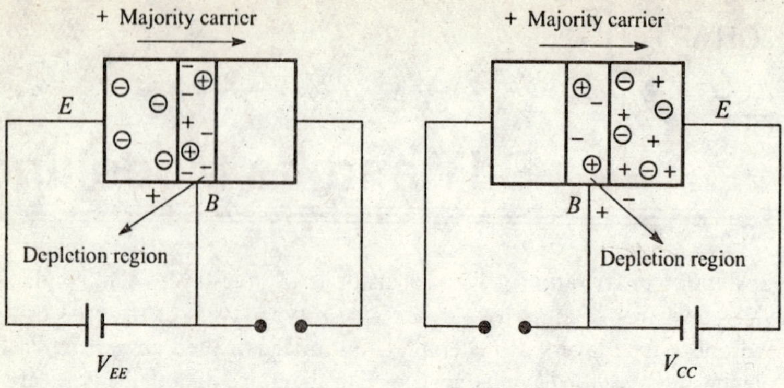

(*a*) Forward biased junction of a *pnp* transistor

(*b*) Reverse biased junction of a *pnp* transistor

Fig. 2.2

Applying Kirchoff's law of current to the transistor, we obtain:

$$\boxed{I_E = I_C + I_B}$$

The minority current component is called the leakage current and is given by the symbol I_{CO} (I_C current with emitter terminal open). The collector current, therefore, is determined in total as:

$$\boxed{I_C = I_{C\,\text{majority}} + I_{C\,\text{minority}}}$$

For general purpose transistors, I_C is measured in milliamperes, while I_{CO} is measured in microamperes or nanoamperes. I_{CO} is temperature sensitive and can severely affect the stability of a system at high temperature if not considered properly.

Operation of BJT (Common base BJT): The different operations of BJT are as follows:

(*i*) ***Linear-region operation***
Base-emitter junction forward biased.
Base-collector junction reverse biased.

(*ii*) ***Cut-off-region operation***
Base-emitter junction reverse biased.
Base-collector junction reverse biased.

(*iii*) ***Saturation-region operation***
Base-emitter junction forward biased.
Base-collector junction forward biased.

Types of configurations: There are basically three types of configurations:

(*i*) Common base configuration.

(*ii*) Common emitter configuration.

(*iii*) Common collector configuration.

(*i*) ***Common base configuration:*** The symbol of common base BJT of *npn* type is shown below:

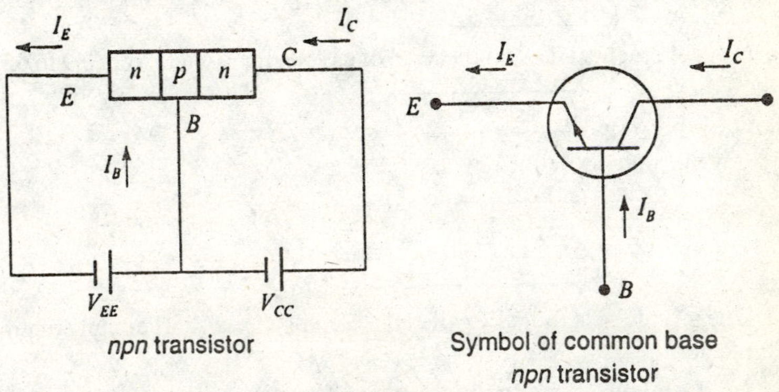

Fig. 2.3

The arrow in the graphic symbol defines the direction of emitter current (conventional flow) through the device.

Input characteristics of the common base silicon transistor is shown below:

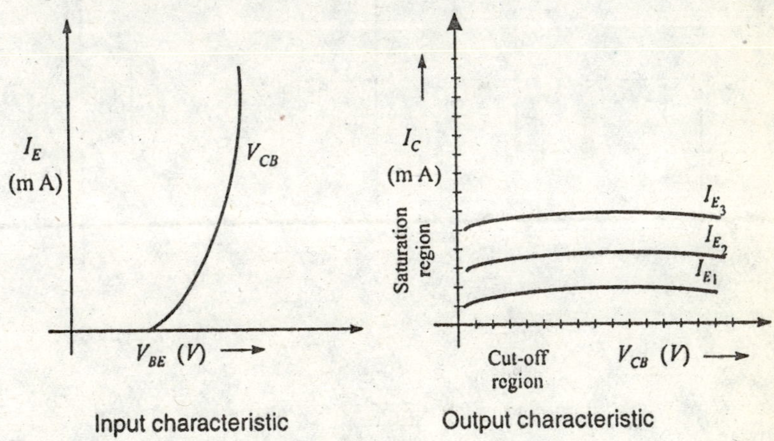

Fig. 2.4

The output characteristic is drawn between the output voltage V_{CB} and the output current I_C at constant I_E. On an average basis, the base-to-emitter voltage (threshold voltage) of an operating transistor is 0.7 V.

i.e. $\boxed{V_{BE} = 0.7\text{ V}}$ ⇒ for silicon transistor

Alpha (α): In the d.c. mode the levels of I_C and I_E due to the majority carriers are related by a quantity called alpha.

i.e. $\boxed{\alpha_{d.c.} = \frac{I_C}{I_E}}$, since $I_C \cong I_E$. The value of α is nearly equal to unity for practical devices the level of a typically extends from 0.9 to 0.998.

$$\boxed{I_C = \alpha I_E + I_{CBO}}$$

An a.c. alpha is defined by

$$\boxed{\alpha_{a.c.} = \frac{\Delta I_C}{\Delta I_E}\bigg|_{V_{CE} = \text{Constt}}}$$

The a.c. alpha is formally called the common base short circuit amplification factor.

(*ii*) ***Common emitter configuration:*** The symbol and rotation are shown below:

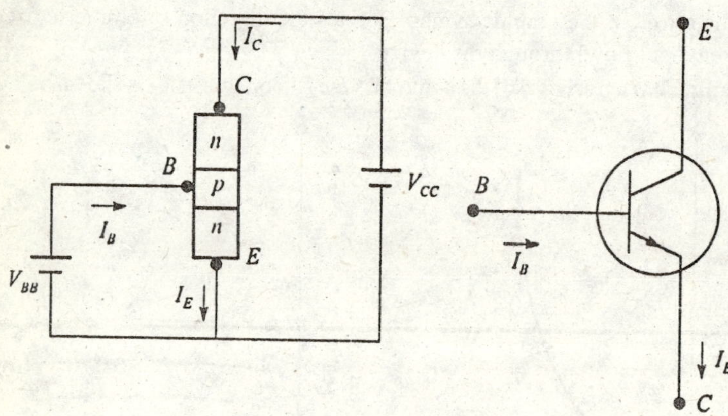

Fig. 2.5

The emitter current is $\boxed{I_E = I_C + I_B}$

and The collector current is $\boxed{I_C = \alpha I_E}$

The collector characteristic and the base characteristic of the *CE* configuration is shown below:

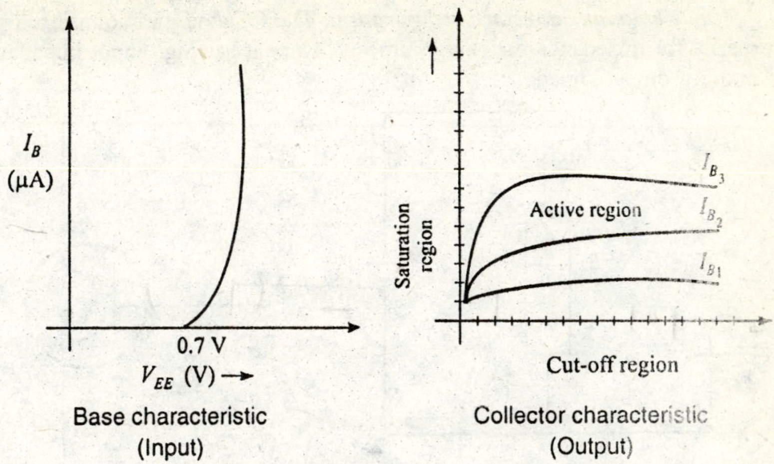

Fig. 2.6

$$\boxed{I_{CEO} = \left.\frac{I_{CBO}}{1-\alpha}\right|_{I_B = 0\mu A}}$$

Beta (β): In the d.c. -mode the levels of I_C and I_B are related by a quantity called beta and defined by the following equation.

$\boxed{\beta_{\text{d.c.}} = \frac{I_C}{I_B}}$ Range of 'β' is from 50 to 400. $\boxed{\beta_{\text{d.c.}} \simeq h_{FE}}$

$\boxed{\beta_{\text{a.c.}} = \left.\frac{\Delta I_C}{\Delta I_B}\right|_{V_{CE} = \text{Constt}}}$ The $\beta_{\text{a.c.}}$ is common emitter forward current amplification factor.

Relation between α and β

(*i*) $\boxed{\alpha = \frac{\beta}{\beta+1}}$ (*ii*) $\boxed{\beta = \frac{\alpha}{1-\alpha}}$

(*iii*) $\boxed{I_{CEO} = \frac{I_{CBO}}{1-\alpha}}$ (*iv*) $\boxed{I_{CEO} \cong \beta\, I_{CBO}}$

(*v*) $\boxed{I_C = \beta\, I_B}$ (*vi*) $\boxed{I_E = (\beta+1)\, I_B}$

(*iii*) ***Common collector configuration:*** The *CC* configuration is used primarily for impedance-matching purposes, since it has high input impedance and low output impedance.

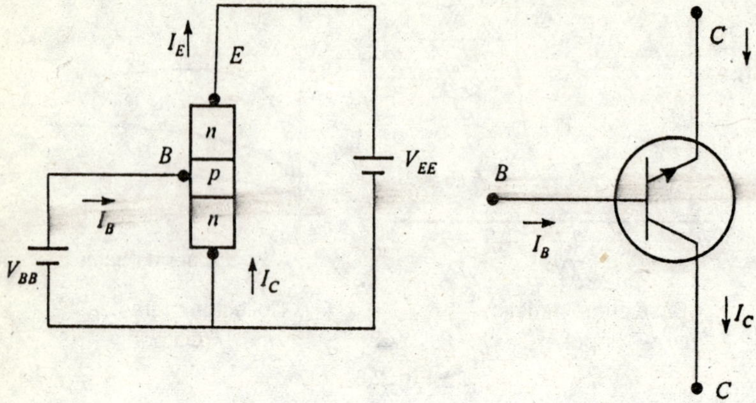

npn transistor
CC configuration

Symbol

Fig. 2.7

Problem 2.1. If the emitter current of a transistor is 8 mA and I_B is $\frac{1}{100}$ of I_C, determine the level of I_C and I_B.

Solution. The emitter curren of a BJT is the summation of the collector current and the base current, i.e.

$$I_E = I_B + I_C \qquad \text{Now} \quad I_B = \frac{1}{100} I_C \Rightarrow I_C = 100 I_B.$$

$$\text{or} \quad I_E = I_B + 100 I_B \qquad \text{and} \quad I_C = 100 I_B$$

$$\Rightarrow \quad I_B = \frac{I_E}{101} \qquad \Rightarrow \quad I_C = 100 \times 79\ \mu\text{A}$$

$$\Rightarrow \quad I_B = \frac{8\text{ mA}}{101} = 79\ \mu\text{A} \quad \Rightarrow \quad I_C = 7.9\text{ mA}$$

Problem 2.2. (*a*) Given an $\alpha_{d.c.}$ of 0.998, determine I_C, if $I_E = 4$ mA.

(*b*) Determine $\alpha_{d.c.}$, if $I_E = 2.8$ mA and $I_B = 20\ \mu$A.

(*c*) Find I_E, if $I_B = 40\ \mu$A and $\alpha_{d.c.} = 0.98$.

Solution. For a bipolar function transistor

$$\alpha_{d.c.} = \frac{I_C}{I_E} \quad \text{and} \quad I_E = I_B + I_C$$

(*a*) $I_C = \alpha_{d.c.} \times I_E = 0.998 \times 4 \times 10^{-3} = 3.992$ mA

(b) $I_E = 2.8$ mA, $I_B = 0.02$ mA, therefore, $I_C = I_E - I_B$.

$$\Rightarrow \quad I_C = (2.8 - 0.02) \text{ mA} \quad \text{and} \quad \alpha_{d.c.} = \frac{I_C}{I_E} = \frac{2.78 \text{ mA}}{2.8 \text{ mA}}$$

$$= 2.78 \text{ mA} \qquad = 0.9928$$

(c) $I_B = 40\ \mu\text{A} = 0.04$ mA

$$\alpha_{d.c.} = 0.98 \qquad I_E = I_C + I_B$$

$$\alpha_{d.c.} = \frac{I_C}{I_E} \qquad \Rightarrow \qquad I_C = I_E - I_B$$

$$\Rightarrow \quad 0.98 = \frac{I_E - I_B}{I_E} \quad \text{or} \quad \frac{I_B}{I_E} = 1 - 0.98$$

$$\Rightarrow \quad 0.98 = 1 - \frac{I_B}{I_E} \quad \Rightarrow \quad I_E = \frac{I_B}{0.02} = \frac{0.04}{0.02}$$

$$\Rightarrow \quad I_E = 2 \text{ mA.}$$

Problem 2.3. Calculate the voltage gain $(A_V = V_L/V_C)$ for the network shown, if $V_i = 500$ mV and $R = 1\text{k}\Omega$.

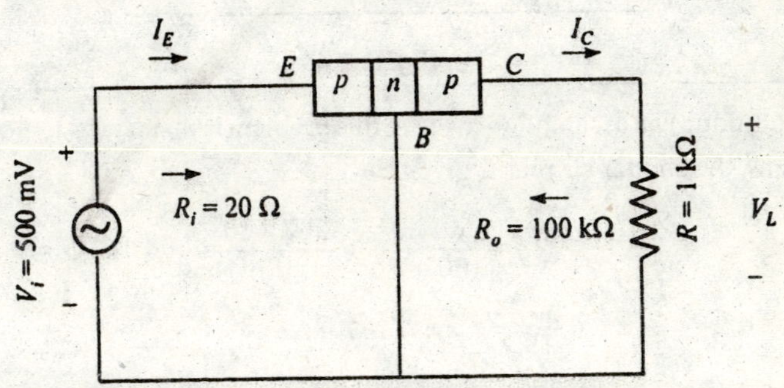

Fig. P. 2.3

Solution. From the figure we have, assume $\alpha_{d.c.} = 1$.

i.e. $\quad I_E = I_C$.

Now $\quad I_E = \frac{V_i}{R_i} = \frac{500 \text{ mV}}{20} = 25 \text{ mA}$

Hence $\quad I_C = I_E = 25$ mA

The load voltage $\quad V_L = I_C R = 25 \times 10^{-3} \times 1 \times 10^3$

$\quad V_L = 25$ V

Since the resistance $R_0 = 100\,\text{k}\Omega$, which is very large than $R = 1\,\text{k}\Omega$ hence, it is considered as an open circuit equivalent.

Thus, the voltage gain is given as

$$A_V = \frac{V_L}{V_i} = \frac{25\text{ V}}{300\text{ mV}} = \frac{25 \times 10^3}{500} = 50.$$

Problem 2.4. Calculate the voltage gain ($A_V = V_L/V_i$) for the network shown, if the source has an internal resistance of 100 Ω in series with V_i, where $V_i = 500$ mV and R_L is equal to 1 kΩ.

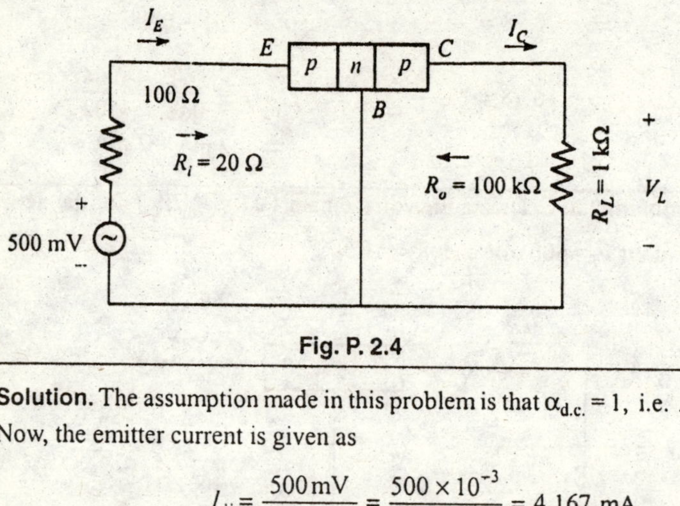

Fig. P. 2.4

Solution. The assumption made in this problem is that $\alpha_{d.c.} = 1$, i.e. $I_B = I_C$

Now, the emitter current is given as

$$I_E = \frac{500\,\text{mV}}{100 + 20} = \frac{500 \times 10^{-3}}{120} = 4.167\text{ mA}.$$

$$\therefore \qquad I_C = I_E = 4.167\text{ mA}$$

The load voltage $\quad V_L = I_C\, R_L \quad \begin{Bmatrix} \text{as } R_0 \gg R_L \\ R_0 \parallel R_L = R_L \end{Bmatrix}$

$$\Rightarrow \qquad V_L = 4.167 \times 10^{-3} \times 1 \times 10^3$$

$$\Rightarrow \qquad V_L = 4.167\text{ V}$$

Thus, the voltage gain $A_V = V_L/V_i$.

$$\Rightarrow \qquad A_V = \frac{4.167\text{ V}}{500\text{ mV}} = \frac{4.167 \times 10^3}{500} = 8.33$$

Problem 2.5. Using the characteristics shown below determine.

(*a*) The value of I_C corresponding to $V_{BE} = +\ 750$ mV and $V_{CE} = +\ 5$ V.

(*b*) Find the value of V_{CE} and V_{BE} corresponding to $I_C = 3$ mA and $I_B = 30\ \mu$A.

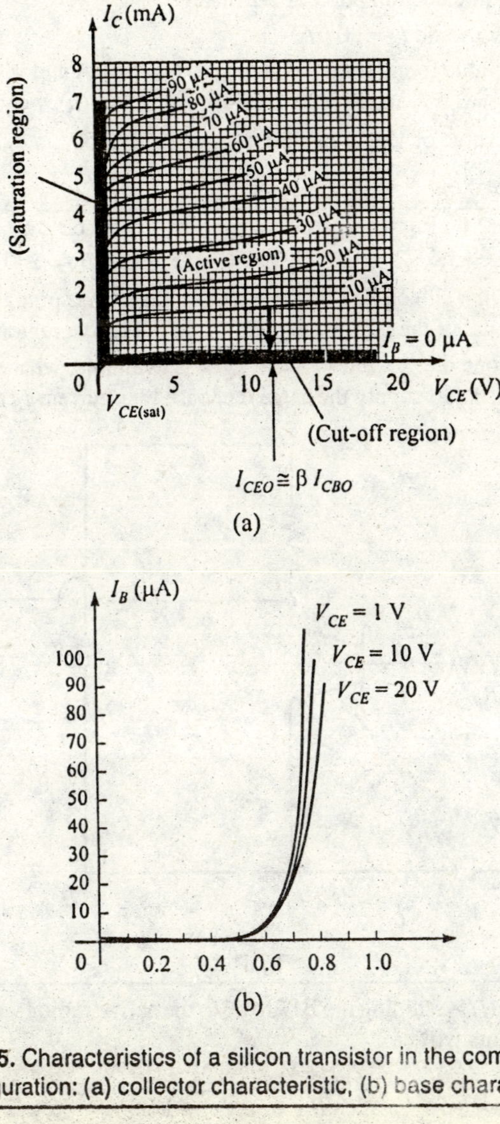

Fig. P. 2.5. Characteristics of a silicon transistor in the common emitter configuration: (a) collector characteristic, (b) base characteristic.

Solution. (*a*) From the graph, we have

$$V_{BE} = 0.7 \text{ V and } V_{CE} = 5 \text{ V}.$$

The base current corresponding to V_{BE} = 0.7 V is I_B = 20 μA. [from Fig. P. 2.5 (*b*)].

Thus, from the figure at the collector current for I_B = 20 μA, and V_{CE} = 5V is found at an intersection point of 2.2 mA.

(*b*) I_C = 3 mA and I_B = 30 μA.

V_{BE} is the value found from the graph for I_B = 30 μA and V_{CE}, so first find out V_{CE}, for I_C = 3 mA and I_B = 30 μA from Fig. P. 2.5 (*a*), we get V_{CE} = 2.5 V.

Thus, the value of V_{BE} at I_B = 30 μA and V_{CE} = 2.5 V is V_{BE} = 0.72 V.

Hence, V_{CE} = 2.5 V and V_B = 0.72 V.

Problem 2.6. For the circuit diagram given below, values are : V_{BB} = 1V, R_B = 5K, R_L = 650 Ω, V_{CC} = + 15 V and B_F = 200.

(*i*) Find the *npn* silicon BJT currents and voltages. Assuming and an idealized d.c. model of the BJT, show that it is in the active region.

(*ii*) If only one d.c. source V_{CC} = + 15 V is available, what should be the new value of R_B to supply the same d.c. base bias current I_B as in (*i*)?

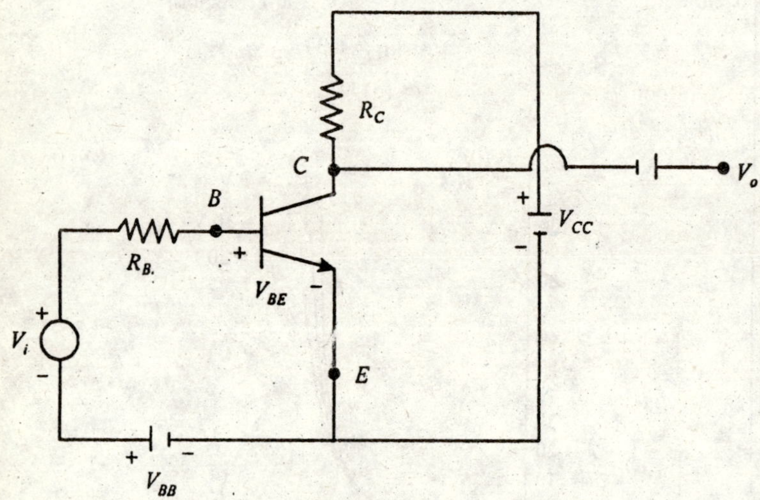

Fig. P. 2.6

Solution. (*i*) Assuming the BJT to be in the active region with V_{BE} = 0.7 V, the BJT currents will be

$$I_B = \frac{V_{BB} - V_{BE}}{R_B} = \frac{1 \text{ V} - 0.7 \text{ V}}{5 \text{ k}\Omega} = \frac{0.3 \text{ V}}{5 \text{ k}\Omega}$$

or $I_B = 60\ \mu A$

Now $I_C = B_F \times I_B = 200 \times 60 \times 10^{-6} = 12000 \times 10^{-6}$

or $I_C = 12\ mA$

Hence $I_E = I_B + I_C = (B_F + 1)\ I_B = 201 \times 60 \times 10^{-6}$

or $I_E \approx 12\ mA$

Now $V_{CE} = V_{CC} - I_C\ R_C = 15\ V - (12\ mA)\ (650\ \Omega)$

or $V_{CE} = 15\ V - 7.8\ V = 7.2\ V$

and $V_{CB} = V_{CE} - V_{BE} = 7.2\ V - 0.7\ V$

or $V_{CB} = 6.5\ V$

Thus, the BJT is in the normal operating (active) region.

(*ii*) Now, only one d.c. supply $V_{CC} = +\ 15\ V$

Also given $I_B = 60\ \mu A$

from the relation $I_B = \dfrac{V_{CC} - V_{BE}}{R_B}$

we get $R_B = \dfrac{V_{CC} - V_{BE}}{I_B}$

or $R_B = \dfrac{15\ V - 0.7\ V}{60 \times 10^{-6}} = \dfrac{14.3\ V}{60 \times 10^{-6}} = 238 \times 10^3$

or, Required value of R_B for the same base current I_B is $R_B = 238\ k\Omega$.

We observe that with only one d.c. source, for the base current I_B to remain same, R_B should be increased many times.

Problem 2.7. For the given common base configuration in the diagram draw the a.c. equivalent using the notation for the transistor model.

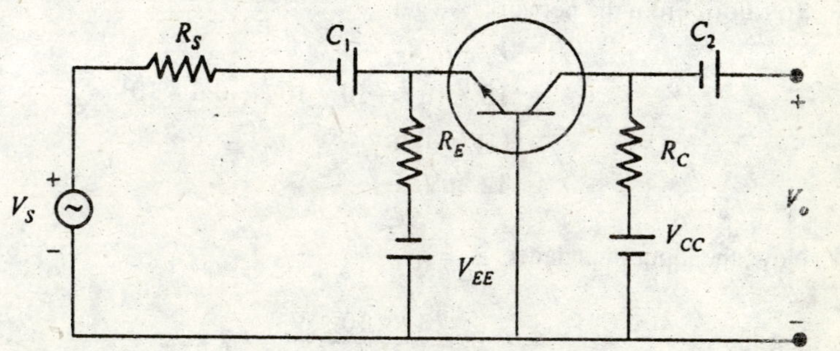

Fig. P. 2.7

Solution. The a.c. equivalent can be drawn using the notation of a transistor model. The capacitors are equated to a short-circuit, and the voltage sources are also equal to short-circuit. Thus, we get

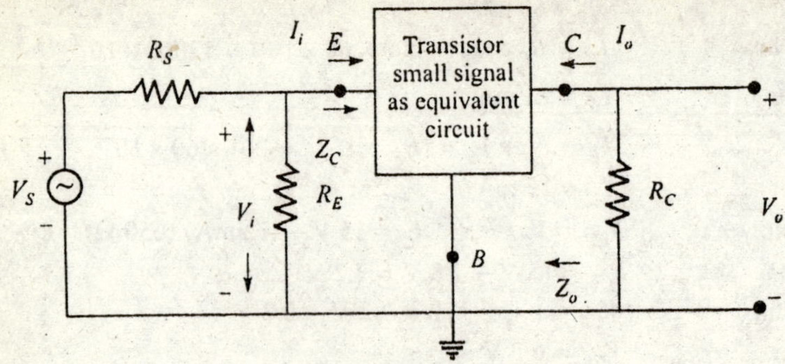

Fig. 2.7 (a)

Problem 2.8. Given the BJT configuration determine.

(*a*) V_i (*b*) Z_i (*c*) A_{VNL} (*d*) A_{VS}

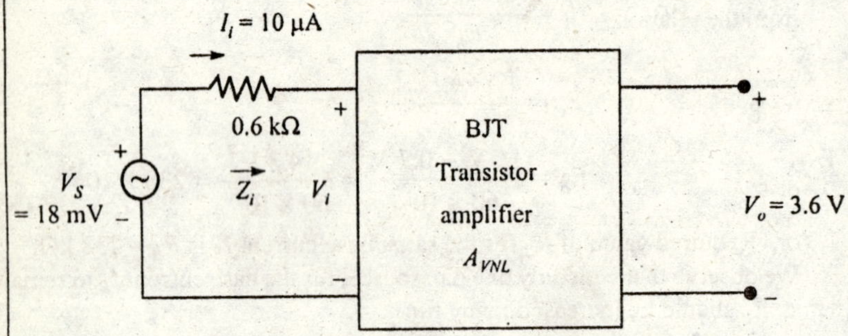

Fig. P. 2.8

Solution. From the network, we get

$$\begin{aligned} V_i &= V_s - I_i R_S \\ &= 18 \times 10^{-3} - 10 \times 10^{-6} \times 0.6 \times 10^3 \\ &= (18 - 6) \text{ mV} \\ &= 12 \text{ mV} \end{aligned}$$

Now, the input impedance $Z_i = \dfrac{V_i}{I_i}$

$$\Rightarrow \qquad Z_i = \frac{12 \times 10^{-3} \text{ V}}{10 \times 10^{-6} \text{ A}} = 1.2 \text{ k}\Omega$$

To calculate gain

$$A_{VNL} = \frac{V_o}{V_i} = \frac{3.6 \text{ V}}{12 \text{ mV}} = 300$$

and $$A_{VS} = \frac{V_o}{V_s} = \frac{3.6\text{ V}}{18\text{ mV}} = 200$$

Thus, we have

(*a*) $V_i = 12$ mV (*b*) $Z_i = 1.2$ kΩ (*c*) $A_{VNL} = 300$ (*d*) $A_{VS} = 200$

Problem 2.9. For the network shown in Fig. P. 2.9 determine:

(*a*) Z_0 if $V = 600$ mV, $R_{\text{sense}} = 10$ kΩ and $I_0 = 10$ μA.

(*b*) Using the Z_0 obtained in part (*a*), determine I_L for the configuration of Fig. P. 2.9 (*b*) if $R_L = 2.2$ kΩ and $I_{\text{amplifier}} = 6$ mA.

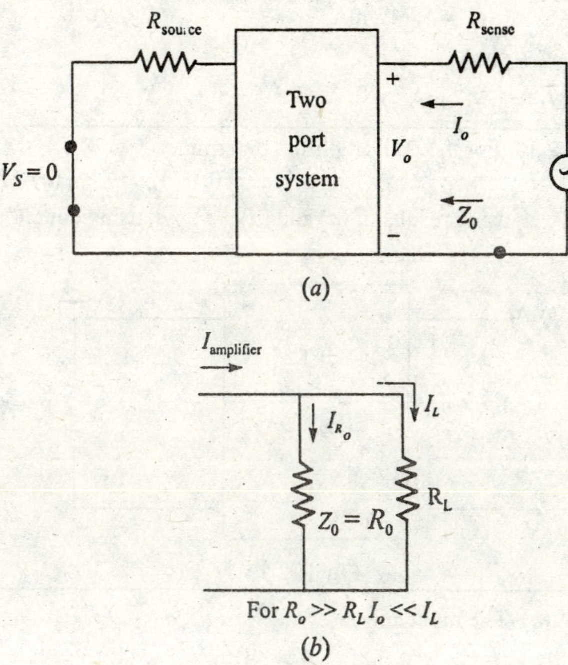

Fig. P. 2.9

Solution. Given $I_0 = 10$ μA, $R_{\text{sense}} = 10$ kΩ, $V = 600$ mV.

The voltage $V_0 = I_0 Z_0 = V - I_0 R_{\text{sense}}$

$$\Rightarrow \quad Z_0 = \frac{V - I_0 R_{\text{sense}}}{I_0}$$

$$\Rightarrow \quad Z_0 = \frac{600\text{ mV} - (10\,\mu\text{A} \times 10\text{ k}\Omega)}{10\,\mu\text{A}}$$

$$\Rightarrow \quad Z_0 = \frac{600\text{ mV} - 100\text{ mV}}{10\,\mu\text{A}}$$

$$\Rightarrow \qquad Z_0 = \frac{500 \text{ mV}}{10 \,\mu\text{A}} = 50 \text{ k}\Omega$$

(*b*) From the Fig. P. 2.9 (b)

$$\text{Current through } R_L,\ I_L = \frac{Z_0 \times I_{amplifier}}{R_L + Z_0}$$

$$\Rightarrow \qquad I_L = \frac{50 \text{ k}\Omega}{2.2 \text{ k}\Omega + 50 \text{k}\Omega} \times 6 \text{ mA}$$

$$\Rightarrow \qquad I_L = 0.9578 \times 6 \text{ mA}$$

$$\Rightarrow \qquad I_L = 5.747 \text{ mA}$$

Thus (*a*) $Z_0 = 50$ kΩ

(*b*) $I_L = 5.747$ mA

Problem 2.10. For the BJT amplifier, determine:

(*a*) I_i (*b*) Z_i (*c*) V_o (*d*) I_o

(*e*) A_i using results of (*a*) and (*d*) (*f*) A_i using equation.

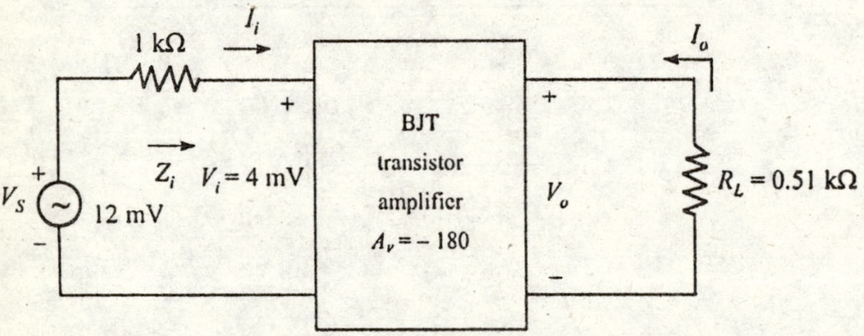

Fig. P. 2.10

Solution. (*a*) The input current

$$I_i = \frac{V_s - V_i}{1 \text{ k}\Omega} = \frac{(12 - 4) \text{ mV}}{1 \text{ k}\Omega} = 8 \mu\text{A}$$

(*b*) The input inpedance

$$Z_i = \frac{V_i}{I_i} = \frac{4 \text{ mV}}{8 \,\mu\text{A}} = 0.5 \text{ k}\Omega$$

(*c*) The output voltage

$$V_o = A_v V_i = -180 \times 4 \text{ mV} = -720 \text{ mV}$$

(*d*) The output current

$$I_o = \frac{V_o}{R_L} = \frac{720 \text{ mV}}{0.51 \text{ k}\Omega} = 1411.76 \,\mu\text{A} = 1.412 \text{ mA}$$

(*e*) The current gain

$$A_i = \frac{I_o}{I_i} = \frac{1411.76\ \mu\text{A}}{8\ \mu\text{A}} = 176.47$$

(*f*) The current gain can also be calculated by using the equation as

$$A_i = -A_v\ \frac{Z_i}{R_L} = +\,180 \times \frac{0.5\ \text{k}\Omega}{0.5\ \text{k}\Omega} = 176.47$$

Problem 2.11. For the common-base configuration, an a.c. signal of 10 mV is applied, resulting in an emitter current of 0.5 mA. If $\alpha = 0.98$, determine:

(*a*) Z_i (*b*) V_o if $R_L = 1.2\ \text{k}\Omega$

(*c*) $A_V = V_o / V_i$ (*d*) Z_0 with $r_0 = \infty\,\Omega$

(*e*) $A_i = I_o / I_i$ (*f*) I_b

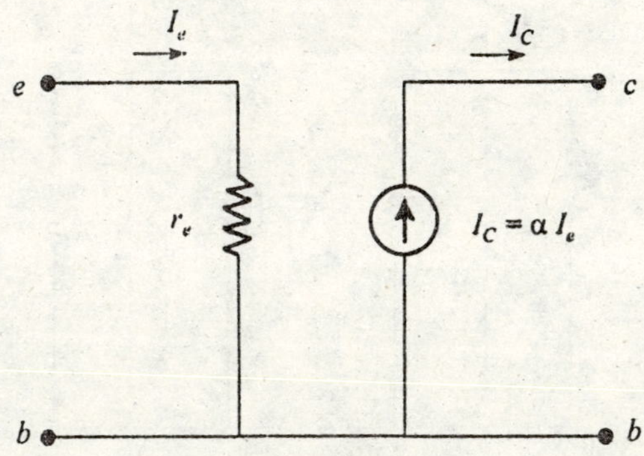

Fig. P. 2.11

Solution. (*a*) The input inpedance

$$Z_i = \frac{V_i}{I_i} = \frac{V_i}{I_C} \Rightarrow Z_i = \frac{10\ \text{mV}}{0.5\ \text{mA}} = 20\ \Omega$$

(*b*) The output voltage

$$V_o = I_C R_L = \alpha I_e R_L$$

$$= 0.98 \times (0.5\ \text{mA}) \times (1.2\ \text{k}\Omega) = 0.588\ \text{V}$$

(*c*) The voltage gain

$$A_v = \frac{V_o}{V_i} = \frac{0.588\ \text{V}}{10\ \text{mA}} = 58.8$$

(*d*) The output impedance

$$Z_0 = r_0 = \infty\ \Omega$$

(*e*) The current gain

$$A_i = \frac{I_o}{I_i} = -\alpha = -0.98$$

(*f*) $I_b = I_e - I_c = I_e - \alpha I_e = (1 - \alpha)\ I_c = (1 - 0.98) \times 0.5$ mA

$$\Rightarrow \qquad I_b = 0.02 \times 0.5 \text{ mA}$$

$$= 0.01 \text{ mA} = 10\ \mu\text{A}$$

Probelm 2.12. For the common base configuration, the emitter current is 3.2 mA and $\alpha = 0.99$. Determine the following, if the applied voltage is 48 *mV* and the load is 2.2 kΩ.

(*a*) r_e (*b*) Z_i (*c*) I_c (*d*) V_o

(*e*) A_v (*f*) I_b

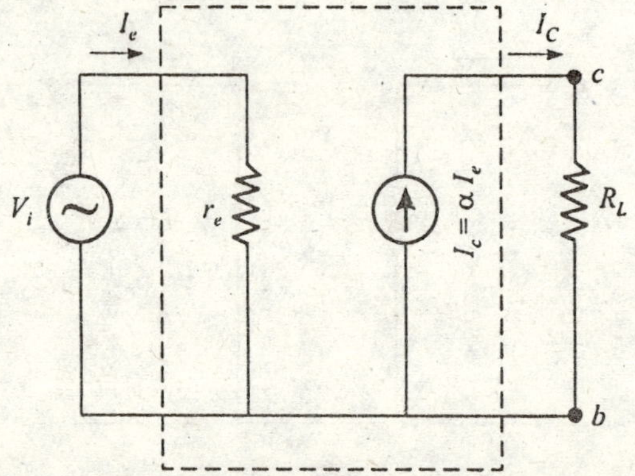

Fig. P. 2.12

Solution. $r_e = \frac{V_i}{I_c} = \frac{48 \text{ mA}}{3.2 \text{ mA}} = 15\ \Omega$

(*b*) The input impedance

$$Z_i = r_e = 15\ \Omega$$

(*c*) The collector current

$$I_c = \alpha\ I_c = 0.99 \times 3.2 \text{ mA} = 3.168 \text{ mA}$$

(*d*) The output voltage

$$V_o = I_C R_L = (3.168 \text{ mA}) \times (2.2 \text{ k}\Omega)$$

$$\Rightarrow \qquad V_o = 6.9696 \text{ V}$$

(*e*) The voltage gain

$$A_V = \frac{V_o}{V_i} = \frac{6.9696 \text{ V}}{48 \text{ mV}} = 145.2$$

(*f*) The base current

$$I_b = I_e - I_c$$
$$= I_e - \alpha I_e$$
$$= (1 - \alpha) I_e$$
$$= (1 - 0.99) \times (3.2 \text{ mA}) = 32 \text{ μA}$$

Probelm 2.13. Using the model shown below, determine the following for a common emitter amplifier if $\beta = 80$, I_E (d.c.) = 2 mA and $r_0 = 40$ kΩ.

(*a*) Z_i (*b*) I_b (*c*) $A_i = I_0/I_i = I_L/I_b$

(*d*) A_v if $R_L = 1.2$ kΩ if $R_L = 1.2$ kΩ

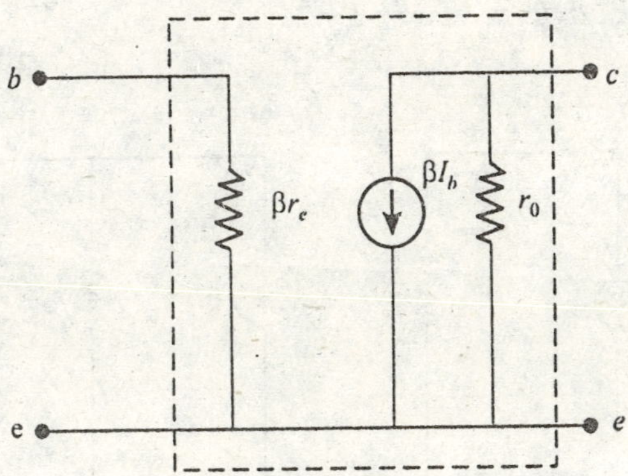

Fig. P. 2.13

Solution. $r_e = \frac{26 \text{ mV}}{I_E} = \frac{26 \text{ mV}}{2 \text{ mA}} = 13\ \Omega$

(*a*) The input impedance

$$Z_i = \beta r_e = 80 \times 13 = 1004 \text{ k}\Omega$$

(*b*) The base current

$$I_b = I_e/_{(1+\beta)} = \frac{2 \text{ mA}}{(80+1)} = \frac{2 \text{ mA}}{81}$$

$\Rightarrow$ $I_b = 24.7$ μA

(*c*) The current gain

$$A_i = \frac{I_o}{I_i} = \frac{I_c}{I_b} = \frac{\beta I_b}{I_b} = \beta = 80$$

(*d*) The voltage gain

$$A_v = V_o/V_i$$

$$\Rightarrow \quad A_v = \frac{I_o R_L}{I_i R_i} = \frac{I_o R_L}{I_i Z_i} = A_i \times \frac{R_L}{Z_i}$$

$$\Rightarrow \quad A_v = 80 \times \frac{1.2 \text{ k}\Omega}{1.04 \text{ k}\Omega} = 92.3076$$

$$\Rightarrow \quad A_v = 92.3$$

Probelm 2.14. The input impedance to a common emitter transistor amplifier is 1.2 kΩ with $\beta = 140$, $r_0 = 50$ kΩ and $R_L = 2.7$ kΩ. Determine:

(*a*) r_e (*b*) I_b if $V_i = 30$ mV (*c*) I_c

(*d*) $A_i = I_o/I_i$ (*e*) $A_v = V_o/V_i$

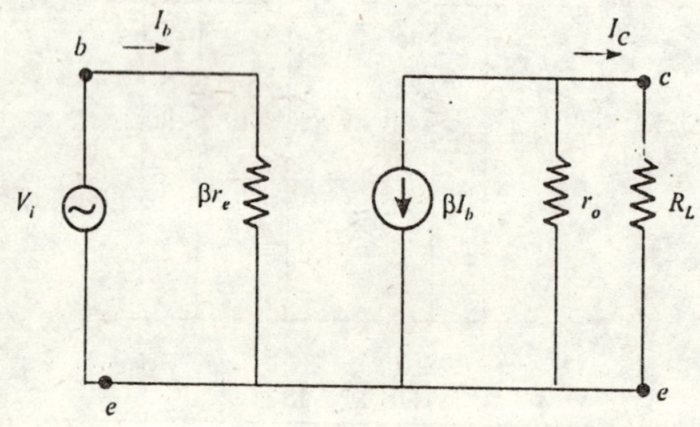

Fig. P. 2.14

Solution. (*a*)

$$r_e = \frac{Z_i}{\beta} = \frac{1.2 \text{ k}\Omega}{140} = 8.5714 \ \Omega$$

(*b*) $$I_b = \frac{V_i}{Z_i} = \frac{30 \text{ mV}}{1.2 \text{ k}\Omega} = 25 \ \mu\text{A}$$

(*c*) The collector current

$$I_c = \beta I_b = 140 \times 25 \ \mu\text{A} = 3.5 \text{ mA}$$

(*d*) The current gain

$$A_i = \frac{I_0}{I_i} = \frac{r_0 \times I_c}{(r_0 + R_L)\, I_b} = \beta \times \frac{r_0}{(R_L + r_0)}$$

$$= 140 \times \frac{50}{(2.7 + 50)} = 132.83$$

(*e*) The voltage gain

$$A_v = \frac{V_o}{V_i} = \frac{-I_o R_L}{V_i} = \frac{-I_o R_L}{I_i R_i} = -A_i \frac{R_L}{Z_i}$$

$$= \frac{-132.83 \times 2.7 \text{ k}\Omega}{1.2 \text{ k}\Omega} = -298.86$$

Problem 2.15. An *npn* silicon BJT is used in the unstabilized *CE* circuit shown below. The device and circuit parameters are:
$V_{CC} = V_{BB} = 20$ V, $R_L = 5$ K, $R_B = 965$ K, $V_{BE} = 0.7$ V, $B_F = 50$. Find the operating point of the BJT. Assume I_{co} to be 10 μA, a negligible value. Assume $V_i = 0$ V.

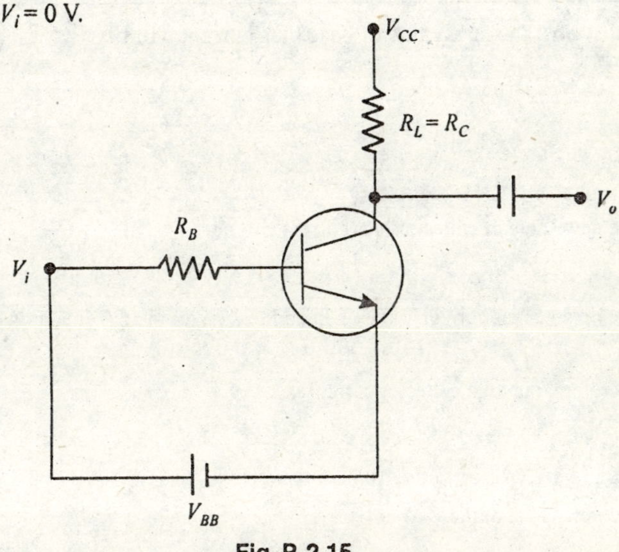

Fig. P. 2.15

Solution. The operating point is obtained on follows:

$$I_{BQ} = \frac{V_{CC} - V_{BE}}{R_B} = \frac{20 \text{ V} - 0.7 \text{ V}}{965 \text{ k}\Omega} = 0.02 \text{ mA}$$

or $$I_{BQ} = 20 \ \mu\text{A}$$

$$I_{CQ} = B_F \times I_{BQ} = 50 \times 20 \ \mu\text{A} = 1 \text{ mA}$$

$$V_{CEQ} = V_{CC} - I_{CQ} \times R_L = 20 \ V - 1 \text{ mA} \times 5 \text{ k}\Omega$$

or $$V_{CEQ} = 15 \text{ V}$$

Problem 2.16. In Problem 2.15, the BJT is replaced by another with $B_{F1} = 100$. Find the new operating point of the BJT.

Solution. The base bias circuit maintains the I_{BQ} at its 0.02 mA value. Hence, the new values of I_{CQ} and V_{CEQ} will be

$$I_{CQ1} = B_{F1} \cdot I_{BQ} = 100 \times 0.02 \text{ mA} = 2 \text{ mA}$$

and $$V_{CEQ_1} = V_{CC} - I_{CQ_1} \times R_L = 20 \text{ V} - 2 \text{ mA} \times 5 \text{ K}$$

or $$V_{CEQ1} = 20 \text{ V} - 10 \text{ V} = 10 \text{ V}$$

Thus, the Q-point shifts considerably from its original location.

Problem 2.17. In last problem, an ambient temperature rise from 25° C to 85° C changes the BJT parameters as follows: $V_{BE_2} = 0.5$ V, $B_{F_2} = 150$ and $I_{CO_2} = 500$ mA. Find the new operating point.

Solution. The new value of base bias current will be

$$I_{BQ_2} = \frac{V_{CC} - V_{BE2}}{R_B} = \frac{20 \text{ V} - 0.5 \text{ V}}{965 \text{ k}\Omega}$$

or $$I_{BQ_2} = \frac{19.5 \text{ V}}{965 \text{ k}\Omega} = 0.02 \text{ mA}$$

The new value of collector bias current will be

$$I_{CQ_2} = B_{F_2} \times I_{BQ_2} = 150 \times 0.02 \text{ mA} = 3 \text{ mA}$$

The reverse bias current, which will add to I_{CQ_2} for the *CE* circuit, is given by

$$B_{F_2} I_{CQ_2} = 150 \times 500 \text{ } \mu\text{A} = 0.075 \text{ mA}$$

Therefore, the total collector current is

$$I_{CQ_3} = I_{CQ_2} + B_{F_2} \times I_{CO_2}$$

or $$I_{CQ_3} = 3 \text{ mA} + 0.075 \text{ mA} = 3.075 \text{ mA}$$

The new value of the V_{CEQ} is

$$V_{CEQ_3} = V_{CC} - I_{CQ_3} R_L = 20 \text{ V} - 3.075 \text{ mA} \times 5 \text{ K}$$

or $$V_{CEQ_3} = 20 \text{ V} - 15.375 \text{ V} = 4.625 \text{ V}$$

Thus, the *Q*-point shifts to the extremity of the active region which might be unacceptable.

Probelm 2.18. The following circuit shows the stabilized *CE* circuit. The circuit parameters are : $V_{CC} = 20$ V, $R_L = 5$ kΩ, $R_C = 1$ kΩ, $R_1 = 90$ kΩ and $R_2 = 10$ kΩ. The parameter variations of the BJT over an ambient temperature range of 25 to 85° C are:

$B = 50$ to 150, $V_{BE} = 0.70$ V to 0.5 V, $I_{Co} = 10$ to 500 mA. Calculate the extreme variation in I_{CQ}.

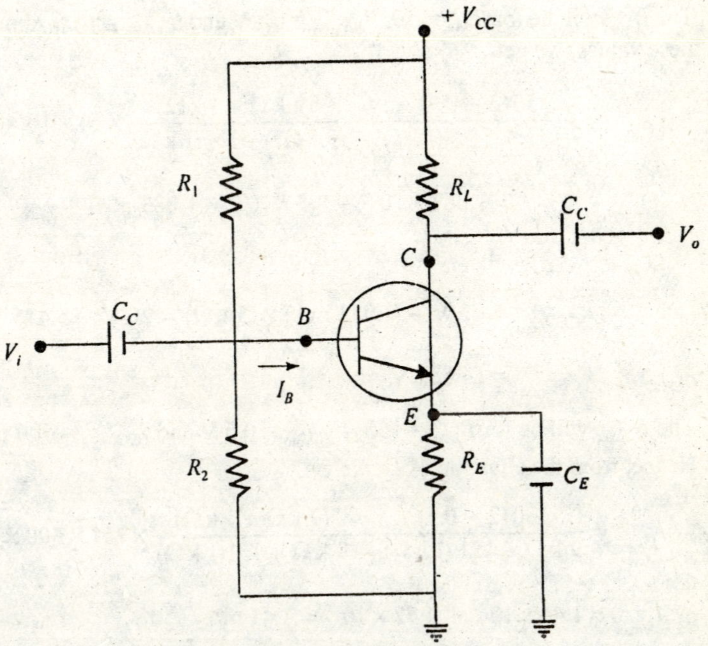

Fig. P. 2.18

Solution. For the voltage-divider bias, we have

$$V_{th} = \frac{V_{CC} \times R_2}{R_1 + R_2} = (20\text{ V})\,\frac{10\text{ k}\Omega}{10\text{ k}\Omega + 90\text{ k}\Omega}$$

or $$V_{th} = \frac{20 \times 10}{100}\text{ V} = 2\text{ V}$$

$$R_{th} = R_1 \parallel R_2 = (10\text{ k}\Omega) \parallel (90\text{ k}\Omega) = \frac{10\text{ k}\Omega \times 90\text{ k}\Omega}{100\text{ k}\Omega}$$

or, $$R_{th} = \frac{900(\text{k}\Omega)^2}{100\text{ k}\Omega} = 9\text{ k}\Omega$$

From the equations

$$V_{th} - V_{BE} = R_{th}\, I_B + (I_B + I_C)\, R_E$$

and $$I_B = \frac{I_C}{\beta} - \frac{(\beta + 1)}{\beta}\, I_{co},$$ we get the equation of I_C as

$$I_C = \frac{(V_{th} - V_{BE})\beta}{R_{th} + (\beta + 1)R_E} + \frac{(R_{th} + R_E)}{R_{th} + (\beta + 1)\, R_E}(\beta + 1)\, I_{Co}$$

The $I_{C_{min}}$ will be for $B_{min} = 50$, $V_{BE\,max} = 0.7$ V and $I_{CO_{min}} = 10$ mA, putting all these values, we get

$$I_{C_{min}} = \frac{50(2-0.7)}{9\text{ k}\Omega + 51 \times (1\text{ k}\Omega)} + \frac{(9\text{ k}\Omega + 1\text{ k}\Omega)}{9\text{ k}\Omega + 51 \times (1\text{ k}\Omega)} \times 51 \times 10 \times 10^{-9}$$

or $$I_{C_{min}} = 1.083 \times 10^{-3} + 8.5 \times 10^{-8} \approx 1.083\text{ mA}$$

$$V_{CEQ_{max}} = V_{CC} - I_{C_{min}} \times R_L$$

or $$V_{CEQ_{max}} = 20\text{ V} - (1.083\text{ mA}) \times (5\text{ k}\Omega) = 20\text{ V} - 5.415\text{ V}$$

or $$V_{CEQ_{max}} = 14.585\text{ V}$$

The $I_{C\,max}$ will be for $B_{max} = 150$, $V_{BE\,min} = 0.5$ V and $I_{CO\,max} = 500$ mA, Hence, from the equation of I_C, we get

$$I_{C_{max}} = \frac{150(2-0.5)}{9\text{ k}\Omega + 151\,(1\text{ k}\Omega)} + \frac{(9\text{ k}\Omega + 1\text{ k}\Omega)}{9\text{ k}\Omega + 151(1\text{ k}\Omega)} \times 151 \times 500 \times 10^{-9}$$

or $$I_{C_{max}} = 1.41 \times 10^{-3} + 4.72 \times 10^{-6} \approx 1.41\text{ mA}$$

$$V_{CEQ_{min}} = V_{CC} - I_{C_{max}} \times R_L = 20\text{ V} - (1.41\text{ mA}) \times (5\text{ k}\Omega)$$

$$= 20\text{ V} - 7.05\text{ V}$$

or $$V_{CEQ_{min}} = 12.95\text{ V}$$

Note that for *Si* BJT, the contribution of I_{co} term is negligible further, I_C has shifted only by 0.397 mA, i.e. about 29.3 percent for these extreme temperature variations of more than 60° C and for three times variation in β.

Problem 2.19. A power transistor which has θ_{jc} of 1.0° C/W is mounted on a heatsink of thermal resistance 3.0° C/W. The maximum allowable junction temperature is 90° C and the thermal resistance between the case and heatsink is 1.0° C/W. How much power can be dissipated when the ambient temperature is 25° C? What is the rating of transistor?

Solution. $$P_D = \frac{T_j - T_A}{\theta_{jA}};\ \theta_{jA} = \theta_{jc} + \theta_{cs} + \theta_{sA} = 1 + 3 + 1 = 5$$

$$= \frac{90-25}{5} = \frac{65}{5} = 13 \text{ W}.$$

Rating of transistor = 13 × 2 = 26 W (factor of safety 2), i.e. the power that can be dissipated = 13 W, and the rating of transistor = 26 W.

Probelm 2.20. A transistor in common emitter configuration employes a 4 kΩ load and V_{CC} = 13 V. Estimate the maximum peak-to-peak value of the input signal for linear operation, if β = 100. Given: V_{knee} = 1 V and a change of 1 V in V_{BE} causes a change of 5 mA in collector current.

Solution. $r_e = \frac{\Delta V_{BE}}{\Delta I_C}$, assuming $\Delta I_C \approx \Delta I_e$

or $r_e = \frac{1}{5 \times 10^{-3}} = 200\ \Omega$

Now, For *CE*, $r_{in} = \beta r_e = 100 \times 200 = 20000\ \Omega$

$$A_V = \frac{\beta i_b R_C}{\beta i_b r_e} = \frac{R_C}{r_e} = \frac{4000}{200} = 20$$

V_{knee} = 1 V, so that maximum swing of collector = 13 – 1 = 12 V

$$V_{CQ} = \frac{12}{2} = 6 \text{ V}$$

$$V_{CQ} = V_{CC} - I_{CQ} R_C - V_{knee}$$

$$I_{CQ} = \frac{13-6-1}{4 \times 10^3} = 1.5 \text{ mA}$$

Maximum swing below *Q* point can be 6 V only. Maximum swing above *Q* point ($I_{CQ} R_C$) = 6 V. For distortionless swing in collector the swing has been restricted to 6 V.

V_o (max) maximum collector swing = 12 V ($p-p$)

$$\therefore\ V_{in}(\max p-p) = \frac{V_o(\max\ p-p)}{A_V}$$

$$= \frac{12}{20} = 600 \text{ mV} (p-p)$$

Problem 2.21. Determine P_D capability of a transistor fitted with a heatsink of thermal resistance $\theta_{HA} = 9°$ C/W. Given: $T_A = 50°$ C, $T_f = 200°$ C, $\theta_{jC} = 7°$ C/W.

Solution. $$P_D = \frac{T_f - T_A}{\theta_{jA}},$$

Where $$\theta_{jA} = \theta_{HA} + \theta_{jc} = 16°\text{C/W}$$

$$\Rightarrow \quad P_D = \frac{(200 - 50)°\text{C}}{16°\text{ C/W}} = 9.4 \text{ W}.$$

CHAPTER

3

BJT and Its Biasing

The operating point defines, where the transistor will operate on its characteristic curves under d.c. conditions. For linear (minimum distortion) amplication, the d.c. operating point should not be too close to the maximum power, voltage or current rating and should avoid the regions of saturation and cut-off.

Biasing: The term biasing appearing in the title of this chapter is an all inclusive term for the application of d.c. voltages to establish a fixed level of current and voltage since the operating point on the characteristics, it is also called the *quiescent point* (Q-point).

There are basically four types of biasing in BJT:

(*i*) Fixed biased (*ii*) Emitter biased

(*iii*) Feedback biased (*iv*) Voltage divide biased

(*i*) ***Fixed biased***: The fixed biased circuit is drawn below:

General equations used are:

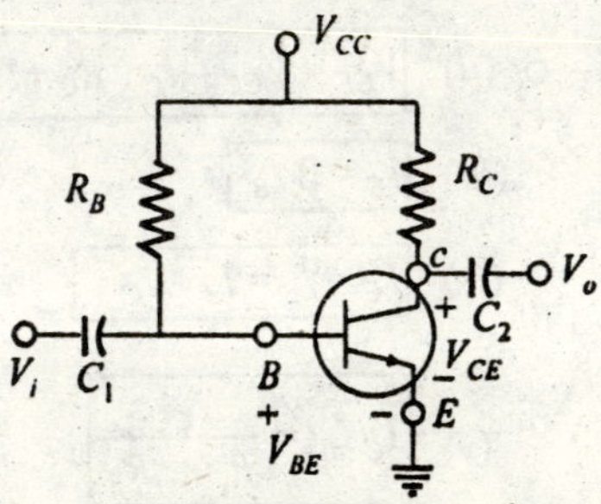

Fig. 3.1 (a)

(*i*) $$I_B = \frac{V_{CC} - V_{BE}}{R_B}$$

(*ii*) $$V_{CE} = V_{CC} - I_C R_C$$

(*iii*) $$I_C = \beta I_B$$

(*iv*) $$V_{BE} = V_B - V_E$$

Saturation level: The saturation current I_C occurs when $V_{CE} = 0$, i.e.

$$I_{C\,sat} = \frac{V_{CE}}{R_C}.$$

(*ii*) ***Emitter biased:*** An emitter resistor is used to improve the stability level over that of the fixed bias configuration the diagram is shown below:

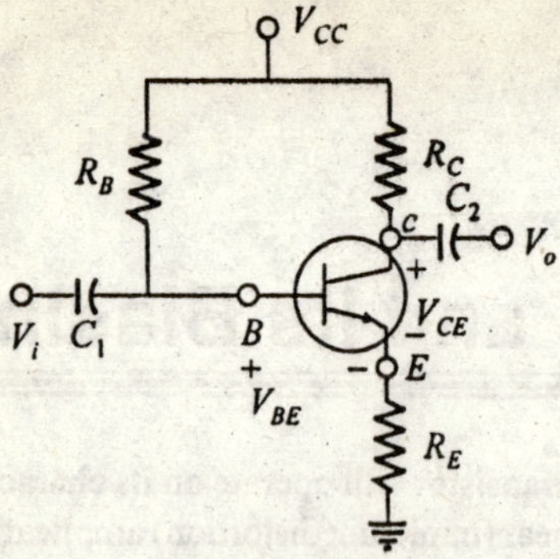

(i) $$I_B = \frac{V_{CC} - V_{BE}}{R_B + (\beta + 1) R_E}$$

(ii) $$V_{CE} = V_{CC} - I_C (R_C + R_E)$$

(iii) $$V_B = V_{CC} - I_B R_B$$

(iv) $$V_C = V_{CC} - I_C R_C$$

(v) $$I_C = \beta I_B$$

Fig. 3.1 (b)

Saturation level: The collector saturation level is the maximum collector current for an emitter collector function to be short circuited, i.e $V_{CE} = 0$.

$$I_{C\,sat} = \frac{V_{CC}}{R_E + R_E}$$

(*iii*) ***Feedback biased circuit:*** This is also called as d.c. bias with voltage feedback. The configuration is drawn below:

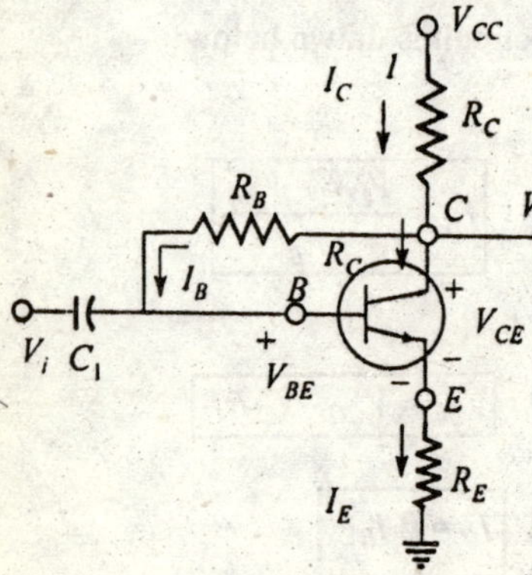

(i) $$I_B = \frac{V_{CC} - V_{BE}}{R_B + \beta(R_E + R_E)}$$

(ii) $$V_{CE} = V_{CC} - I_C (R_C + R_E)$$

(iii) $$I_C = \beta I_B$$

(iv) $$I'_C = I_E = I_B + I_C$$

(v) $$I_{C\,sat} = \frac{V_{CC}}{R_e + R_E}$$

Fig. 3.1 (c)

(iv) ***Voltage-divider bias circuit:*** In the previous bias configuration the bias current I_{CQ} and voltage V_{CEQ} are a function of the current gain (β) of the transistor. However, since β is temperature sensitive the actual value of beta is usually not well defined, it would be desirable to develop a bias circuit that is less dependent or infact independent of the transistor beta. The voltage divider is such an example.

The thevenin equivalent circuit is drawn as:

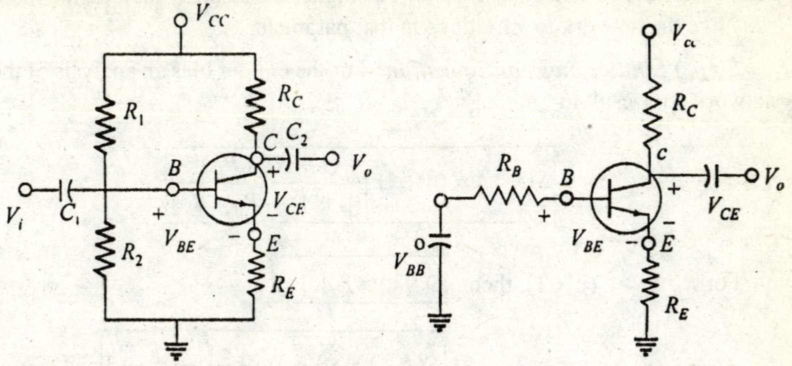

Fig. 3.2

Equation used are:

(*i*) $R_{th} = R_B = R_1 \| R_2 = \dfrac{R_1 R_2}{R_1 + R_2}$ (*ii*) $V_B = V_{th} = \dfrac{R_2 V_{CC}}{R_1 + R_2}$

(*iii*) $I_B = \dfrac{V_{th} - V_{BE}}{R_{th} + (\beta + 1) R_E}$ (*iv*) $V_{CE} = V_{CC} - I_C(R_C + R_E)$

(*v*) $I_{csat} = \dfrac{V_{CC}}{R_C + R_E}$

Bias stabilization: The stability of a network system is a measure of the sensitivity of the network to variations in its parameters. In a transistor the collector current (I_C) is sensitive to each of the following parameters:

(*i*) ***β*:** Increases with the increase in temperature.

(*ii*) $|V_{BE}|$: Decreases about 7.5 mV per degree celcius (°C) increase in temperature.

(*iii*) ***I_{CO}* (*reverse saturation current*):** Doubles in value for every 10°C increase in temprature.

Stability factors: A stability factors, *S*, is defined for each of the parameters affecting bias stability as listed below:

$$\boxed{S(I_{CO}) = \frac{\Delta I_e}{\Delta I_{co}}} \qquad \boxed{S(V_{BE}) = \frac{\Delta I_e}{\Delta V_{BE}}}$$

$$\boxed{S(\beta) = \frac{\Delta I_c}{\Delta \beta}}$$

In each case, Δ signifies the change in that quantity.

The network that are quite stable and relatively insensitive to temperature variations have low stability factors. The higher the stability factors, the more sensitive the network to variations in that parameter.

$S(I_{CO})$ Emitter bias configuration: For the emitter bias an analysis of the network will result in

$$S(I_{CO}) = (\beta + 1)\,\frac{(1 + R_B / R_E)}{[(\beta + 1) + R_B / R_E]}$$

For $R_B/R_E >> (\beta + 1)$, then $S(I_{CO}) = \beta + 1$

For $R_B / R_E << 1$, then $S(I_{CO}) = (\beta + 1) \cdot \dfrac{1}{(\beta + 1)} = \rightarrow 1$

Thus, the emitter bias is quite stable, when the ratio R_B/R_E is as small as possible and the least stable when R_B/R_E ratio approaches $(\beta + 1)$.

For fixed bias configuration $S(I_{CO}) = \beta + 1$

For voltage divider bias $S(I_{CO}) = (\beta + 1)\,\dfrac{1 + R_{th} / R_E}{[(\beta + 1) + R_{th} / R_E]}$

For feedback bias $S(I_{CO}) = (\beta + 1)\,\dfrac{1 + R_B / R_C}{[(\beta + 1) + R_B / R_C]}$

$S(V_{BE})$: $S(V_{BE}) = \dfrac{\Delta I_c}{\Delta V_{BE}}$

For emitter bias

$$S(V_{BE}) = \frac{-\beta}{R_B + (\beta + 1)\, R_E}$$

$$\Rightarrow \qquad S(V_{BE}) = \frac{-\beta / R_E}{R_B / R_E + (\beta + 1)}$$

When, $(\beta + 1) >> R_B / R_E$, then

$$S(V_{BE}) = \frac{-\beta / R_E}{(\beta + 1)} \cong \frac{-\beta / R_E}{\beta} = \frac{-1}{R_E}$$

For fixed bias configuration, $R_E = 0\ \Omega$

$$\boxed{S(V_{BE}) = \frac{-\beta}{R_B}}$$

For voltage divider bias

$$\boxed{S(V_{BE}) = \frac{-\beta}{R_{th} + (\beta+1)R_E}}$$

For feedback bias configuration

$$\boxed{S(V_{BE}) = \frac{-\beta}{R_B + (\beta+1)R_C}}$$

S* (β) *For emitter bias configuration

$$\boxed{S(\beta) = \frac{I_{C_1}\,(1 + R_B/R_E)}{\beta_1(1+\beta_2 + R_B/R_E)}}$$

The notations I_{C_1} and β_1 is used to define their values under one set of network conditions, while the notations β_2 is used to define the new value of beta as established by such causes as temperature change, variation in β for the same transistor or a change in transistor.

For fixed bias configuration, $R_E = 0$

$$\boxed{S_{(\beta)} = \frac{I_{C_1}}{\beta_1}}$$

For voltage divider bias

$$\boxed{S(\beta) = \frac{I_{C_1}\,(1 + R_{th}/R_E)}{\beta_1\,(1+\beta_2 + R_{th}/R_E)}}$$

For feedback bias with, $R_E = 0\ \Omega$

$$\boxed{S(\beta) = \frac{I_{C_1}(R_B + R_C)}{\beta_1\left((R_B + R_C(1+\beta_2)\right)}}$$

Thus, the total effect on the collector current can be determined using the following equation:

$$\boxed{\Delta I_C = S(I_{CO})\,.\,\Delta I_{CO} + S(V_{BE})\,.\,\Delta V_{BE} + S(\beta)\,.\,\Delta\beta}$$

Problem 3.1. In the self bias circuit shown below, a silicon transistor is used with $\beta = 50$, $V_{BE} = 0.6$ V, $V_{CC} = 1.8$ V and $R_C = 4.3$ kΩ. It is desired to establish quiescent point at $I_C = 1.5$ mA, $V_{CE} = 10$ V and stability factor of $S \leq 4$, find R_e, R_1 and R_2.

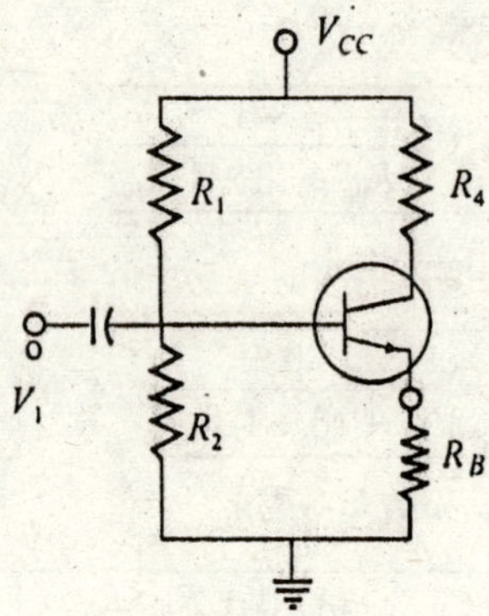

Fig. P. 3.1

(IES 1993)

Solution. It is given that

$$\beta = 50,\ V_{BE} = 0.6\text{ V},\ V_{CC} = 1.8\text{ V}$$
$$R_C = R_4 = 4.3\text{ k}\Omega,\ I_C = 1.5\text{ mA},\ V_{CE} = 10\text{ V}$$
$$S \leq 4 \quad R_e,\ R_1 \text{ and } R_2 = ?$$

We can easily reduce the above circuit as shown.

$$V_B = \frac{V_{CC} \times R_2}{R_1 + R_2}$$

and $R_B = R_1 \parallel R_2 = \dfrac{R_1\ R_2}{R_1 + R_2}$

Now, $R_B = \dfrac{R_1\ R_2}{R_1 + R_2}$

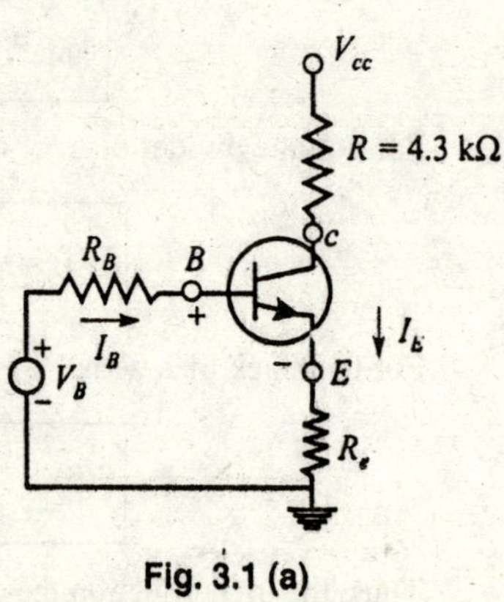

Fig. 3.1 (a)

Apply KVL in the base emitter loop junction

$$V_B = R_e I_E + V_{BE} + R_B I_B$$

$$\Rightarrow \quad V_B = R_e (I_B + I_C) + V_{BE} + R_B I_B$$

$$\Rightarrow \quad V_B = (R_e (b + 1) + R_B) I_B + V_{BE} \qquad \left\{\begin{array}{l} \because\ I_C = \beta I_B \\ \text{and } I_E = I_B + I_C \end{array}\right\}$$

$$\Rightarrow \quad \frac{V_{CC}\ R_2}{R_1 + R_2} = [R_e\ (\beta + 1) + R_B)\ I_B + V_{BE}$$

Put the values

$$\Rightarrow \quad \frac{18\ R_2}{R_1 + R_2} = \left(R_e . 51 + \frac{R_1\ R_2}{R_1 + R_2} \right) I_B + 0.6$$

$$\Rightarrow \quad \left(18\ \frac{R_B}{R_1} \right) = (51\ R_e + R_B)\ I_B + 0.6 \qquad ...(1)$$

We have $I_B = \frac{I_C}{\beta} = \frac{1.5}{50} = 0.03$ mA

$$\Rightarrow \quad \left(18\ \frac{R_B}{R_1} \right) = (51\ R_e + R_B)\ 0.03 + 0.6 \qquad ...(2)$$

Now, apply KVL in the emitter collector loop

$$V_{CC} = R_e I_E + 10 + I_C R_C$$

$$\Rightarrow \quad 18 = R_e (\beta + 1) I_B + 10 + \beta I_B 4.3$$

$$\Rightarrow \quad 8 = (51\ R_e + 215)\ I_B$$

$$\Rightarrow \quad 8 = (51\ R_e + 215)\ 0.03$$

$$\Rightarrow \quad 51\ R_e + 215 = \frac{8}{0.03}$$

$$\Rightarrow \quad 51\ R_e + 215 = 266.67$$

$$\Rightarrow \quad 51\ R_e = 51.6667$$

$$\Rightarrow \quad R_e = 1.013\ \text{k}\Omega$$

Now, stability $S \leq 4$

$$\Rightarrow \quad (1+\beta)\frac{\left[1 + \frac{R_B}{R_e}\right]}{(1+\beta) + \frac{R_B}{R_e}} \leq 4$$

$$\Rightarrow \quad \frac{51\left[1 + \frac{R_B}{R_e}\right]}{(51 + R_B / R_e)} \leq 4$$

$$\Rightarrow \quad \frac{51}{4}\left[1 + \frac{R_B}{R_e}\right] \leq 51 + \frac{R_B}{R_e}$$

$$\Rightarrow \quad \left(\frac{51}{4} - 1\right)\frac{R_B}{R_e} \leq 51 - \frac{51}{4}$$

$$\Rightarrow \quad \left(\frac{47}{4}\right)\frac{R_B}{R_e} \leq \frac{204 - 51}{4}$$

$$\Rightarrow \qquad 47\frac{R_B}{R_e} \le 153$$

$$\Rightarrow \qquad \frac{R_B}{R_e} \le \frac{153}{47}$$

$$\Rightarrow \qquad R_B \le 3.255 \times 1.013$$

$$\Rightarrow \qquad R_B \le 3.2976\ \text{k}\Omega$$

$$\therefore \qquad R_B = 3.3\ \text{k}\Omega$$

Put R_B equation (2), we get

$$\Rightarrow \qquad 18 \times \frac{3.3}{R_1} = \left[51\,(1.013) + 3.3\right] 0.03 + 0.6$$

$$\Rightarrow \qquad \frac{18 \times 3.3}{R_1} = 2.249$$

$$\Rightarrow \qquad R_1 = \frac{18 \times 3.3}{2.25} = 26.4\ \text{k}\Omega$$

Now $$R_B = \frac{R_1\,R_2}{R_1 + R_2}$$

$$\Rightarrow \qquad 3.3 = \frac{26.4 \times R_2}{(26.4 + R_2)}$$

$$\Rightarrow \qquad 87.12 + 3.3R_2 = 26.4R_2$$

$$\Rightarrow \qquad R_2 = \frac{87.12}{23.1}$$

$$\Rightarrow \qquad R_2 = 3.772\ \text{k}\Omega$$

Hence
$$R_e = 1.013\ \text{k}\Omega$$
$$R_1 = 26.4\ \text{k}\Omega$$
$$R_2 = 3.772\ \text{k}\Omega$$

Problem 3.2. For the fixed bias configuration of the figure, determine:

(a) I_{BQ}

(b) V_{CEQ}

(c) I_{CQ}

(d) V_C

(e) V_B

(f) V_E

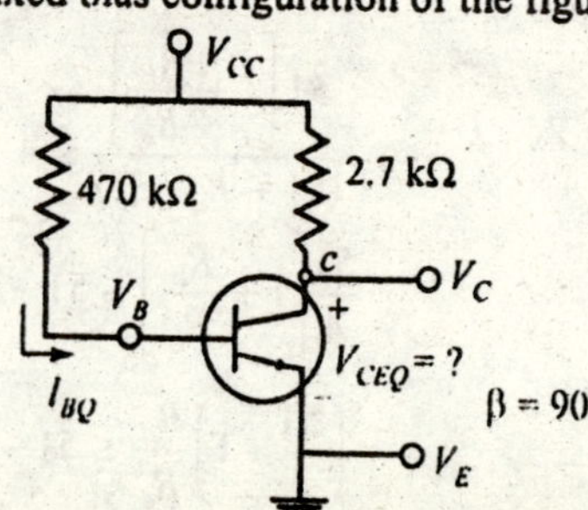

Fig. P. 3.2

Solution. Apply KVL in the base emitter loop

$\Rightarrow \quad 16 = 470\text{K}\, I_{BQ} + V_{BE}$

$\Rightarrow \quad 16 = 470\text{ K}\, I_{BQ} + 0.7$

$\Rightarrow \quad 15.3 = 470\text{ K}\, I_{BQ}$

$\Rightarrow \quad I_{BQ} = \dfrac{15.3}{470\text{ K}} = 0.03255\text{ mA}$

$\Rightarrow \quad I_{BQ} = 0.03255\text{ mA}$

$\Rightarrow \quad I_{BQ} = 32.55\ \mu\text{A}$

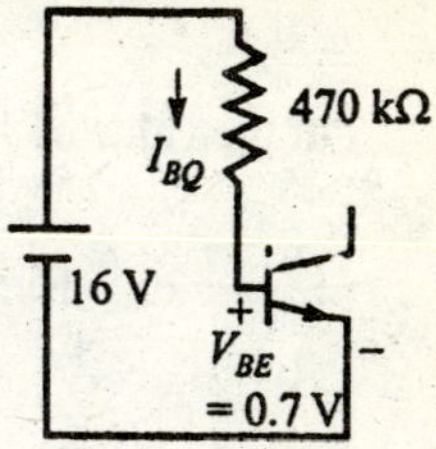

Fig. 3.2 (a)

Now, again apply KVL in the collector emitter loop

$16 = V_{CEQ} + 2.7\text{ K}\, I_C \qquad I_C = \beta\, I_B = 90 \times 32.55$

$I_C = 2.93\text{ mA}$

$\Rightarrow \quad V_{CEQ} = 16 - 2.7 \times 2.93$

$\Rightarrow \quad V_{CEQ} = 8.09\text{ V}$

Now $\quad V_{CE} = V_C - V_E$

$\Rightarrow \quad V_E = 0 \qquad$ [$\because$ Emitter terminal is grounded]

$\therefore \quad V_C = V_{CE} = 8.09$

$V_{BE} = V_B - V_E$

$\therefore \quad V_B = V_{BE} = 0.7\text{ V}$

Hence

(*a*) $I_{BQ} = 32.55\ \mu\text{A}$ (*b*) $I_{CQ} = 2.93\text{ mA}$ (*c*) $V_{CEQ} = 8.09\text{ V}$

(*d*) $V_C = 8.09\text{ V}$ (*e*) $V_B = 0.7\text{ V}$ (*f*) $V_E = 0$

Problem 3.3. Given the information in the figure find out the following:

(*a*) I_C (*b*) V_{CC} (*c*) β (*d*) R_B

Solution. Apply KVL in the base emitter loop

$V_{CC} = I_B R_B + V_{BE}$

$V_{CC} = 20 \times 10^{-6} R_B + 0.7 \qquad \ldots (1)$

$I_C = I_E - I_B = (4 - 0.02) \times 10^{-3} = 3.98\text{ mA}$

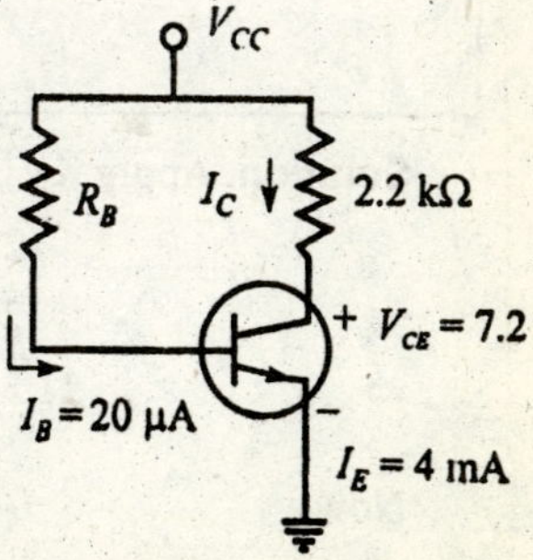

Fig. P. 3.3

Apply KVL in the co-emitter loop

$V_{CC} = V_{CE} + I_C \times 2.2\text{ K}$

$\Rightarrow \quad V_{CC} = 7.2 + (I_E - I_B)\, 2.2\text{ K}$

$\Rightarrow \quad V_{CC} = 7.2 + (4 - 0.02) \times 10^{-3} \times 2.2 \times 10^{3}$

$$\Rightarrow \qquad V_{CC} = 7.2 + 8.756$$

$$\Rightarrow \qquad V_{CC} = 15.956 \text{ V}$$

Put the value of V_{CC} in equation (1)

$$V_{CC} = 20 \times 10^{-6} R_B + 0.7$$

$$\Rightarrow \qquad 20 \times 10^{-6} R_B = 15.956 - 0.7$$

$$\Rightarrow \qquad R_B = \frac{15.256}{20 \times 10^{-6}}$$

$$\Rightarrow \qquad R_B = 726.8 \text{ k}\Omega$$

Now

$$I_C = \beta I_B$$

$$\beta = \frac{I_C}{I_B} = \frac{3.98 \text{ mA}}{0.2 \text{ mA}} = 199$$

Hence

(*a*) $I_C = 3.98$ mA (*b*) $V_{CC} = 15.956$ V (*c*) $\beta = 199$

(*d*) $R_B = 762.8$ kΩ

Problem 3.4. Given the information provided in the figure determine:

(*a*) R_C (*b*) R_E (*c*) R_B (*d*) V_{CE} (*e*) V_B

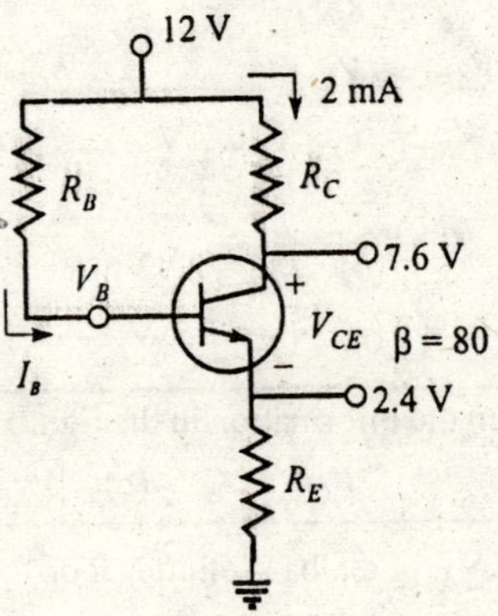

Fig. P. 3.4

Solution. Apply KVL in the emitter base junction loop

$$12 = I_B R_B + V_{BE} + (I_B + I_C) R_E$$

$$\Rightarrow \qquad 12 = I_B R_B + 0.7 + (I_B + \beta I_E) R_E$$

$$\Rightarrow \qquad 11.3 = I_B [R_B + (\beta + 1) R_E] \qquad \{\because I_C = \beta I_B\}$$

Now

$$I_C = 2 \text{ mA}, I_B = \frac{I_C}{\beta} = \frac{2}{80} = 25 \, \mu\text{A}$$

$$\Rightarrow \qquad 11.3 = 25 \times 10^{-6} [R_B + 81 R_E]$$

$$\Rightarrow \quad [R_B + 81R_E] = \frac{11.3 \times 10^6}{25}$$

$$\Rightarrow \quad R_B + 81\,R_E = 425 \times 10^3 \qquad \text{... (1)}$$

$$\Rightarrow \quad V_{CE} = 7.6 - 2.4 = 5.2 \text{ V}$$

Apply KVL in the collector emitter loop

$$12 = V_{CE} + I_C R_C + I_E R_E$$

$$\Rightarrow \quad 12 = V_{CE} + 2 \times 10^{-3} R_C + (\beta + 1) I_B R_E$$

$$\Rightarrow \quad 12 = V_{CE} + 2 \times 10^{-3} R_C + 81 \times 25 \times 10^{-6} R_E$$

$$\Rightarrow \quad 12 = V_{CE} + 2 \times 10^{-3} R_C + 2.025 \times 10^{-3} R_E$$

$$\Rightarrow \quad 12 = V_{CE} + (2\,R_C + 2.025\,R_E) \times 10^{-3}$$

$$\Rightarrow \quad 12 = 5.2 + (2\,R_C + 2.025\,R_E) \times 10^{-3}$$

$$\Rightarrow \quad 2\,R_C + 2.025\,R_E = 6.8 \times 10^{3} \qquad \text{... (2)}$$

From the figure, we have

$$R_C = \frac{12 - 7.6}{2 \times 10^{-3}} = 2.2 \text{ k}\Omega$$

Putting R_C in equation (2)

$$\Rightarrow \quad 2 \times 2.2 \times 10^3 + 2.025\,R_E = 6.8 \times 10^3$$

$$\Rightarrow \quad R_E = \frac{(6.8 - 4.4)}{2.025} \times 10^3$$

$$\Rightarrow \quad R_E = 1.185 \text{ k}\Omega$$

$$V_{BE} = V_B - V_E$$

$$0.7 = V_B - 2.4$$

$$V_B = 3.1 \text{ V}$$

Put the values in equation (1)

$$\Rightarrow \quad 11.3 = 25 \times 10^{-6} [R_B + 81 \times 1.185 \times 10^3]$$

$$\Rightarrow \quad 452 \times 10^{-3} = R_B + 95.985 \times 10^3$$

$$\Rightarrow \quad R_B = (452 - 96) \times 10^3$$

$$\Rightarrow \quad R_B = 356 \text{ k}\Omega$$

Hence

(*a*) $R_C = 2.2$ kΩ (*b*) $R_E = 1.185$ kΩ (*c*) $R_B = 356$ kΩ

(*d*) $V_{CE} = 5.2$ V (*e*) $V_B = 3.1$ V

Problem 3.5. For the voltage divider bias configuration, determine:

(*a*) I_C

(*b*) V_E

(*c*) V_B

(*d*) R_1

(*e*) I_B

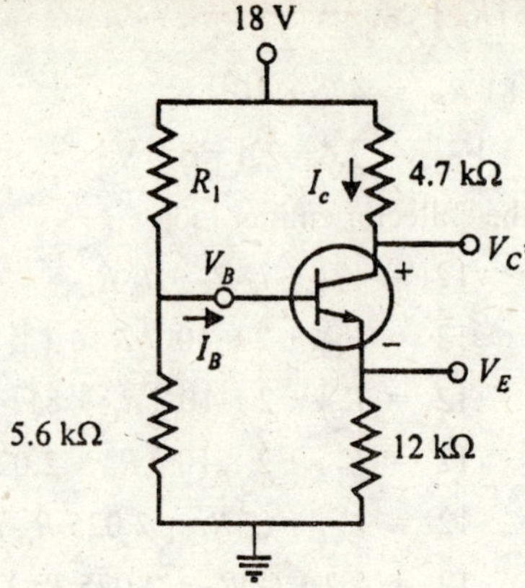

Fig. P. 3.5

Solution. First reduce the above circuit in an easier circuit as shown below:

$$V_{BB} = \frac{V_C\ R_2}{R_1 + R_2}$$

$$\Rightarrow \qquad V_{BB} = \frac{18 \times 5.6}{(5.6 + R_1)} = \frac{100.8}{(5.6 + R_1)}$$

$$R_B = R_1 \| R_2 = \frac{R_1\ R_2}{R_1 + R_2} = \frac{5.6\ R_1}{(5.6 + R_1)}$$

The circuit reduce as

$$I_C = \frac{18 - 12}{4.7}$$

$$= \frac{6}{4.7}\ \text{mA}$$

$$= 1.28\ \text{mA}$$

$$I_B = \frac{I_C}{99} = \frac{1.28}{99} = 12.9\ \mu\text{A}$$

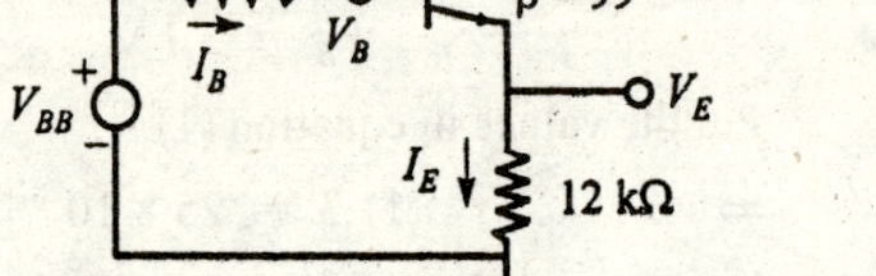

Fig. 3.5 (a)

Take KVL in the emitter collector region (loop)

$$18 = 4.7\ \text{K}\ I_C + (\beta + 1)\ I_B\ 1.2 \times 10^3 + V_{CE}$$

$$\Rightarrow \qquad 18 = 4.7 \times 1.28 + 100 \times 12.9 \times 10^{-3} \times 1.2 + V_{CE}$$

$$\Rightarrow \qquad V_{CE} = 18 - (6.016 + 1.5515)$$

$$\Rightarrow \qquad V_{CE} = 10.4325$$

$$\Rightarrow \quad V_C - V_E = 10.432$$

$$\Rightarrow \quad V_E = V_C - 10.432$$

$$= 12 - 10.432$$

$$= 1.5675 \text{ V}$$

$$= 1.57 \text{ V}$$

Take KVL in the base emitter loop

$$V_{BB} = I_B R_B + V_{BE} + (\beta + 1) R_E I_B$$

$$\Rightarrow \quad V_{BB} = I_B R_B + 0.7 + 1.57$$

$$\Rightarrow \quad \frac{100.8}{5.6 + R_1} = 0.0129 \; R_B + 2.27$$

$$\Rightarrow \quad \frac{100.8}{5.6 + R_1} = \frac{0.0129 \times 5.6 \; R_1}{5.6 + R_1} + 2.27$$

$$\Rightarrow \quad 100.8 = 0.07224 \; R_1 + 2.27 \; R_1 + 12.712$$

$$\Rightarrow \quad 88.088 = 2.3422 \; R_1$$

$$\Rightarrow \quad R_1 = \frac{88.088}{2.3422} = 37.609 \text{ k}\Omega$$

Now $\quad V_{BE} = 0.7$

$$\Rightarrow \quad V_B - V_E = 0.7$$

$$\Rightarrow \quad V_B = V_E + 0.7 = 1.57 + 0.7$$

$$\Rightarrow \quad V_B = 2.27 \text{ V}$$

Hence

(*a*) $I_C = 1.28$ mA (*b*) $V_E = 1.5$ V (*c*) $V_B = 2.27$ V

(*d*) $R_1 = 37.609$ kΩ (*e*) $I_B = 12.9$ μA

Problem 3.6. For the collector feedback configuration shown in Fig. P. 3.6, determine:

(*a*) I_B (*b*) I_C (*c*) V_C

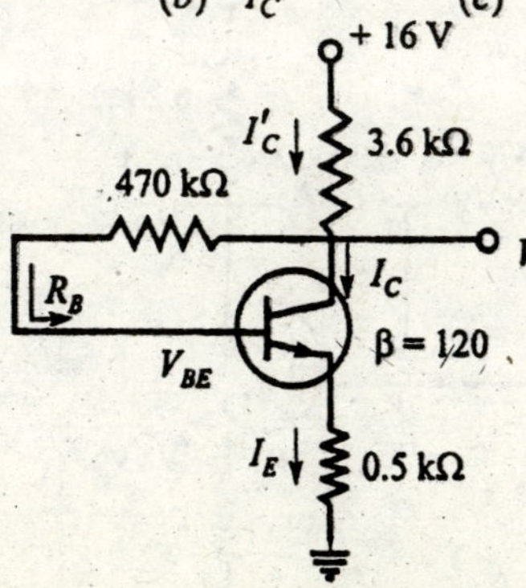

Fig. P. 3.6

Solution. $I'_C = I_C + I_B$

Take KVL in emitter base loop

$$16 = I'_C\, 3.6 + I_B \times 470 + V_{BE} + (\beta + 1)\, I_B\, 0.51$$

$$\Rightarrow \quad 16 = (I_C + I_B)\, 3.6 + 470\, I_B + 0.7 + (\beta + 1)\, I_B\, 0.51$$

$$\Rightarrow \quad 15.3 = (\beta + 1)(3.6 + 0.51)\, I_B + 470\, I_B$$

$$\Rightarrow \quad 15.3 = [(121)(4.11) + 470]\, I_B$$

$$\Rightarrow \quad 15.3 = 967.3\, I_B$$

$$\Rightarrow \quad I_B = \frac{15.3}{967.31} \text{ mA}$$

$$\Rightarrow \quad I_B = 15.82\ \mu\text{A}$$

$$I_C = \beta\, I_B = 120 \times 15.82$$

$$I_C = 1.898 \text{ mA}$$

$$\frac{16 - V_C}{3.6 \text{ K}} = I'_C$$

$$\Rightarrow \quad 16 - V_C = (I_C + I_B)\, 3.6 \text{ K}$$

$$\Rightarrow \quad 16 - V_C = (\beta + 1)\, I_B \times 3.6 \text{ K}$$

$$\Rightarrow \quad 16 - V_C = 121 \times 15.82 \times 10^{-3} \times 3.6$$

$$\Rightarrow \quad 16 - V_C = 6.891$$

$$V_C = 16 - 6.891$$

$$\Rightarrow \quad V_C = 9.1088 \simeq 9.11 \text{ V}$$

Hence

$$I_B = 15.82\ \mu\text{A}$$

$$I_E = 1.898 \text{ mA}$$

$$V_E = 9.11 \text{ V}$$

Problem 3.7. For the voltage feedback network shown in Fig. P. 3.7, determine:

(*a*) I_C (*b*) V_C (*c*) V_E (*d*) V_{CE} (*e*) I_B

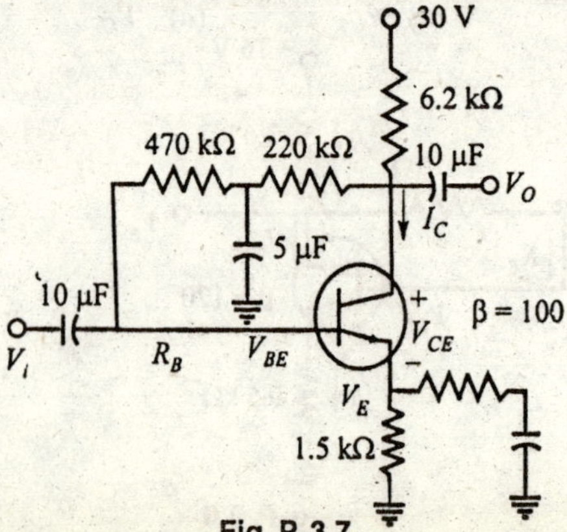

Fig. P. 3.7

Solution. For $\omega = 0$ all the capacitor acts as an open circuit, and the circuit reduce to

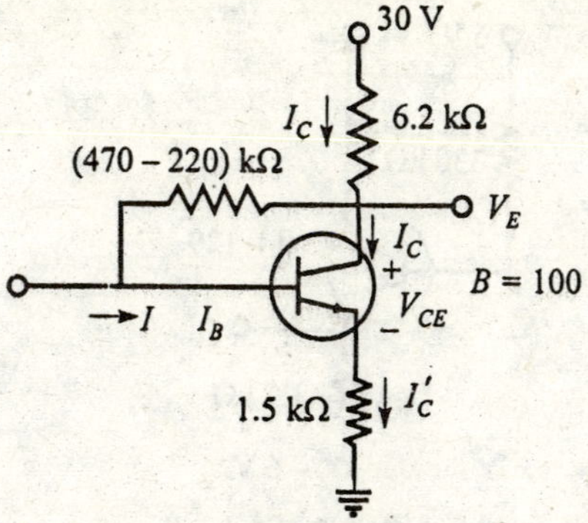

Fig. 3.7 (a)

Apply KVL at emitter base loop

$$30 = 1.5\,I'_C + V_{BE} + 690\,I_B + 6.2\,I'_C$$

$$\Rightarrow \quad 30 = 7.7\,I'_C + 0.7 + 690\,I_B$$

$$\Rightarrow \quad 30 = 7.7\,(\beta + 1)\,I_B + 0.7 + 690\,I_B$$

$$\Rightarrow \quad 29.3 = (7.7 \times 101 + 690)\,I_B$$

$$\Rightarrow \quad I_B\,(777.7 + 690) = 29.3$$

$$\Rightarrow \quad I_B = \frac{29.3}{1467.7}\text{ mA}$$

$$\Rightarrow \quad I_B = 0.01996\text{ mA}$$

Hence

$$I_C = \beta\,I_B = 1.996\text{ mA} \simeq 2\text{ mA}$$

$$I'_C = (\beta + 1)\,I_B = 101 \times 0.01996 = 2.0163\text{ mA}$$

$$\therefore \quad V_E = I'_C \times 1.5\text{ K}$$

$$\Rightarrow \quad V_E = 2.0163 \times 1.5$$

$$\Rightarrow \quad V_E = 3.024\text{ V}$$

$$V_C = 30 - I'_C\ 6.2\text{ K} = 30 - 2.0163 \times 6.2 = 17.5\text{ V}$$

$$V_{CE} = V_C - V_E = 17.5 - 3.024 = 14.475\text{ V}$$

$\therefore$ (*a*) $I_C = 2$ mA (*b*) $V_C = 17.5$ V (*c*) $V_E = 3.024$ V

(*d*) $V_{CE} = 14.475$ (*e*) $I_B = 0.01996$ mA

Problem 3.8. Determine the level of V_E and I_E for the network shown in Fig. P. 3.8.

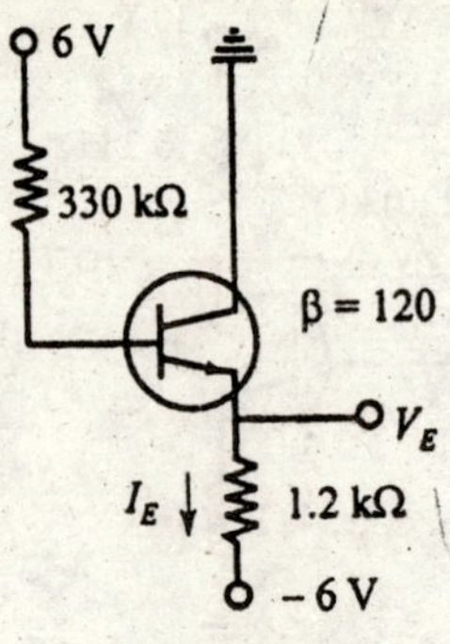

Fig. P. 3.8

Solution. Taking the KVL in the emitter base loop

$$6 - I_B\, 330\text{ K} - V_{BE} - 1.2\text{ K}\, I_E + 6 = 0$$

$$\Rightarrow \quad 12 - 0.7 = I_B\, 330\text{ K} + 1.2\text{ K}\,(\beta + 1)\, I_B$$

$$\Rightarrow \quad 11.3 = [\,330\text{ K} + 1.2 \times 121\text{ K}\,]\, I_B$$

$$I_B = \frac{11.3}{475.2}\text{ mA}$$

$$\Rightarrow \quad I_B = 0.02378\text{ mA} \qquad \Rightarrow I_B = 23.78\ \mu\text{A}$$

$$\therefore \quad I_E = (\beta + 1)\, I_B = 121 \times I_B$$

$$\Rightarrow \quad I_E = 2.877\text{ mA} \qquad \Rightarrow I_E \approx 2.88\text{ mA}$$

$$V_E = -6 + I_E \times 1.2\text{ K} \qquad \Rightarrow V_E = -6 + 2.88 \times 1.2$$

$$\Rightarrow \quad V_E = -6 + 3.453 \qquad \Rightarrow V_E = -2.547\text{ V}$$

Hence $\quad I_E = 2.88\text{ mA},\ V_E = -2.547\text{ V}$

Problem 3.9. For the network of figure determine:

(*a*) I_E (*b*) V_C (*c*) V_{CE}

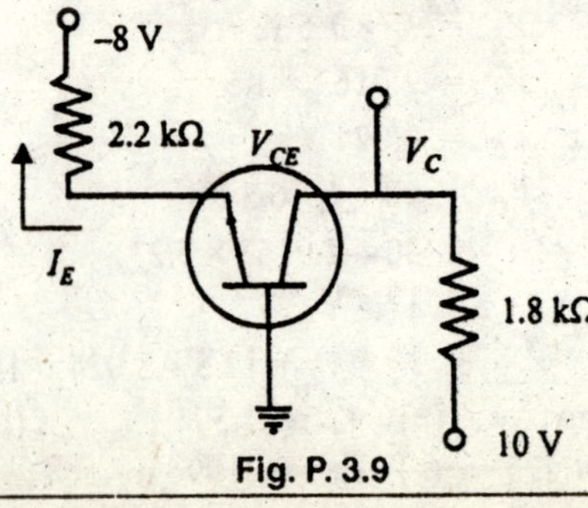

Fig. P. 3.9

Solution. Apply KVL in emitter base loop

$$-8 + 2.2\,I_E + V_{BE} = 0$$

$$-8 + 2.2\,I_E + 0.7 = 0$$

$$\Rightarrow \quad 2.2\,I_E = +7.3$$

$$\Rightarrow \quad I_E = \frac{7.3}{2.2}\text{ mA}$$

$$\Rightarrow \quad I_E = 3.3182\text{ mA}$$

$$V_C = 10 - I_E \times 1.8$$

$$= 10 - 1.8 \times 3.3182$$

$$= 4.027\text{ V}$$

Take KVL in emitter collector loop

$$-8 + 2.2\,I_E + V_{CE} + 1.28\,I_E - 10 = 0$$

$$\Rightarrow \quad 18 - 4\,I_E = V_{CE}$$

$$\Rightarrow \quad V_{CE} = 18 - 4 \times 3.3182$$

$$\Rightarrow \quad V_{CE} = 18 - 13.2728$$

$$\Rightarrow \quad V_{CE} = 4.7272\text{ V}$$

$$\Rightarrow \quad V_{CE} = 4.73\text{ V}$$

Hence

(*a*) $I_E = 3.32$ mA (*b*) $V_C = 4.027$ V (*c*) $V_{CE} = 4.73$ V

Problem 3.10. Determine V_C and I_B for the network below, (PNP):

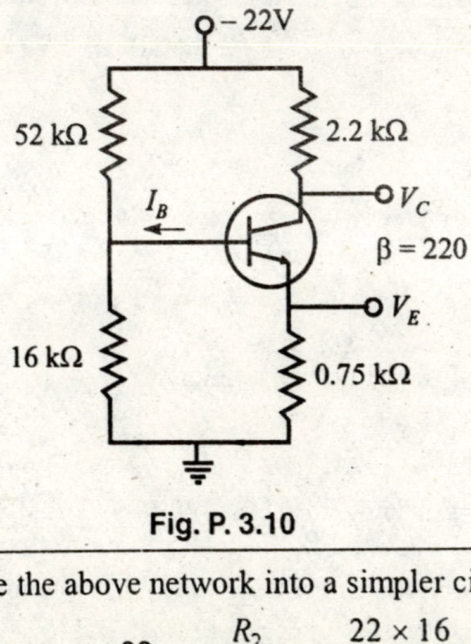

Fig. P. 3.10

Solution. Reduce the above network into a simpler circuit as below:

$$V_B = 22 \times \frac{R_2}{R_1 + R_2} = \frac{22 \times 16}{98} = 3.592\text{ V}$$

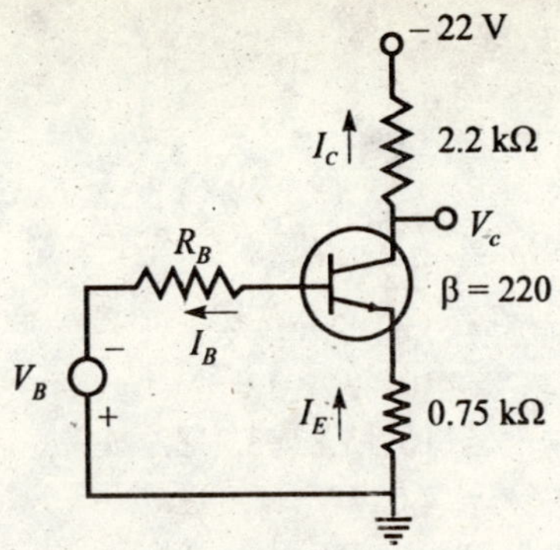

Fig. 3.10 (a)

$$R_B = R_1 \parallel R_2 = \frac{16 \times 82}{98} = 13.39\,\text{k}\Omega$$

Apply KVL in base emitter loop

$$-3.592 + I_B R_B + V_{BE} + 0.75\,(\beta + 1)\,I_B = 0$$

$$\Rightarrow \quad (221 \times 0.75 + 13.39)\,I_B = 3.592 - 0.7$$

$$\Rightarrow \quad 179.14\,I_B = 2.892$$

$$\Rightarrow \quad I_B = 16.14\ \mu\text{A}$$

$$I_C = \beta I_B$$

$$= 220 \times 16.14\ \mu\text{A}$$

$$= 3.5516\ \text{mA}$$

$$\frac{V_C + 22}{2.2} = I_C$$

$$\Rightarrow \quad V_C = -22 + 3.5576 \times 2.2$$

$$\Rightarrow \quad V_C = -22 + 7.813$$

$$\Rightarrow \quad V_C = -14.186\ \text{V}$$

Problem 3.11. For the network determine:

(*a*) $S(I_{CO})$

(*b*) $S(V_{BE})$

(*c*) $S(\beta)$ using T_1 as the temperature, at which the parameter values are specified and $\beta(T_2)$ as 25% more than $\beta(T_1)$.

(*d*) Determine the net change in I_C, if a change in operating conditions result in I_{CO} increasing from 0.2 to 10 μA, V_{BE} drops from 0.7 to 0.5 V and β increase 25%.

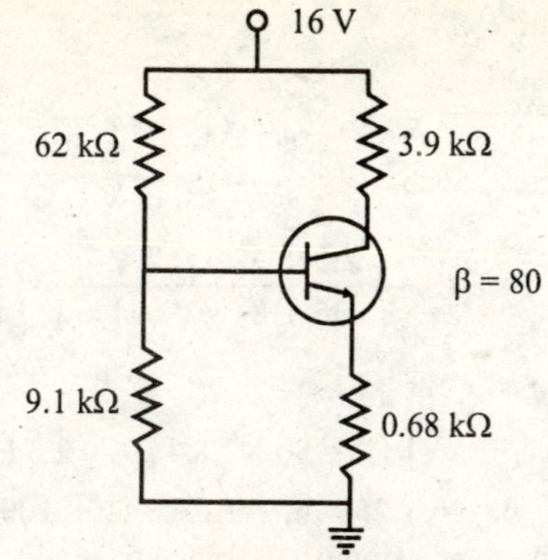

Fig. F. 3.11

Solution. $$S(I_{CO}) = (\beta + 1)\frac{[1 + R_{th}/R_E]}{(\beta + 1) + (R_{th}/R_E)}$$

Where $R_{th} = R_1 \parallel R_2$

$$\Rightarrow \quad R_{th} = \frac{62 \times 9.1}{71.1} = 7.935 \text{ k}\Omega$$

(*a*) $$S(I_{CO}) = 81\frac{\left[1 + \dfrac{7.935}{0.68}\right]}{81 + \left(\dfrac{7.935}{0.68}\right)} = \frac{81\,[1 + 11.669]}{81 + 11.669}$$

$$= \frac{81 \times 12.669}{92.669} = 11.073$$

(*b*) $$S(V_{BE}) = \frac{-\beta}{R_{th} + (\beta + 1)R_E}$$

$$= \frac{-80}{(7.935 + 81 \times 0.68) \times 10^3} = \frac{-80}{63.015 \times 10^3}$$

$$= -1.269 \times 10^{-3} \text{ mho}$$

$$= -1.27 \times 10^{-3}\, S$$

(*c*) $$S(\beta) = \frac{\Delta I_E}{\Delta \beta} = \frac{I_{C_1}(1 + R_{th}/R_E)}{\beta_1(1 + \beta_2 + R_{th}/R_E)}$$

Now $$\frac{R_{th}}{R_E} = \frac{7.935 \text{ k}\Omega}{0.68 \text{ k}\Omega} = 11.67$$

Now, calculation of I_{C_1}

$$I_{B_1} = \frac{V_{th} - V_{BE}}{R_{th} + (\beta + 1) R_E}, \text{ where } V_{th} = (16 \text{ V}) \times \frac{9.1 \text{ k}\Omega}{9.1 \text{ k}\Omega + 62 \text{ k}\Omega}$$

$$= 2.048 \text{ V}$$

Therefore $\quad I_{B_1} = \dfrac{2.048 \text{ V} - 0.7 \text{ V}}{(7.935 + 81 \times 0.68) \text{ k}\Omega} = \dfrac{1.348 \text{ V}}{63.015 \text{ k}\Omega}$

or $\quad I_{B_1} = 21.4 \text{ μA}$

and hence $\quad I_{C_1} = \beta I_{B_1} = 80 \times 21.4 \text{ μA} = 1.711 \text{ mA}$

Given $\quad \beta_2 = 1.25 \times b_1 = 1.25 \times 80 = 100$

Therefore, from the expression of $S(\beta)$, we have

$$S(\beta) = \frac{(1.711 \text{ mA})(1 + 11.67)}{80(1 + 100 + 11.67)}$$

or $\quad S(\beta) = \dfrac{21.7 \times 10^{-3}}{9013.6} = 2.408 \times 10^{-6}$

$$S(\beta) \approx 2.41 \times 10^{-6}$$

(*d*) Given $\quad \Delta I_{CO} = (10 - 0.2) \text{ μA} = 9.8 \text{ μA}$

$$\Delta V_{BE} = (0.7 - 0.5) \text{ V} = 0.2 \text{ V}$$

and $\quad \Delta\beta = 0.25 \times 80 = 20$

For the net change of I_C, we have the relation

$$\Delta I_C = S(I_{CO})\, \Delta I_{CO} + S(V_{BE})\, \Delta V_{BE} + S(\beta)\, \Delta\beta$$

or $\quad \Delta I_C = 11.073 \times 9.8 \times 10^{-6} + \left| (-1.27) \times 10^{-3} \right| \times 0.2 + 2.41 \times 10^{-6} \times 20$

$$= 10^{-6} (108.515 + 254 + 48.2)$$

$$= 10^{-6} \times 410.8 \cong 0.411 \text{ mA}$$

Problem 3.12. Define the transistor inverter of figure to operate with a saturation current of 8 mA using a transistor with a beta of 100. Use a level of I_B equal to 120% of I_B max and standard resistor values.

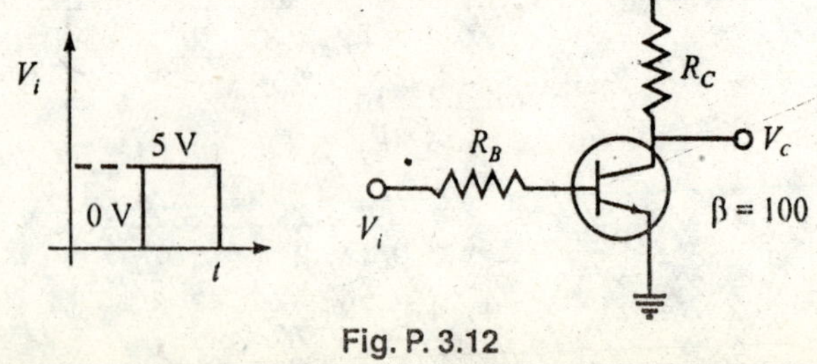

Fig. P. 3.12

Solution. The output wave form is given as

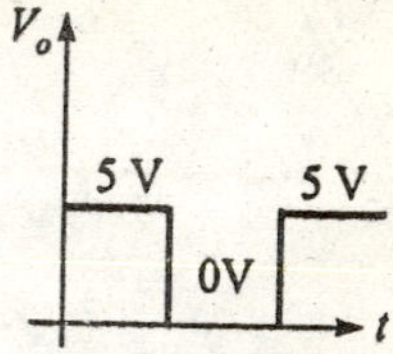

Fig. 3.12 (a)

Given

$$I_{c\,sat} = 8 \text{ mA}$$

$$\beta = 100$$

$$I_B = 120\% \text{ of } I_B \text{ max}$$

$$I_B = 1.2\, I_B \text{ max}$$

$$R_B = R_C = ?$$

$$R_C = \frac{V_{CC}}{I_{Csat}} \qquad (\because V_{CE\,sat} = 0)$$

$$\Rightarrow \quad R_C = \frac{5}{8} = 0.625 \text{ k}\Omega$$

$$\Rightarrow \quad R_C = 625 \text{ k}\Omega$$

$$I_{B\max} \frac{I_{Csat}}{\beta} = \frac{8}{100} = 80 \ \mu\text{A}$$

$$I_B = 1.2 \times 80 \ \mu\text{A} = 96 \ \mu\text{A}$$

$$R_B = \frac{V_i - V_{BE}}{I_B} = \frac{5 - 0.7}{96 \times 10^{-6}} = 44.79 \text{ k}\Omega$$

Problem 3.13. Determine the range possible values for V_C for the network of figure using 1 mΩ potentiometer.

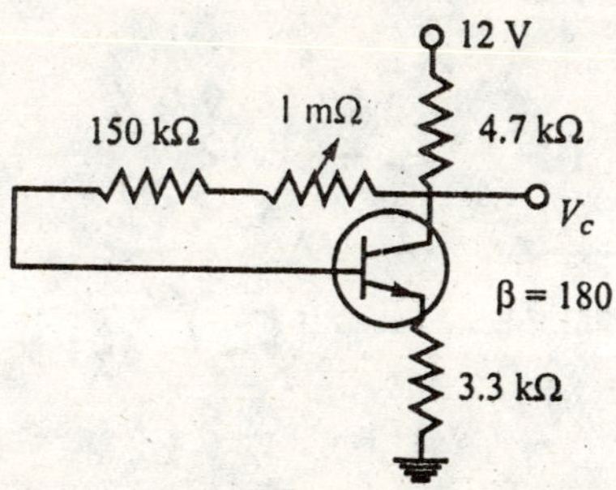

Fig. P. 3.13

Solution. The range of possible values for V_C is obtained (*i*) by short-circuit 1mΩ and (*ii*) by keeping the full 1 mΩ resistor.

(*i*) Short-circuit 1 mΩ resistor

Apply KVL in emitter base loop

$$12 = 3.3\,(\beta + 1)\, I_B + 0.7 + 150\, I_B + 4.7\, \beta\, I_B$$

$$\Rightarrow \quad 12 = (181 \times 3.3 + 150 + 4.7 + 180)\, I_B + 0.7$$

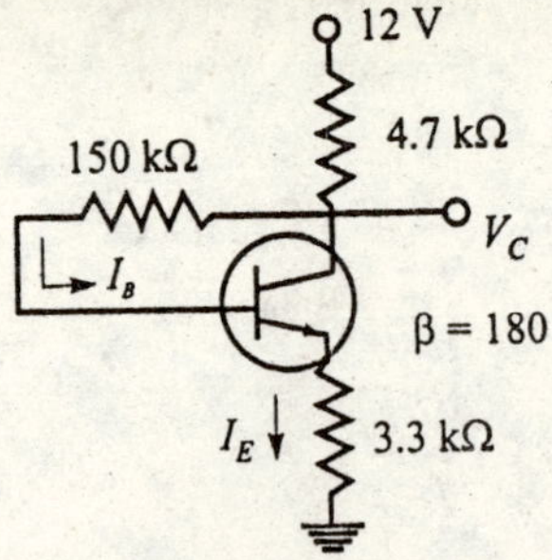

Fig. 3.13 (a)

$$\Rightarrow \quad 11.3 = (1593.3)\, I_B$$

$$\Rightarrow \quad I_B = 7.092 \text{ mA}$$

$$I_C = \frac{12 - V_C}{4.7} \qquad \therefore \quad I_C = \beta\, I_B$$

$$I_C = 180 \times 7.092$$

$$= 1.276 \text{ mA}$$

$$\Rightarrow \quad V_C = 12 - I_C\, 4.7$$

$$= 12 - 180 \times 4.7 \times 7.092$$

$$= 12 - 6$$

$$= 6 \text{ V}$$

(*ii*) When 1 mΩ is present.

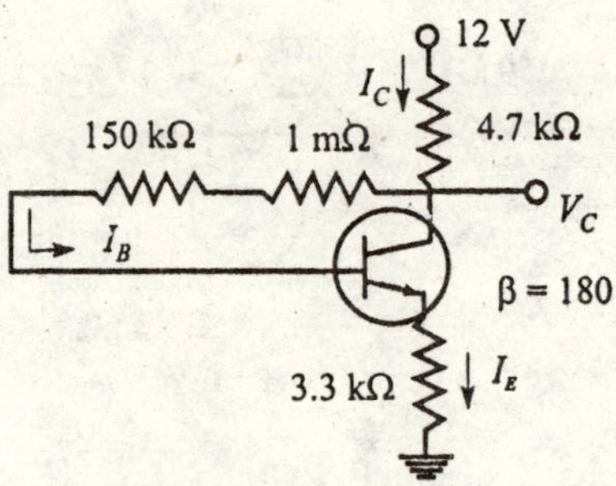

Fig. 3.13 (b)

Apply KVL in emitter base loop

$$12 = I_E\, 3.3 \text{ K} + V_{BE} + 1150 \text{ K}\, I_B + 4.7 \text{ K}\, I_C$$

$$\Rightarrow \quad 11.3 = (\beta + 1)\, I_B\, 3.3 \text{ K} + 1150 \text{ K}\, I_B + \beta\, 4.7 \text{ K}\, I_B$$

$$\Rightarrow \quad 11.3 = (181 \times 3.3 \text{ K} + 1150 \text{ K} + 180 \times 4.7 \text{ K})\, I_B$$

$$\Rightarrow \quad 11.3 = (597.3 \text{ K} + 1150 \text{ K} + 846 \text{ K})\, I_B$$

$$\Rightarrow \quad I_B = \frac{11.3}{2593.3 \text{ K}} = 4.357 \ \mu\text{A}$$

$$I_C = \beta I_B = 0.784 \text{ mA}$$

$$V_C = 12 - I_C \times 4.7 \text{ K}$$

$$= 12 - 3.686 = 8.3136 \text{ V}$$

Hence, the range of voltage V_C is between 6 V to 8.313 V.

Problem 3.14. In the figure shown below, has a Si transistor used has $\beta \geq 30$ and $I_{CBO} = 100$ A. Determine:

(*i*) The values of V_o for $V_i = 12$ V and $R = 20$ kΩ and show that the transistor is in saturation.

(*ii*) The minimum values of R for the transistor to remain in the active region for $V_i = 12$ V.

(*iii*) The values of V_o for $V_i = 1$ V and $R = 15$ kΩ.

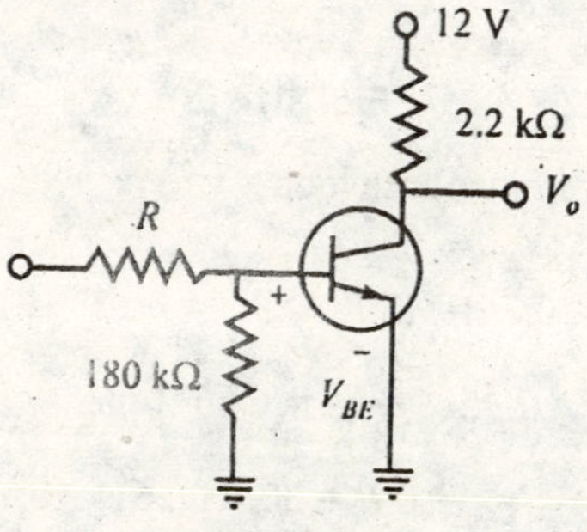

Fig. P. 3.14

Solution. (*i*) When $V_i = 12$ V , $R = 20$ kΩ then , R and 180 kΩ resistor are in parallel, i.e. voltage divider bias configuration.

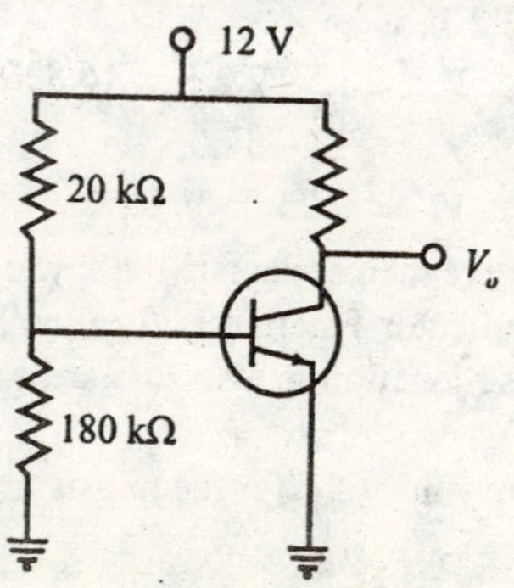

Fig. 3.14 (a)

Reduce the above circuit in a simpler form.

$$V_B = 12 \times \frac{180}{200} = 10.8\,\text{V}$$

$$R_B = 20\text{ K} \parallel 180\text{ K} = \frac{20 \times 180}{200} = 18\,\text{k}\Omega$$

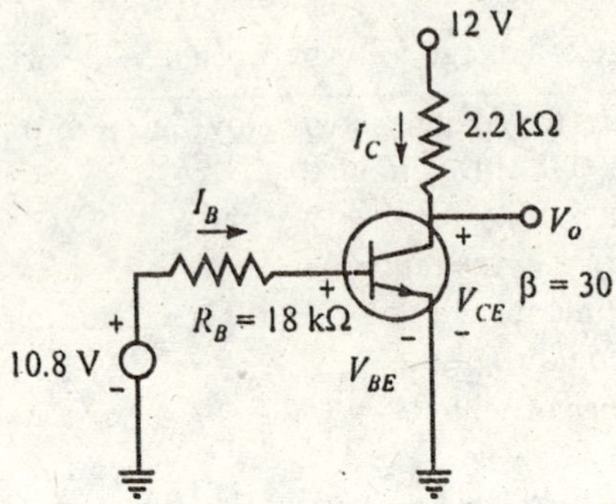

Fig. 3.14 (b)

Take KVL in the emitter base loop

$$10.8 - 18\text{ K}\, I_B - V_{BE} = 0$$

$$10.8 - 0.7 = 18\text{ K}\, I_B$$

$$\Rightarrow \quad I_B = \frac{10.1}{18\text{ K}} = 0.5611\text{ mA}$$

$$I_C = \beta I_B = 30 \times 0.5611$$

$$= 16.833\text{ mA}$$

To calculate V_o

$$\frac{12 - V_o}{2.2\text{ K}} = I_C$$

$$\Rightarrow \quad V_o = 12 - 2.2\text{ K} \times 16.833 \times 10^{-3}$$

$$\Rightarrow \quad V_o = 12 - 37.03$$

$$\Rightarrow \quad V_o = -25.03\text{ V}$$

Since, V_o is negative then collector is at lower potential then emitter and hence, the emitter collector function is forward bias since the transistor is *npn*. The emitter base junction is also forward bias and the transistor is in saturation.

(*ii*) The minimum value of R for the transistor to remain in active region when $V_i = 12$ V

We get $$I_{csat} = \frac{12}{2.2\text{ K}} \qquad \{\because V_{CE} = 0\}$$

$$= 5.4545\text{ mA}$$

$$I_B = \frac{I_{c\,sat}}{\beta_{max}} = \frac{5.4545}{30} = 0.1818 \text{ mA}$$

$$R_B = \frac{R \times 180}{180 + R} \text{ and } V_B = \frac{12 \times 180}{(R + 180)}$$

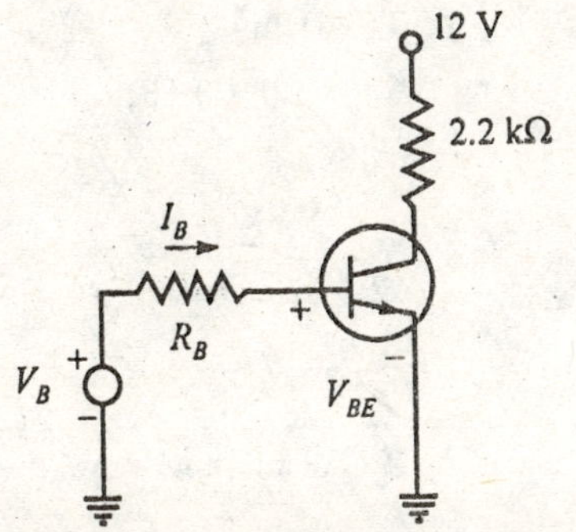

Fig. 3.14 (c)

Take KVL in emitter base loop

$$V_B = I_B R_B + V_{BE}$$

$$\Rightarrow \quad \frac{12 \times 180}{(R + 180)} = \frac{0.1818 \times 180R}{R + 180} + 0.7$$

$$\Rightarrow \quad 2160 = 32.724\,R + 0.7\,R + 126$$

$$\Rightarrow \quad 33.424\,R = 2034$$

$$\Rightarrow \quad R_{min} = 60.854 \text{ k}\Omega$$

(*iii*) $V_i = 1$ V, $R = 15$ kΩ

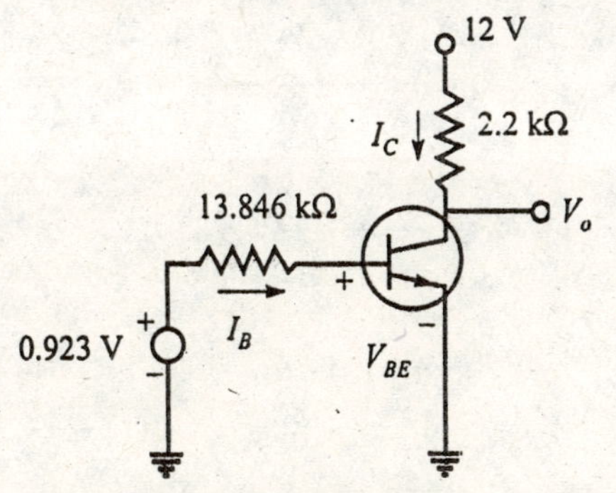

Fig. 3.14 (d)

$$R_B = 15\text{ K} \parallel 180\text{ K}, \quad V_B = \frac{180 \times 1}{180 + 15} \text{ V} = 0.923 \text{ V}$$

$$= \frac{15 \times 180}{195} = 13.846 \text{ k}\Omega$$

In emitter base loop

$$0.923 = 13.846\, I_B + 0.7$$

$$\Rightarrow \quad I_B = 0.0161 \text{ mA}$$

$$I_C = \beta\, I_B = 30 \times 0.161$$

$$= 0.4833 \text{ mA}$$

Now

$$V_o = V_C - 2.2 \text{ K}\, I_C$$

$$= 12 - 2.2 \times 0.4833$$

$$= 10.936 \text{ V}$$

Problem 3.15. Determine ICQ at a temperature of 100° C, if $I_{CQ} = 2$ mA at 25° C. Use $\beta_1 = 50$ and $\beta_2 = 80$ and a resistance ratio $\frac{R_B}{R_E}$ of 20.

Solution. From the problem, it is evident that out of three variations possible only β is varying. Neither V_{BE} nor I_{CO} is varying, hence, we will use the relation of sensitivity regarding beta, i.e of S_β.

$$S_\beta = \frac{I_{C_1}\left(1 + \frac{R_B}{R_E}\right)}{\beta_1\left(1 + \beta_2 + \frac{R_B}{R_E}\right)}$$

or

$$S_\beta = \frac{\left(2 \times 10^{-3}\right)(1 + 20)}{50\,(1 + 80 + 20)} = \frac{42 \times 10^{-3}}{5050}$$

or

$$S_\beta = \frac{42 \times 10^{-6}}{5.050} = 8.32 \times 10^{-6}$$

and

$$\Delta I_C = S(\beta)\, \Delta\beta$$

or,

$$\Delta B = 80 - 50 = 30$$

hence

$$\Delta I_C = 8.32 \times 10^{-6} \times 30$$

$$= 250 \times 10^{-6}$$

or

$$\Delta I_C = 0.25 \text{ mA}$$

Therefore, the collector current changed from 2 mA at room temperature to 2.25 mA at 100°C, representing a 12.5 % increase.

Problem 3.16. Given the information of a fixed biased circuit, determine:

(*a*) I_C (*b*) R_C (*c*) R_B (*d*) V_{CE}

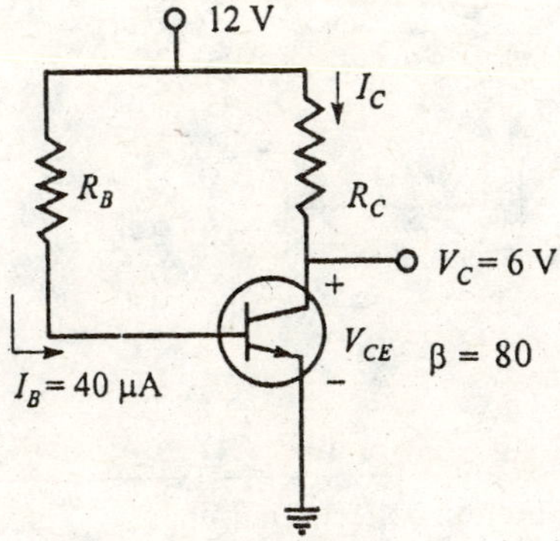

Fig. P. 3.16

Solution. It is an fixed biased circuit having $V_{BE} = 0.7$ V, taking KVL in the base emitter loop.

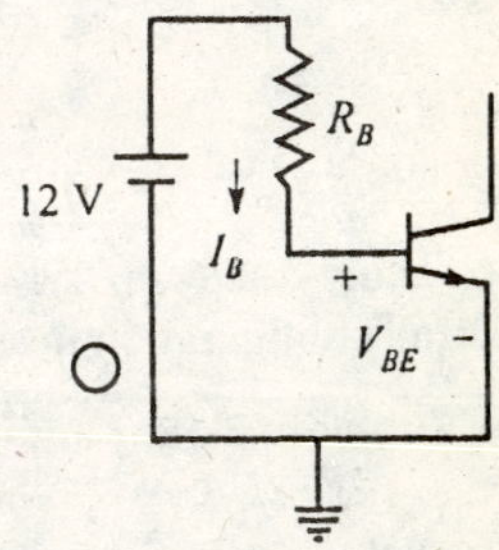

Fig. 3.16 (a)

$$12 = I_B R_B + V_{BE}$$

$$\Rightarrow \quad 12 = 40 \times 10^{-6} \times R_B + 0.7$$

$$\Rightarrow \quad R_B = \frac{12 - 0.7}{40 \times 10^{-6}}$$

$$\Rightarrow \quad R_B = 0.2825 \times 10^{-6}\ \Omega$$

$$\Rightarrow \quad R_B = 283.5\ \text{k}\Omega$$

$$V_{CE} = V_C - V_E = V_C - 0 \quad \text{(since the emitter is grounded)}$$

$$\Rightarrow \quad V_{CE} = V_C = 6\ \text{V}$$

Now $\quad I_C = \beta I_B = 80 \times 40 \times 10^{-6} = 3.2$ mA

From the circuit, we have

$$R_C = \frac{V_{CC} - V_C}{I_C} = \frac{12 - 6}{3.2 \times 10^{-3}} = 1.875\ \text{k}\Omega$$

Thus, we have

(*a*) $I_C = 3.2$ mA (*b*) $R_C = 1.875$ kΩ

(*c*) $R_B = 282.5$ kΩ (*d*) $V_{CE} = 6$ V

Problem 3.17. Find the saturation current ($I_{c\ sat}$) for the fixed bias configuration shown below.

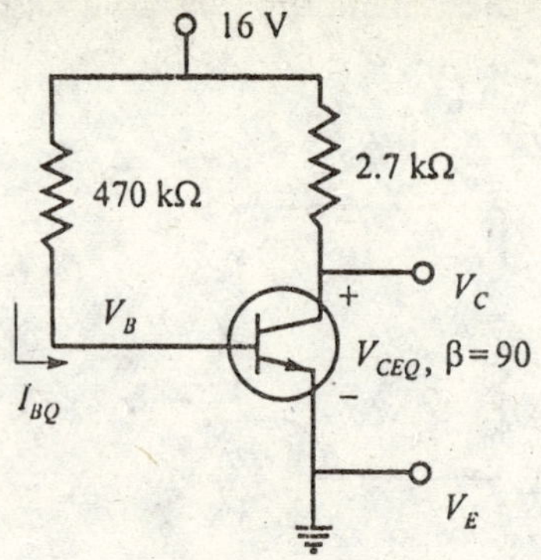

Fig. P. 3.17

Solution. The saturation collector current ($I_{c\ sat}$) is the maximum current that flows, when $V_{CE} = 0$ that is the collector emitter terminals are short circuit.

Thus $$I_{c\ sat} = \frac{V_{CC}}{R_C}$$

$$\Rightarrow \quad I_{c\ sat} = \frac{16\,\text{V}}{2.7\,\text{k}\Omega}$$

$$\Rightarrow \quad I_{c\ sat} = 5.926\ \text{mA}$$

Thus, 5.296 mA is the minimum collector current to drive the fixed biased configuration in saturation region.

Problem 3.18. For the emitter stabilized bias circuit determine:

(*a*) I_{BQ} (*b*) I_{CQ} (*c*) V_{CEQ} (*d*) V_C
(*e*) V_B (*f*) V_E

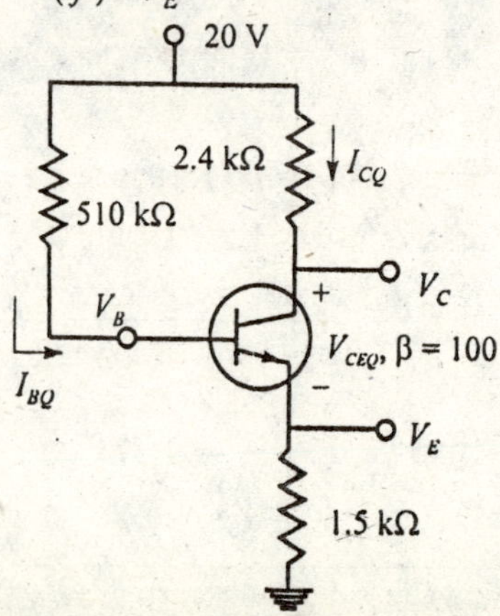

Fig. P. 3.18

Solution. Take KVL in the base emitter loop

$$20 = 510 \times I_B + V_{BE} + (I_B + I_C)\, 1.5$$

$$\Rightarrow \quad 20 - 0.7 = 510\, I_B + (\beta + 1)\, I_B \times 1.5$$

$$\Rightarrow \quad 19.3 = I_B\, 510 + 101 \times 1.5 \times I_B$$

$$\Rightarrow \quad I_B = \frac{19.3}{510 + 151.5} \text{ mA}$$

$$\Rightarrow \quad I_B = 0.0292 \text{ mA}$$

$$I_{BQ} = 29.2\ \mu\text{A}$$

$$I_{CQ} = \beta\, I_{BQ} = 100 \times 0.0292 = 2.92 \text{ mA}$$

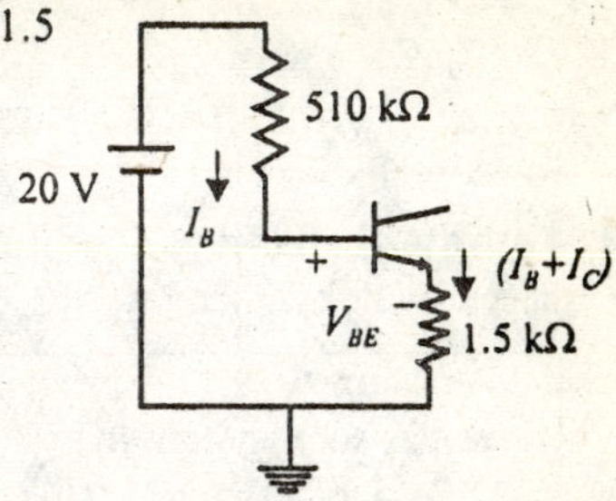

Fig. 3.18 (a)

$$I_E = (\beta + 1)\, I_{BQ} = 101 \times 0.0292 = 2.949 \text{ mA}$$

Thus $\quad V_E = I_E R_E = 2.949 \times 1.5 = 4.424$ V

$$V_B = V_{BE} + V_E = 0.7 + 4.424 = 5.124 \text{ V}$$

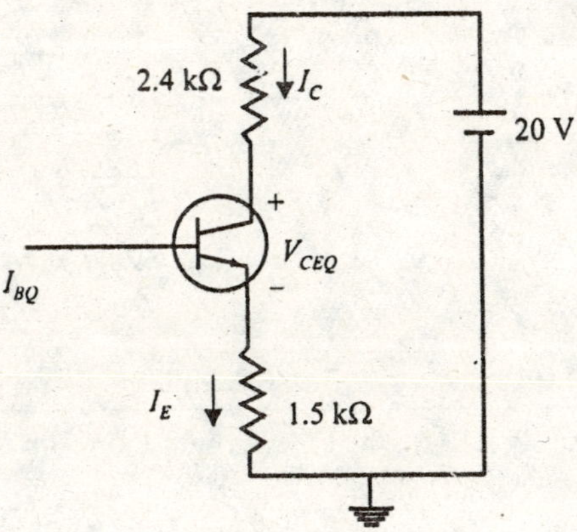

Fig. 3.18 (b)

Taking KVL in the emitter collector loop, we have

$$20 = 1.5 \times I_E + V_{CEQ} + 2.4 \times I_C$$

$$\Rightarrow \quad 20 = 1.5 \times 101 I_B + 2.4 \times 100\, I_B + V_{CEQ}$$

$$\Rightarrow \quad V_{CEQ} = -(1.5 \times 101 + 2.4 \times 100)\, I_B + 20$$

$$\Rightarrow \quad V_{CEQ} = 20 - 391.5 \times I_B$$

$$\Rightarrow \quad V_{CEQ} = 20 - 11.432$$

$$\Rightarrow \quad V_{CEQ} = 8.568 \text{ V}$$

Thus $\quad V_{CEQ} = V_C - V_E$

$$\Rightarrow \quad V_C = V_{CEQ} + V_E$$

$$\Rightarrow \quad V_C = 8.568 + 4.424$$

$$\Rightarrow \quad V_C = 12.992 \simeq 13 \text{ V}$$

Thus, we have

(*a*) $I_{BQ} = 29.2\ \mu A$ (*b*) $I_{CQ} = 2.92$ mA (*c*) $V_{CEQ} = 8.568$ V

(*d*) $V_C = 13$ V (*e*) $V_B = 5.124$ V (*f*) $V_E = 4.424$ V

Note. In this type of problem the resistor R_E is reflected back to the input base circuit by a factor of $(\beta + 1)$.

Problem 3.19. For emitter stabilized bias circuit, the information are given as such, determine

(*a*) β (*b*) V_{CC} (*c*) R_B

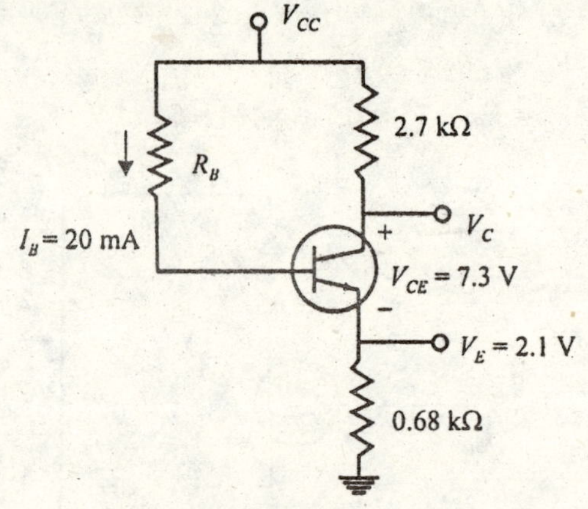

Fig. P. 3.19

Solution. Apply KVL in the base emitter loop with resistor R_E reflected back by a factor of $(\beta + 1)$.

$$V_{CC} = I_B R_B + V_{BE} + I_B R_B (\beta + 1)$$

$$\Rightarrow \quad V_{CC} = I_B [R_B + R_E(\beta + 1)] + V_{CE}$$

$$\Rightarrow \quad V_{CC} = 20 \times 10^{-6} [R_B + R_E(\beta + 1)] + V_{BE} \quad \text{... (1)}$$

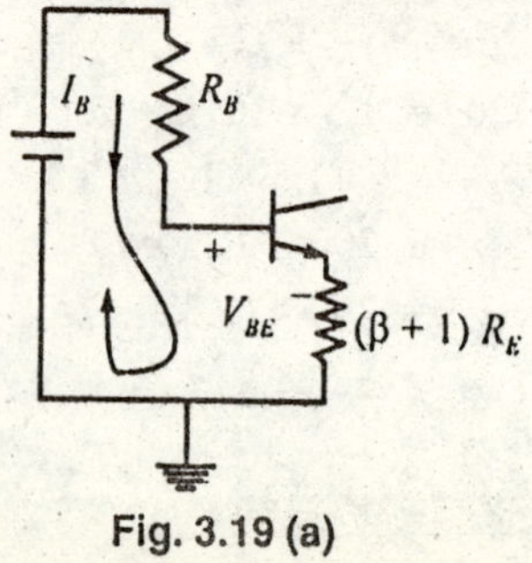

Fig. 3.19 (a)

From the circuit, we have

$$I_E = \frac{V_E}{R_E} = \frac{2.1}{0.68} = 3.088 \text{ mA}$$

$$(\beta + 1)\, I_B = I_E \qquad \text{and} \qquad I_C = \beta\, I_B = \frac{\beta\, R_E}{(\beta + 1)}$$

$$I_B = \frac{I_E}{(\beta + 1)} \qquad \Rightarrow \qquad I_C = \frac{\beta}{(\beta + 1)}\, I_E$$

Now $V_C = V_{CE} + V_E = 7.3 + 2.1 = 9.4$ V

Again, apply KVL in the collector emitter loop

$$V_{CC} = I_C R_C + V_{CE} + (\beta + 1)\, I_B R_E$$

$$\Rightarrow \quad V_{CC} = \beta\, I_B R_C + V_{CE} + (\beta + 1)\, I_B R_E$$

$$\Rightarrow \quad V_{CC} = \frac{\beta\, I_E}{(\beta + 1)} R_C + V_{CE} + I_B\, R_E \qquad \text{... (2)}$$

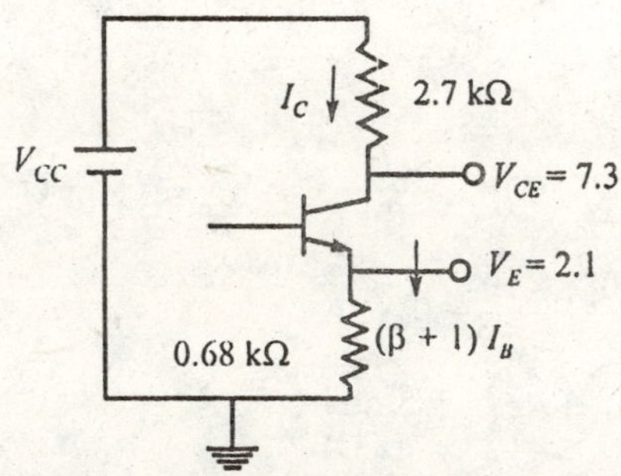

Fig. 3.19 (b)

From equation (1) and (2), we have

$$I_B\,[\,R_B + R_E(\beta + 1)\,] + V_{BE} = \frac{\beta\, I_E}{(\beta + 1)} R_C + V_{CE} + I_E\, R_E$$

$$\Rightarrow \quad I_B R_B + I_E R_E + V_{BE} = \beta\, I_B\, .\, R_C + V_{CE} + I_E R_E$$

$$\Rightarrow \quad I_B R_B - \beta\, I_B R_C = V_{CE} - V_{BE}$$

$$\Rightarrow \quad I_B (R_B - \beta\, R_C) = V_{CE} - V_{BE}$$

Put the values

$$\Rightarrow \quad \frac{I_E}{(\beta + 1)} (R_B - \beta \times 2.7) = 7.3 - 0.7$$

$$\Rightarrow \quad \frac{3.088}{(\beta + 1)} \left(R_B - \beta \times 2.7\right) = 6.6$$

For easier calculation take $\beta + 1 \simeq \beta$

$$\Rightarrow \quad 3.088 \left(\frac{R_B}{\beta} - 2.7\right) = 6.6$$

$$\Rightarrow \qquad \left(\frac{R_B}{\beta} - 2.7\right)\text{K} = 2.137\ \text{K}$$

$$\Rightarrow \qquad \frac{R_B}{\beta} - 2.7 = 20.137$$

$$\Rightarrow \qquad \frac{R_B}{\beta} = 4.837$$

Thus $\qquad R_B = 4.837\ \beta$

From the circuit, we get

$$\Rightarrow \qquad V_{CC} = I_C R_C + V_C$$

$$V_{CC} = I_E + V_C \qquad [\because I = I_C \cong I_E]$$

$$\Rightarrow \qquad V_{CC} = 3.088 \times 2.7 + 9.4$$

$$\Rightarrow \qquad V_{CC} = 17.74\ \text{V}$$

$$I_E = \frac{V_{CC} - V_C}{R_C} = 3.088\ \text{mA}$$

$$I_B = \frac{I_E}{\beta+1} \Rightarrow \frac{R_E}{I_B} = (\beta+1)$$

$$\Rightarrow \qquad \frac{3.088 \times 10^{-3}}{20 \times 10^{-6}} (\beta+1)$$

$$\Rightarrow \qquad (\beta+1) = 154.4$$

$$\Rightarrow \qquad \beta = 153.4$$

Thus $\qquad R_B = 4.837\ \beta = 4.837 \times 153.4$

$$\Rightarrow \qquad R_B = 741.99 \simeq 742\ \text{k}\Omega$$

Thus

(*a*) $\beta = 153.4$ (*b*) $V_{CC} = 17.74$ (*c*) $R_B = 742\ \text{k}\Omega$

Problem 3.20. Determine the saturation current ($I_{c\ sat}$) for the emitter stabilized biased circuit shown below:

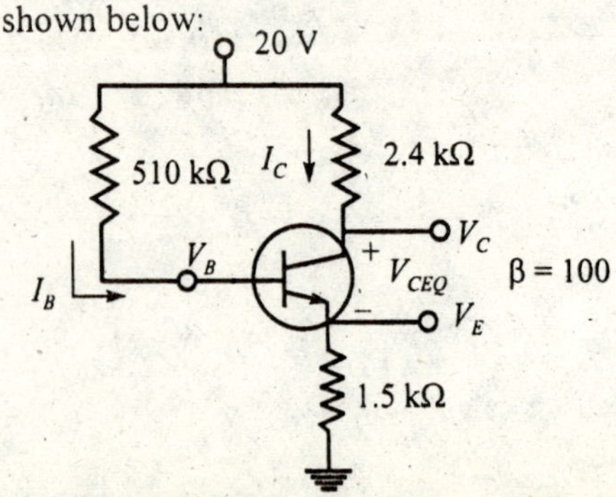

Fig. P. 3.20

Solution. For saturation level in emitter stabilized biased circuit, we have

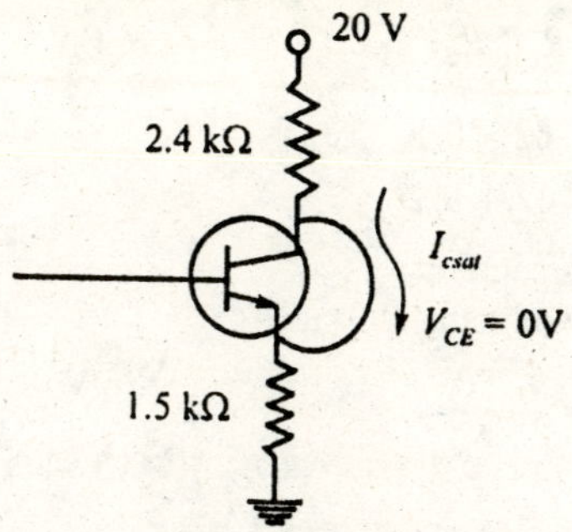

Fig. 3.20 (a)

$$\text{Thus} \qquad I_{c\,sat} = \frac{V_{CC}}{R_E + R_E}$$

$$\Rightarrow \qquad I_{c\,sat} = \frac{20}{(2.4 + 1.5)\text{ K}}$$

$$\Rightarrow \qquad I_{c\,sat} = 5.128 \text{ mA}$$

Thus, we find that the addition of the emitter resistor reduce the collector saturation level below that obtained with a fixed bias configuration ($I_{c\,sat} = \frac{V_{CC}}{R_C}$) having the same collector resistor.

Problem 3.21. For the voltage divider bias configuration, determine:

(*a*) I_{BQ} (*b*) I_{CQ} (*c*) V_{CEQ} (*d*) V_C
(*e*) V_E (*f*) V_B

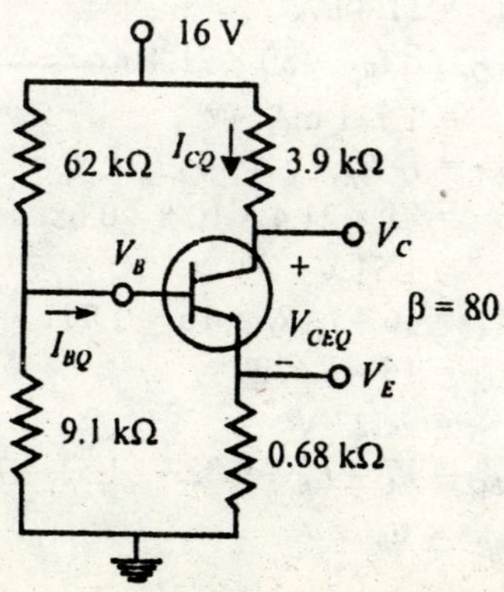

Fig. P. 3.21

Solution. The thevenin resistance and voltage are determined for the network as

$$R_B = R_1 \| R_2 \qquad V_B = \frac{R_2 V_{CC}}{R_1 + R_2}$$

$$= \frac{62 \times 9.1}{62 + 9.1} \qquad = \frac{9.1 \times 16}{(9.1 + 62)}$$

$$= 7.935 \text{ k}\Omega \qquad = 2.048 \text{ V}$$

The circuit is redrawn using R_B and V_B.

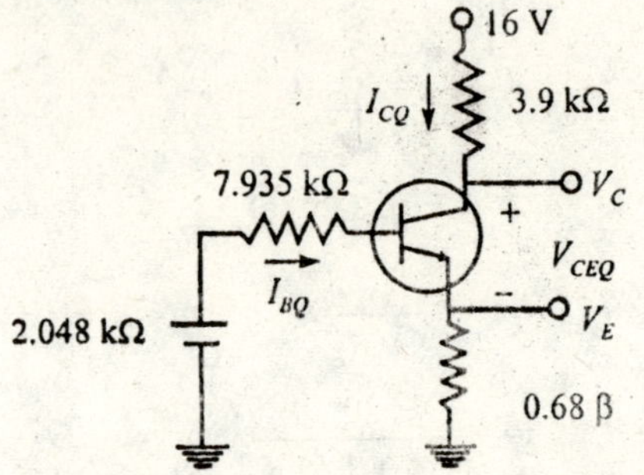

Fig. 3.21 (a)

Given $\beta = 80$

Now take KVL in the base emitter loop

$$2.048 = 0.68 \times (\beta + 1) I_B + V_{BE} + I_B \times 7.935$$

$$\Rightarrow \quad 2.048 - 0.7 = (55.08 + 7.935) I_B$$

$$\Rightarrow \quad I_B = \frac{1.348}{63.015\text{K}}$$

$$\Rightarrow \quad I_B = 0.0214 \text{ mA}$$

$$\Rightarrow \quad I_{BQ} = 21.4 \ \mu\text{A}$$

$$I_{CQ} = \beta I_{BQ} = 80 \times 21.4$$

$$= 1.711 \text{ mA}$$

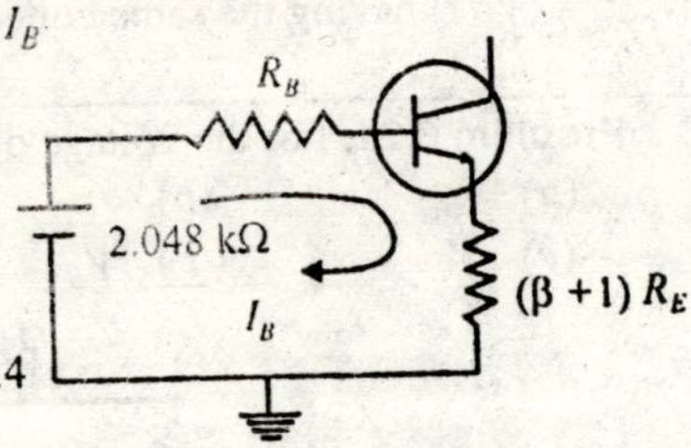

Fig. 3.21 (b)

$$V_E = I_E R_E = (\beta + 1) I_B \times R_E$$

$$= 81 \times 21.4 \times 10^{-6} \times 0.68 \times 10^3$$

$$\Rightarrow \quad = 1.51 \text{ V}$$

$$V_C = 16 - I_C R_C = 16 - 1.711 \times 3.9$$

$$= 16 - 6.6729$$

$$= 9.327 \text{ V}$$

Thus $V_{CEQ} = V_C - V_E = 9.327 - 1.51 = 7.817$ V

Now $V_{BE} = V_B - V_E$

$V_B = V_{BE} + V_E = 0.7 + 1.51 = 2.21$ V

Thus, we get

(a) $I_{BQ} = 21.4$ mA (b) $I_{CQ} = 1.711$ mA (c) $V_{CEQ} = 7.817$ V

(d) $V_C = 9.327$ V (e) $V_E = 1.51$ V (f) $V_B = 2.21$ V

Problem 3.22. Given the information of the voltage divider bias circuit, determine:

(a) I_C (b) V_E (c) V_{CC} (d) V_{CE}

(e) V_B (f) R_1

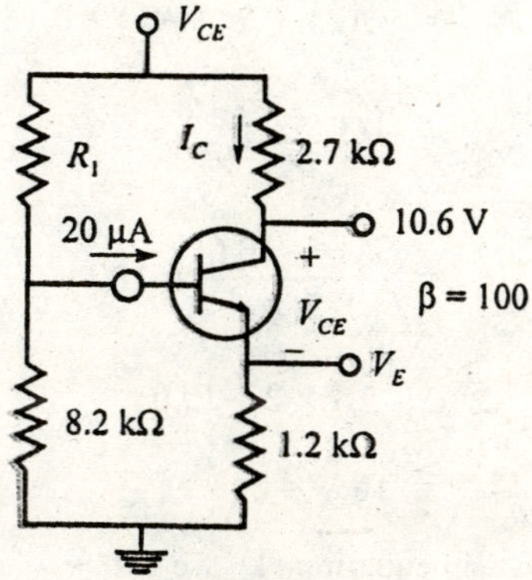

Fig. P. 3.22

Solution. From the figure, we have

$$I_B = 20\ \mu A,\ \beta = 100$$

The thevenin equivalent circuit is drawn.

$$R_B = R_1 \parallel R_2 \quad V_B = \frac{8.2}{R_1 + R_2} \times V_{CC} = \frac{R_1\ R_2}{R_1 + R_2} = \frac{8.2\ R_1}{(R_1 + R_2)}$$

$$V_B = \frac{8.2}{R_1 + R_2} \times V_{CC}$$

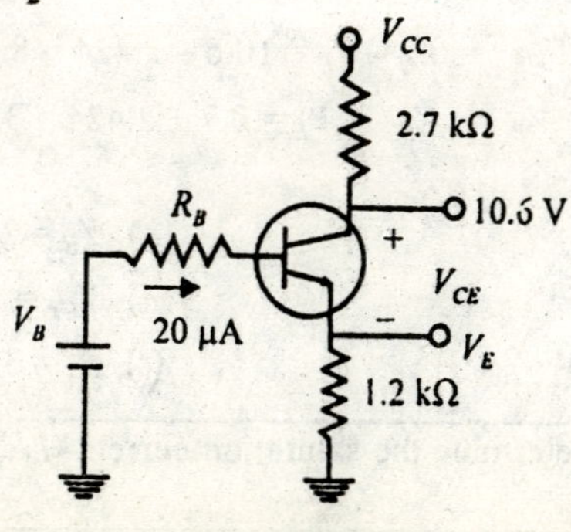

Fig. 3.22 (a)

Apply KVL in the base emitter loop

$$V_B = I_B(\beta + 1)\,1.2 \times 10^3 + V_{BE} + I_B R_B$$

$$\Rightarrow \frac{8.2\,V_{CC}}{(R_1 + R_2)} = 101 \times 20 \times 10^{-6} \times 1.2 \times 10^3 + 0.7 + 20 \times 10^{-6} \times \frac{8.2\,R_1}{(R_1 + R_2)}$$

$$\Rightarrow \frac{8.2\,V_{CC}}{(R_1 + R_2)} = 2.424 + 0.7 + \frac{0.164\,R_1}{R_1 + R_2}$$

$$\Rightarrow 8.2\,V_{CC} = 3.124\,(R_1 + R_2) + 0.164\,R_1 \quad \text{...(1)}$$

Now $\quad I_C = \dfrac{V_{CC} - 10.6}{R_C} \qquad I_C = \beta\, I_B$

$$\Rightarrow I_C = \frac{V_{CC} - 10.6}{2.7} = 100 \times 20 \times 10^{-6}$$

$$V_{CC} = 2.7 \times 2 + 10.6$$

$$\Rightarrow V_{CC} = 16\text{ V} = 2\text{ mA}$$

Put the value of V_{CC} in equation (1), we get

$$\Rightarrow 8.2\,V_{CC} = 3.124\,(R_1 + R_2) + 0.164\,R_1$$

$$\Rightarrow 8.2 \times 16 = 3.124\,(R_1 + 8.2) + 0.164\,R_1$$

$$\Rightarrow 131.2 = 25.617 + 3.288\,R_1$$

$$\Rightarrow R_1 = 32.11\text{ k}\Omega$$

Now
$$V_E = I_E R_E = (\beta + 1)\, I_B \times R_E$$
$$= 101 \times 20 \times 10^{-6} \times 1.2 \times 10^3$$
$$= 2.424\text{ V}$$
$$V_{CE} = V_C - V_E = 10.6 - 2.424 = 8.176\text{ V}$$
$$V_B = V_{BE} + V_E = 0.7 + 2.424 = 3.124\text{ V}$$

Thus, we get

(*a*) $I_C = 2$ mA (*b*) $V_E = 2.424$ V

(*c*) $V_{CC} = 16$ V (*d*) $V_{CE} = 8.176$ V

(*e*) $V_B = 3.124$ V (*f*) $R_1 = 32.11$ kΩ

Problem 3.23. Determine the saturation current ($I_{c\,sat}$) for the voltage divider network.

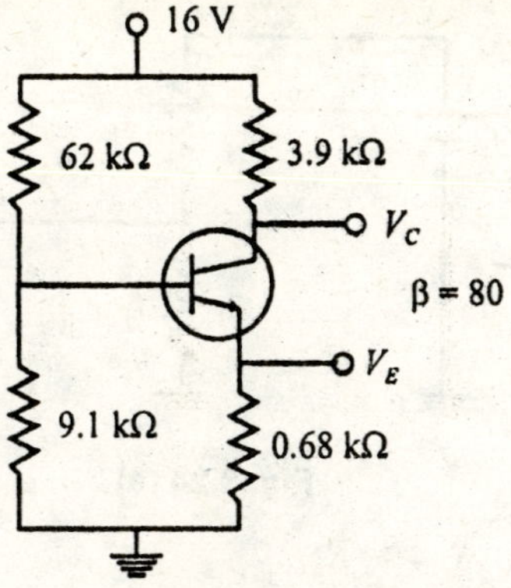

Fig. P. 3.23

Solution. The saturation collector current for the voltage divider configuration is given as

$$I_{c\,sat} = \frac{V_{CC}}{R_C + R_E} = \frac{16}{(3.9 + 0.68)\text{ K}}$$

$$= 3.493 \text{ mA}$$

Thus, we find that the emitter bias configuration and the voltage divider configuration has the same formula for the saturation collector current (approximately).

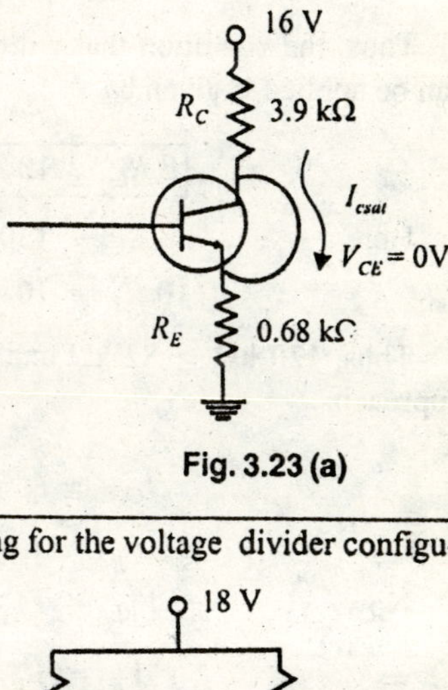

Fig. 3.23 (a)

Problem 3.24. Determine the following for the voltage divider configuration using the approximation approach, if the condition established is satisfied.

(*a*) I_C

(*b*) V_{CE}

(*c*) I_B

(*d*) V_B

(*e*) V_E

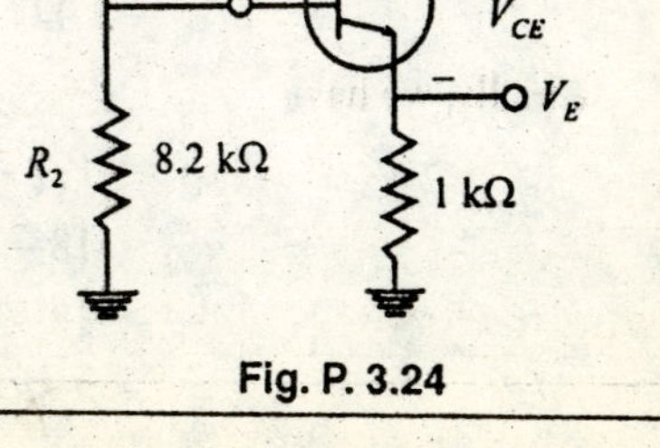

Fig. P. 3.24

Solution. For approximate analysis

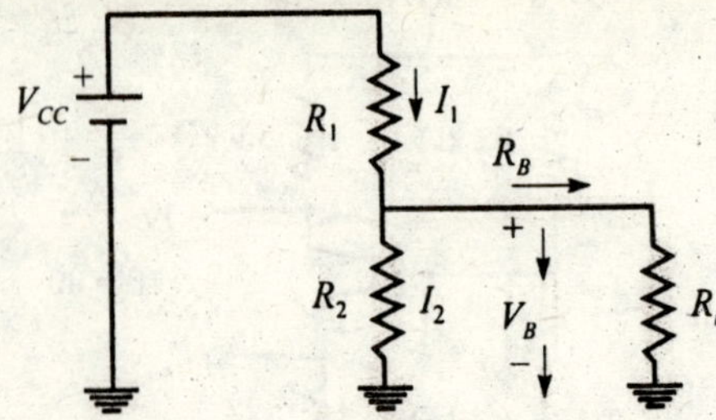

Fig. 3.24 (a)

$$R_i >> R_2$$
$$R_i = (\beta + 1) R_E$$
$$\simeq \beta R_E$$

Thus, the condition that will define whether the approximate approach can be applied is given by

$$\boxed{\beta R_E \geq 10 R_2}$$

Here $\beta R_E = 120 \times 1 = 120 \text{ k}\Omega$

$$10 R_2 = 10 \times 8.2 = 8.2 \text{ k}\Omega$$

Thus, 120 kΩ ≥ 82 kΩ and the condition is satisfied for approximate approach.

$$I_{CQ} \cong I_E \qquad V_B = \frac{R_2 V_{CC}}{R_1 + R_2} = \frac{8.2 \times 18}{8.2 + 39}$$

⇒ $V_B = 3.127$ V

Now $V_{BE} = V_B - V_E$

⇒ $V_E = V_B - V_{BE} \qquad I_{CQ} \cong I_E = \frac{V_E}{R_E}$

⇒ $V_E = 3.127 - 0.7 \qquad I_{CQ} = \frac{2.427}{1} = 2.427 \text{ mA}$

$$= 2.427 \text{ V}$$

Finally, we have

$$V_{CEQ} = V_{CC} - I_{EC} (R_C + R_E)$$
$$= 18 - 2.427 (3.3 + 1)$$
$$= 18 - 10.437$$
$$= 7.563 \text{ V}$$

$$I_B = \frac{I_{CQ}}{\beta} = \frac{2.427}{120} = 0.020 \text{ mA} = 20 \text{ μA}$$

Thus, we have

(*a*) $I_C = 2.427$ mA (*b*) $V_{CE} = 7.563$ V (*c*) $I_B = 20$ μA

(*d*) $V_B = 3.127$ V (*e*) $V_E = 2.427$ V

Problem 3.25. Repeat the above problem using the exact (thevenin) approach and compare solution. Based on the result is the approximate approach a valid analysis technique, if the $\beta R_E \geq 10 R_2$ condition is satisfied?

(*a*) I_C (*b*) V_{CE} (*c*) I_B (*d*) V_B (*e*) V_E

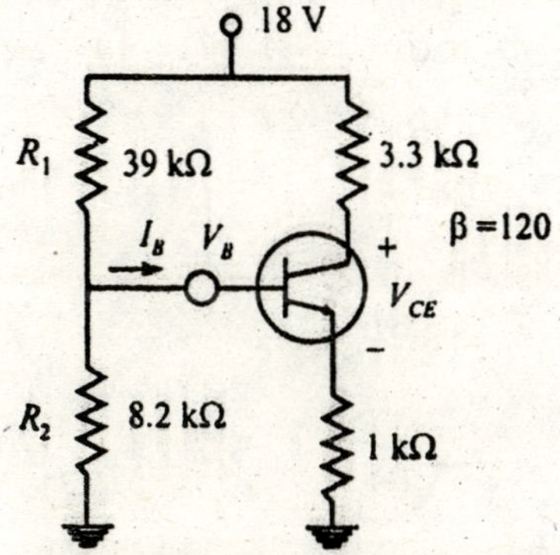

Fig. P. 3.25

Solution. First find out the thevenin equivalent circuit

$$V_B = \frac{R_2}{R_1 + R_2} \times V_{CC} = \frac{8.2 \times 18}{39 + 8.2} = 3.127 \text{ V}$$

$$R_B = R_1 \parallel R_2 = \frac{8.2 \times 39}{8.2 + 39} = 6.775 \text{ kΩ}$$

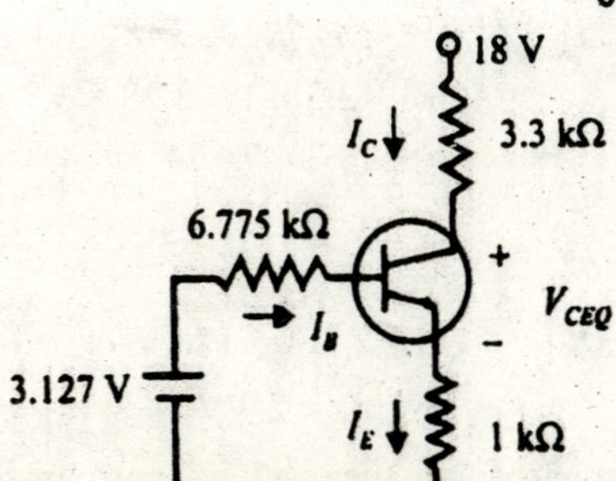

Fig. 3.25 (a)

Now, apply KVL in the base emitter loop

$$3.127 = 6.775\, I_B + V_{BE} + I_E R_E$$

$$\Rightarrow \quad 3.127 - 0.7 = (6.775 + 121)\, I_B$$

$$\Rightarrow \quad I_B = \frac{2.427}{127.775}$$

$$I_E = (\beta + 1)\, I_B \qquad \Rightarrow I_B = 0.01899 \text{ mA}$$

$$\Rightarrow \quad I_E = 121\, I_B \qquad \Rightarrow I_B = 18.99\ \mu\text{A} \simeq 19\ \mu\text{A}$$

Therefore $\quad I_C = \beta\, I_B = 120 \times 19 = 2.28$ mA

$$I_E = (\beta + 1)\, I_B = 121 \times 19 = 2.299 \text{ mA} \simeq 2.3 \text{ mA}$$

$$V_E = I_E R_E = 2.3 \times 1 = 2.3 \text{ V}$$

$$V_B = V_{BE} + V_E = 0.7 + 2.3 = 3 \text{ V}$$

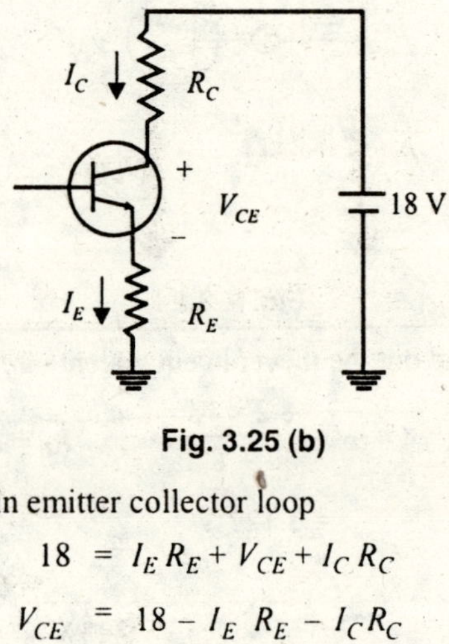

Fig. 3.25 (b)

Applying KVL in emitter collector loop

$$18 = I_E R_E + V_{CE} + I_C R_C$$

$$\Rightarrow \quad V_{CE} = 18 - I_E R_E - I_C R_C$$

$$= 18 - 2.3 \times 1 - 2.28 \times 3.3$$

$$= 18 - 2.3 - 7.524$$

$$= 8.176 \text{ V}$$

Thus, we have

(*a*) $I_C = 2.28$ mA (*b*) $V_{CE} = 8.176$ V (*c*) $I_B = 19\ \mu$A

(*d*) $V_B = 3$ V (*e*) $V_E = 2.3$ V

Hence, when compared with the result of approximate analysis it is satisfied and error is very less. Hence, the approximate analysis method can be used if the desired condition $\beta R_E \geq 10 R_2$ is satisfied.

Problem 3.26. Given $V_B = 4$ V for d.c. biased voltage feedback network determine:

(*a*) V_E (*b*) V_C (*c*) I_C (*d*) V_{CE}

(*e*) I_B (*f*) β

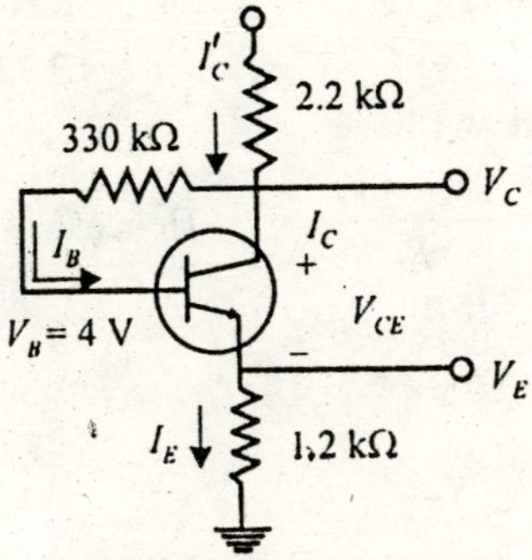

Fig. P. 3.26

Solution. Now, we start form

$$V_E = -V_{BE} + V_B$$

$$= -0.7 + 4 = 3.3 \text{ V}$$

Hence $I_E = \dfrac{V_E}{R_E} = \dfrac{3.3}{1.2\text{K}} = 2.75 \text{ mA}$

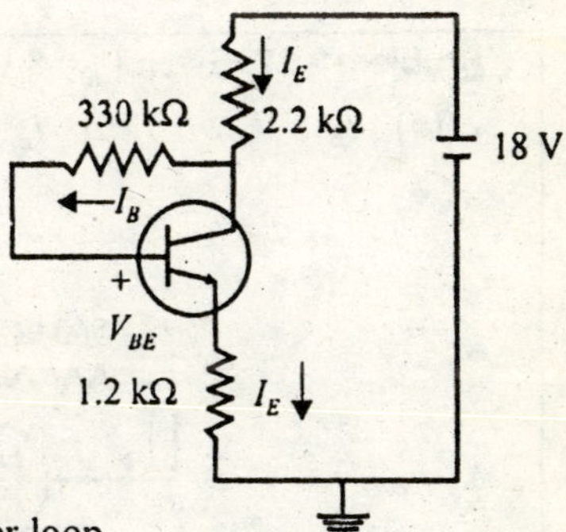

Fig. 3.26 (a)

From the network, we find that

$$I'_C = I_E = I_B + I_E = I_B + \beta\, I_B$$

$$\Rightarrow \quad I_E = (\beta + 1)\, I_B$$

Now, we apply KVL in the base emitter loop

$$18 = 1.2\, I_E + V_{BE} + 330 \times I_B + 2.2\, I_E$$

$$\Rightarrow \quad 18 = (\beta + 1)\, I_B \times 3.4 + 0.7 + 330\, I_B$$

$$\Rightarrow \quad [(\beta + 1)\, 3.4 + 330]\, I_B = 17.3$$

$$\Rightarrow \quad [(\beta + 1)\, 3.4 + 330]\, \frac{I_E}{(\beta + 1)} = 17.3$$

$$\Rightarrow \quad 3.4\, I_E + \frac{330\, I_E}{(\beta + 1)} = 17.3$$

$$\Rightarrow \quad 3.4 + \frac{330}{(\beta + 1)} = \frac{17.3}{I_E} = \frac{17.3}{2.75}$$

$$\Rightarrow \quad 3.4 + \frac{330}{(\beta + 1)} = 6.2909 \qquad \text{Thus,} \quad (\beta + 1)\, I_B = I_E$$

$$\Rightarrow \quad \frac{330}{(\beta + 1)} = 2.891$$

$$\Rightarrow \quad \beta + 1 = 114.15$$

$$\Rightarrow \quad \beta = 113.15$$

$$I_B = \frac{2.75}{114.15}$$

$$= 24.1 \ \mu A$$

$$I_C = \beta \, I_B$$

$$= 2.727 \text{ mA}$$

Now, from network, we have

$$I_E = I'_C = \frac{V_{CC} - V_C}{R_C}$$

$$V_C = V_{CC} - I_E R_C$$

$$= 18 - 2.75 \times 2.2$$

$$= 11.95$$

Hence, $V_{CE} = V_C - V_E$

$$= (11.95 - 3.3) \text{ V}$$

$$= 8.65 \text{ V}$$

So

(*a*) $V_E = 3.3$ V (*b*) $V_C = 11.95$ V (*c*) $I_C = 2.727$ mA

(*d*) $V_{CE} = 8.65$ V (*e*) $I_B = 24.1\ \mu$A (*f*) $\beta = 113.15$

Problem 3.27. Given $V_E = 8$ V for the network shown below, determine:

(*a*) I_B (*b*) I_c (*c*) β (*d*) V_{CE}

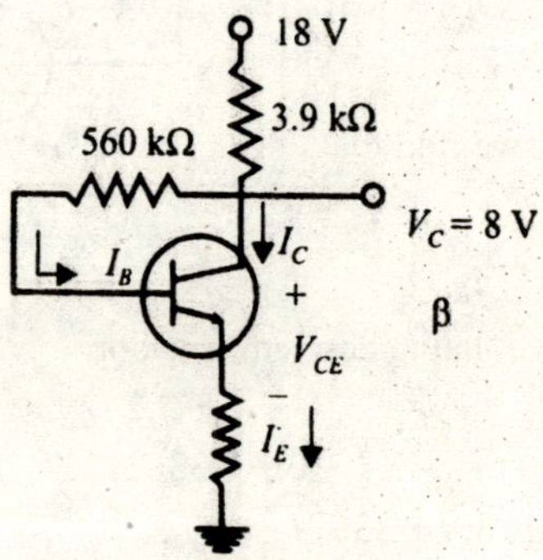

Fig. P. 3.27

Solution. From the network we find that $V_E = 0$. Since the emitter terminal is grounded. Therefore

$$V_{CE} = V_C - V_E = V_C - 0 = V_C = 8 \text{ V}$$

Now, Apply KVL in the base emitter loop

$$18 = V_{BE} + 560\, I_B + 3.9\, I_E$$

$$\Rightarrow \quad 18 - 0.7 = 560\, I_B + 3.9\, (\beta + 1)\, I_B$$

$$\Rightarrow \quad [\ 560 + 3.9\, (\beta + 1)\, I_B = 17.3$$

Now $I_E = (\beta + 1)\, I_B = \dfrac{18 - 8}{3.9\text{ K}} = \dfrac{10}{3.9\text{ K}} = 2.56\text{ mA}$

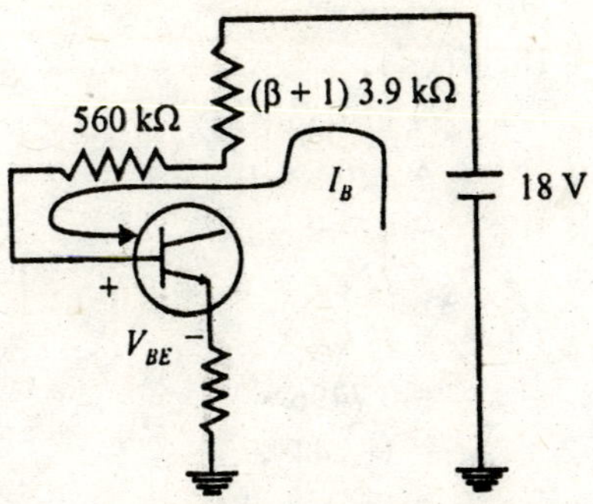

Fig. 3.27 (a)

$\Rightarrow$ $[560 + 3.9\,(\beta + 1)\, I_B = 17.3$ Thus, $(\beta + 1) = \dfrac{2.564 \times 10^{-3}}{13 \times 10^{-6}}$

$\Rightarrow$ $560\, I_B + 3.9 \times 2.564 = 17.3$ $(\beta + 1) = 197.23$

$\Rightarrow$ $560\, I_B = 7.3$ $\beta = 196.23$

$\Rightarrow$ $I_B = 13\ \mu\text{A}$ and $I_C = \beta\, I_B$

$= 2.55\text{ mA}$

Thus

(*a*) $I_B = 13\ \mu\text{A}$ (*b*) $I_C = 2.55\text{ mA}$ (*c*) $\beta = 196.23$

(*d*) $V_{CE} = 8\text{ V}$

Problem 3.28. For the network shown determine, if $\beta = 120$.

(*a*) I_B (*b*) I_C (*c*) V_{CE} (*d*) V_C

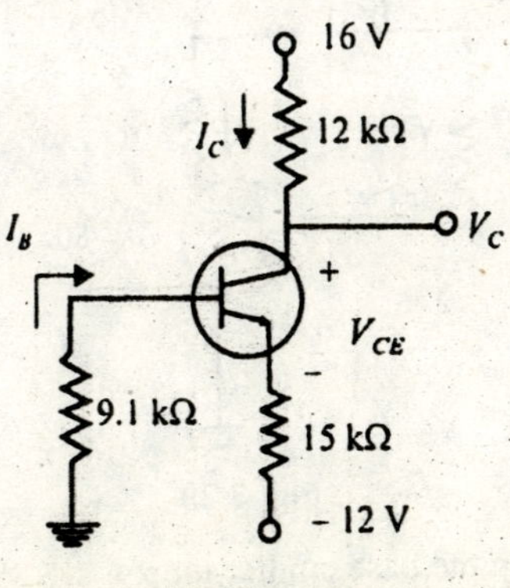

Fig. P. 3.28

Solution. Apply KVL in the base emitter loop

$$-12 + (\beta + 1)\, 15\, I_B + V_{BE} + 9.1\, I_B = 0$$

$\Rightarrow$ $[121 \times 15 + 9.1]\, I_B = 12 - 0.7$

$\Rightarrow$ $I_B \times 1824.1 = 11.3$

$\Rightarrow$ $I_B = 6.195\ \mu A$

So $I_C = \beta\, I_B = 120 \times 6.195 \times 10^{-6}$

$= 0.7496$ mA

and $I_E = I_B + I_C = 0.7496$ mA

$V_E = I_E\, R_E$

$= 0.7496 \times 15$

$= 11.244$ V

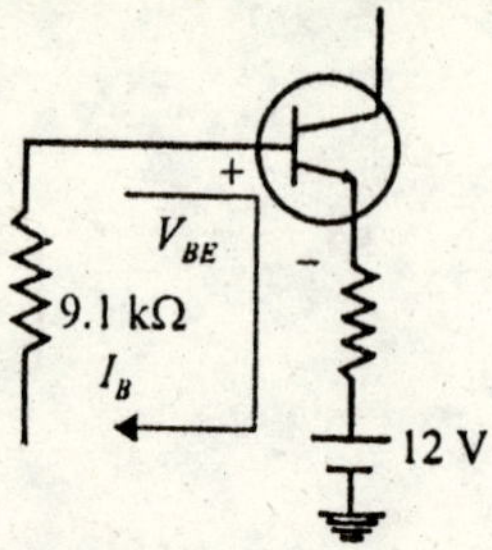

Fig. 3.28 (a)

$V_C = V_{CC} - I_C R_C = 16 - 12 \times 0.743 = 7.084$ V

Therefore $V_{CE} = V_C - V_E = 7.084 - 11.244$

$= -4.16$ V

Therefore, we have

(*a*) $I_B = 6.195\ \mu A$ (*b*) $I_C = 0.743$ mA (*c*) $V_{CE} = -4.16$ V

(*d*) $V_C = 7.084$ V

Problem 3.29. Determine R_E and R_B for a fixed bias configuration if $V_{CC} = 12$ V, $\beta = 80$, $I_{CQ} = 2.5$ mA, and $V_{CEQ} = 6$ V. Use standard values.

Solution. The fixed bias configuration is drawn below:

$V_{CC} = 12$ V $\quad I_{CQ} = 2.5$ mA

$\beta = 80 \quad V_{CEQ} = 6$ V

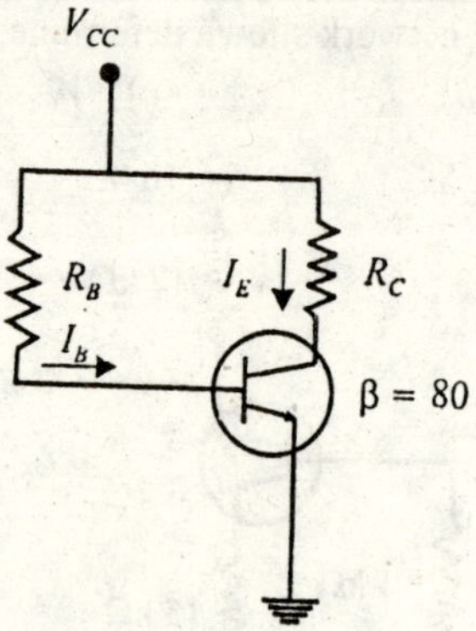

Fig. 3.29

Now, apply KVL in the base emitter loop

$$V_{CC} = V_{BE} + I_B\, R_B$$

$$\Rightarrow \quad I_B R_B = V_{CC} - V_{BE}$$

$$\Rightarrow \quad R_B = \frac{V_{CC} - V_{BE}}{I_B}$$

Now $$I_B = \frac{I_{CQ}}{\beta} = \frac{2.5}{80} = 31.25\ \mu A$$

$$\therefore \quad R_B = \frac{12 - 0.7}{31.25 \times 10^{-6}} = 361.6\ k\Omega \simeq 362\ k\Omega$$

Taking KVL in the collector emitter loop

$$R_C = \frac{V_{CC} - V_{CEQ}}{I_{CQ}} = \frac{12-6}{2.5} = \frac{6}{2.5}\ 2.4\ k\Omega$$

Therefore $R_C \simeq 2.5\ k\Omega$

and $R_B \simeq 362\ k\Omega$

Problem 3.30. Design a voltage divider bias network using supply of 24 V a transistor with a beta of 110 and an operating point of $I_{CQ} = 4$ mA and $V_{CEQ} = 8$ V, choose $V_E = \frac{1}{8} V_{CC}$. Use standard values.

Solution. The voltage divider bias network is shown, where we have to calulate R_1, R_2, R_C and R_E from the given informations:

$\beta = 110$

$I_{CQ} = 4$ mA

$V_{CEQ} = 8$ V

$V_{CC} = 24$ V

$V_E = 3$ V

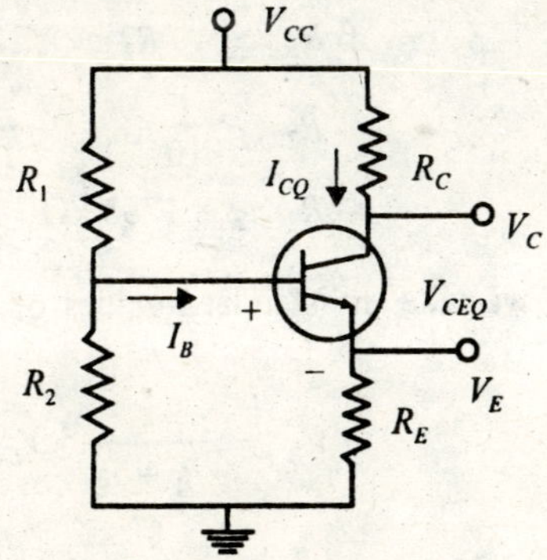

Fig. 3.30 (a)

From the above, we get the reduce thevenin equivalent circuit.

$$V_B = \frac{R_2 V_{CC}}{R_1 + R_2} \quad \text{and} \quad R_B = \frac{R_1 R_2}{R_1 + R_2}$$

$$I_{CQ} = 4\ mA$$

$$\therefore \qquad I_B = \frac{I_C}{\beta} = \frac{4 \times 10^{-3}}{110}$$

$$= 36.36\ \mu A$$

$$I_E = I_B + I_C = (\beta + 1)\, I_B$$

$$= 111 \times 36.36 \times 10^{-6}$$

$$= 4.036 \text{ mA}$$

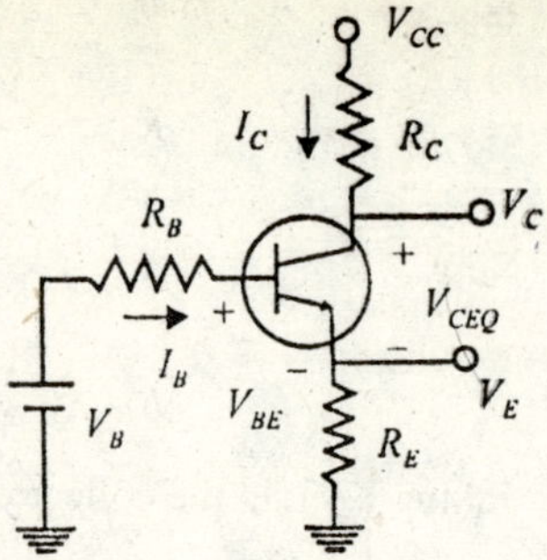

Fig. 3.30 (b)

Therefore $\qquad R_E = \frac{V_E}{I_E} = \frac{3}{4.036}$

$\Rightarrow \qquad R_E = 0.7433 \text{ k}\Omega$

Now $\qquad V_C = V_{CEQ} + V_E = 8 + 3 = 11 \text{ V}$

Hence $\qquad R_C = \frac{V_{CC} - V_C}{I_{CQ}} = \frac{24 - 11}{4 \times 10^{-3}} = 3.25 \text{ k}\Omega$

For the circuit to be operate efficiently, it is assumed that the current through R_1 and R_2 should be approximate equal to and much larger than the base current (at least 10 : 1). This fact and the voltage divider equation for the base voltage provide the two relationships necessary to determine the base resistors, i.e.

$$\beta R_E \geq 10 R_2$$

$$\Rightarrow \qquad R_2 \leq \frac{1}{10} \beta R_E \leq \frac{1}{10} \times 110 \times 0.7433$$

$$\Rightarrow \qquad R_2 \leq 8.1763 \text{ k}\Omega$$

Thus, we take the standard values of $R_2 = 8 \text{ k}\Omega$ and R_1 is calculated from V_B.

$$V_B = \frac{R_2}{R_1 + R_2} \times V_{CC}$$

Here $\qquad V_B = V_{BE} + V_E = 0.7 + 3$, $\quad V_B = 3.7 \text{ V}$

$$3.7 = \frac{8}{R_1 + 8} \times 24$$

$$\Rightarrow \qquad 3.7 R_1 + 29.6 = 192$$

$$\Rightarrow \qquad R_1 = \frac{162.4}{3.7} = 43.89 \text{ k}\Omega$$

Therefore, standard value of $R_1 = 44 \text{ k}\Omega$

Thus, we get the design of voltage divider as

(*a*) $R_E = 0.7433\ k\Omega$ $R_1 = 44\ k\Omega$

(*b*) $R_C = 3.25\ k\Omega$ $R_2 = 8\ k\Omega$

Problem 3.31. Using the information given determine I_B, $I_{B\ max}$ and $I_{c\ sat}$, when $V_i = 10$ V. Determine the collector to emitter resistance at saturation and cut-off.

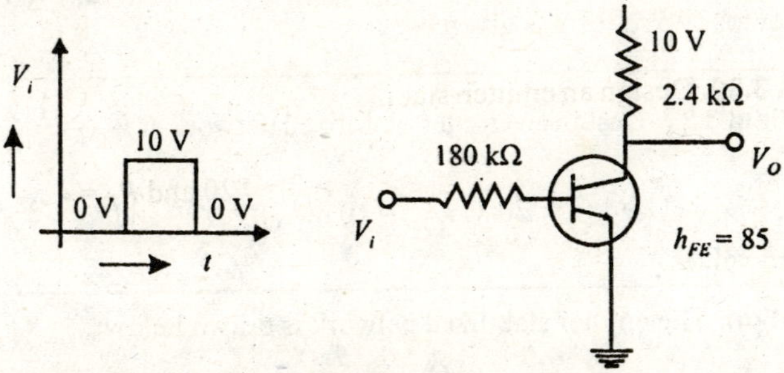

Fig. P. 3.31

Solution. From the network, we have

$$I_{c\,sat} = \frac{V_{CC}}{R_C} \quad [\because V_{CE} = 0] \text{ During saturation } V_{CE} \text{ is short-circuited.}$$

$$= \frac{10}{2.4} = 4.167 \text{ mA}$$

and $$I_B = \frac{V_i - 0.7}{R_B} = \frac{10 - 0.7}{180} = \frac{9.3}{180} = 0.05167 \text{ mA}$$

$$= 51.69\ \mu\text{A}$$

Now $$I_{B\max} = \frac{I_{c\,sat}}{B_{d.c.}} - \frac{4.167}{85} = 49.02\,\mu\text{A}$$

Since, $I_B > I_{B\ max}$ which will drive the transistor into saturation.

During saturation V_{CE} is very low nearly $V_{CE\,sat} = 0.15$ V

$$R_{sat} = \frac{V_{CE\,sat}}{I_{C\,sat}} = \frac{0.15 \text{ V}}{4.167 \text{ mA}} = 35.99\ \Omega = 36\ \Omega$$

Idealy $R_{sat} = 0$ as $V_{CE\,sat} = 0$ (short-circuit).

During cut-off condition ($V_i = 0$)

Ideally $R_{\text{cut-off}} = \frac{V_{CC}}{I_{CEO}}$

$$= \frac{10\text{ V}}{0\text{ mA}} = \infty\ \Omega \text{ (Open circuit)}$$

But practically, $R_{\text{cut-off}} = \frac{10}{10\ \mu\text{A}} = 1000\text{ k}\Omega$

The output wave form is given by

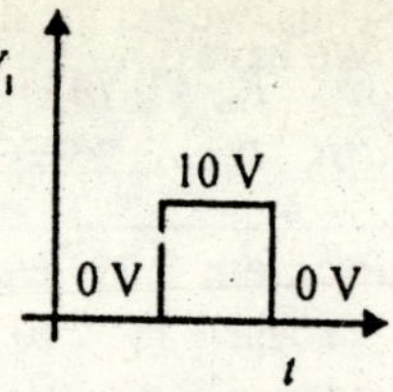

Fig. 3.31 (a)

Problem 3.32. Design an emitter stabilized network at $I_{CQ} = \frac{1}{2} I_{c\,sat}$ and $V_{CEQ} = \frac{1}{2} V_{CC}$. Use $V_{CC} = 20$ V, $I_{c\,sat} = 10$ mA, $\beta = 120$ and $R_E = 4\,R_E$. Use standard values.

Solution. The emitter stabilized network is drawn below:

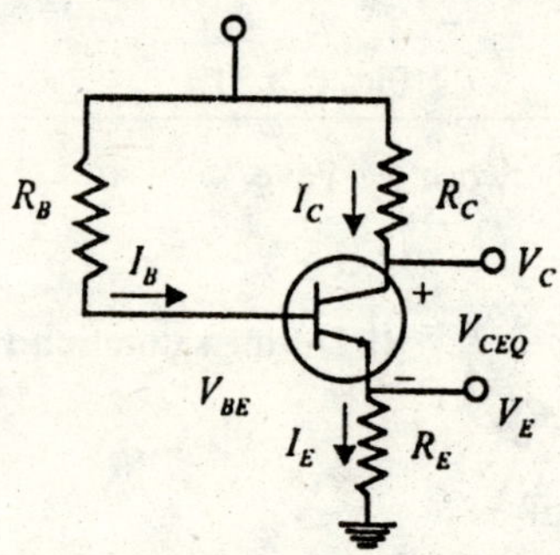

Fig. 3.32 (a)

Given values are

$$V_{CC} = 20\text{ V} \qquad I_{CQ} = \frac{1}{2} I_{sat}$$

$$I_{C\,sat} = 10\text{ mA} \qquad = 5\text{ mA}$$

$$\beta = 120 \qquad V_{CEQ} = \frac{1}{2} V_{CC} = 10\text{ V}$$

$$R_C = 4\,R_E = 10\text{V}$$

$$I_B = \frac{I_{CQ}}{\beta} = \frac{5}{120} = 41.67\ \mu\text{A}$$

$$I_E = (\beta + 1)\ I_B$$

$$= 121 \times 41.67 \times 10^{-6}$$

$$= 5.042\text{ mA}$$

When saturation collector current flows, $V_{CE} = 0$

We have $V_{CC} = I_C R_C + I_E R_E$

$\Rightarrow$ $20 = 5 \times 4 R_E + 5.042 \times R_E$

$20 = 25.042 R_E$

$\Rightarrow$ $R_E = 0.798 \text{ k}\Omega$

Thus $R_C = 4 R_E = 3.194 \text{ k}\Omega$

Now $V_E = I_E R_E = 5.042 \times 0.798 = 4.023 \text{ V}$

Apply KVL in the base emitter loop

$$V_{CC} = I_B R_B + V_{BE} + V_E$$

$\Rightarrow$ $20 = 0.04167 R_B + 0.7 + 4.023$

$\Rightarrow$ $$R_B = \frac{15.277}{0.04167} = 366.62 \text{ k}\Omega$$

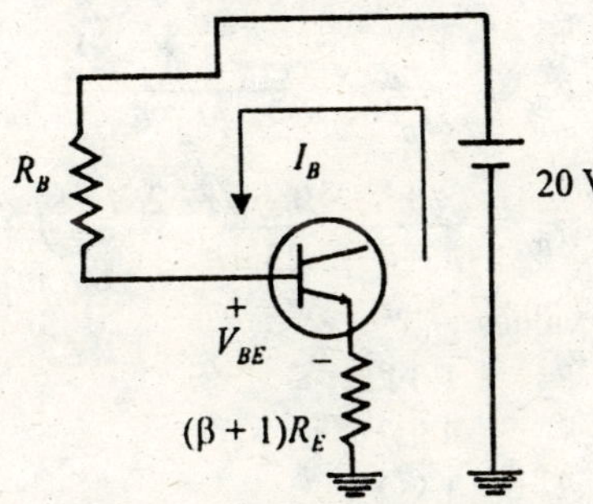

Fig. 3.32 (b)

Thus, we have $R_E = 0.798 \text{ k}\Omega,\ R_C = 3.2 \text{ k}\Omega,\ R_B = 367 \text{ k}\Omega$

Problem 3.33. Determine the resistor values for the network of figure given below for the indicated operating point and supply voltage.

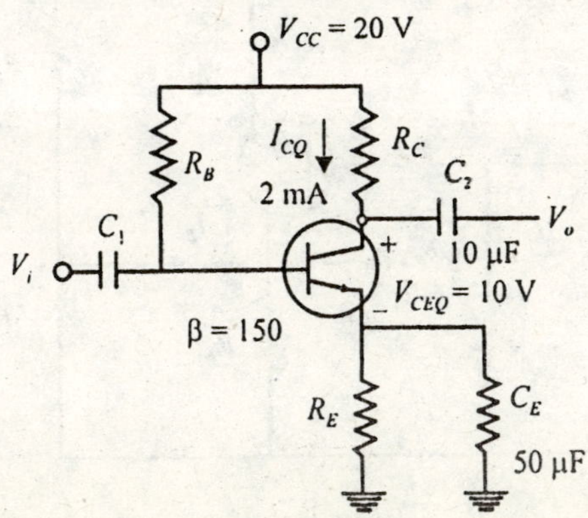

Fig. P. 3.33

Solution.
$$V_E = \frac{1}{10} V_{CC} = \frac{1}{10}(20\text{ V}) = 2\text{ V}$$

$$R_E = \frac{V_E}{I_E} \cong \frac{V_E}{I_C} = \frac{2\text{V}}{2\text{ mA}} = 1\text{ k}\Omega$$

$$R_C = \frac{V_{RC}}{I_C} = \frac{V_{CC} - V_{CE} - V_E}{I_C}$$

$$R_C = \frac{20\text{V} - 10\text{ V} - 2\text{V}}{2\text{ mA}} = \frac{8\text{ V}}{2\text{ mA}} = 4\text{ k}\Omega$$

$$I_B = \frac{I_C}{\beta} = \frac{2\text{ mA}}{150} = 13.33\,\mu\text{A}$$

Therefore
$$R_B = \frac{V_{RB}}{I_B} = \frac{V_{CC} - V_{BE} - V_E}{I_B}$$

or
$$R_B = \frac{20\text{ V} - 0.7\text{ V} - 2\text{ V}}{13.33\,\mu\text{A}} \cong 1.3\text{M}\Omega$$

Thus, we have the values

$$R_B = 1.3\text{ k}\Omega$$
$$R_C = 4\text{ k}\Omega$$
$$R_E = 1\text{ k}\Omega$$

Problem 3.34. An emitter feedback bias network is shown below. With $I_C = 3$ mA, $H_{FE} = 45$, $V_{CC} = 12$ V, and $V_{BE} = 0.5$. Find the resistor values to give a stability factor of 0.05.

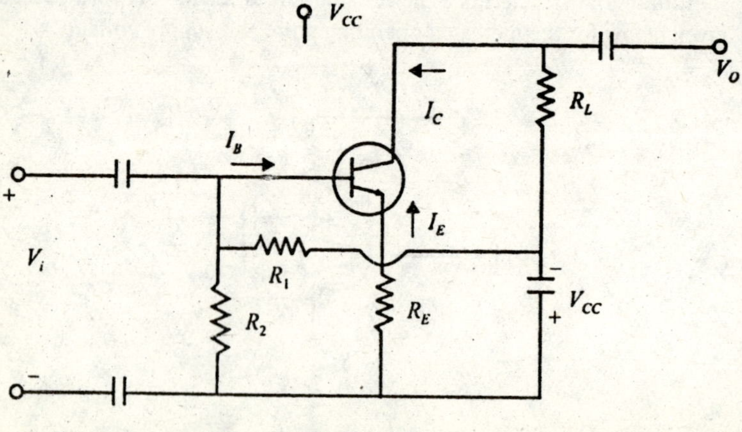

Fig. P. 3.34

Solution. The stability factor is given by

$$S = \frac{R_E}{R_B} = 0.05$$

If I_B is the base current, we have

$$R_L + R_E = \frac{V_{CC}/2}{I_C} = \frac{6}{0.003} = 2000\ \Omega$$

The emitter resistance is normally kept about 20 % of $R_L + R_E$

So that, $R_E = 0.2\ (R_L + R_E)$ $\Rightarrow R_E = 0.2 \times 2000 = 400\ \Omega$

Now, voltage at emitter terminal is given by

$$V_E = I_E R_E = (I_C + I_B)\ R_E = (I_C + \frac{I_C}{\beta})\ R_E$$

or $$V_E = \left(0.003 + \frac{0.003}{45}\right) \times 400 = 1.23\text{ V}$$

$\therefore$ Voltage at the base terminal is

$$V_B = V_E + V_{BE} = 1.23\text{ V} + 0.5\text{ V}$$
$$= 1.73\text{ V}$$

Now $$\frac{V_B}{V_{CC}} = \frac{R_1\ R_2}{R_1\left(R_1 + R_2\right)} = \frac{R_B}{R_1}$$

But $$\frac{R_E}{R_B} = 0.05 \quad \text{or} \quad R_B = \frac{R_E}{0.05}$$

$\therefore$ $$\frac{V_B}{V_{CC}} = \frac{R_E}{0.05\ R_1}$$

$\therefore$ $$R_1 = \frac{V_{CC}\ R_E}{0.05\ V_B} = \frac{12 \times 400}{0.05 \times 1.73}\ \Omega$$

$$55500\ \Omega = 55.5\text{ k}\Omega$$

$$R_L = 2000 - R_E = 2000 - 400 = 1600\ \Omega$$

or $$R_L = 1.6\text{ k}\Omega$$

or $$\frac{V_B}{V_{CC}} = \frac{1.73}{12} = \frac{R_2}{R_1 + R_2} = \frac{R_2}{R_2 + 55500}$$

or $1.73\ (R_2 + 55500) = 12\ R_2$ which yields

$$R_2 = 9350\ \Omega, = 9.35\text{ k}\Omega$$

$$R_1 = 1.6\text{ k}\Omega, \qquad R_E = 0.4\text{ k}\Omega,$$

$$R_1 = 55.5\text{ k}\Omega, \qquad R_2 = 9.35\text{ k}\Omega$$

Problem 3. 35. Calculate the value feedback R_F in the circuit of figure for V_{CC} = 20V, R_c = 5 K, R_e = 100 Ω, V_{CE} = 5 V, V_{BE} = 0.7 V and β = 100. Neglect reverse saturation current.

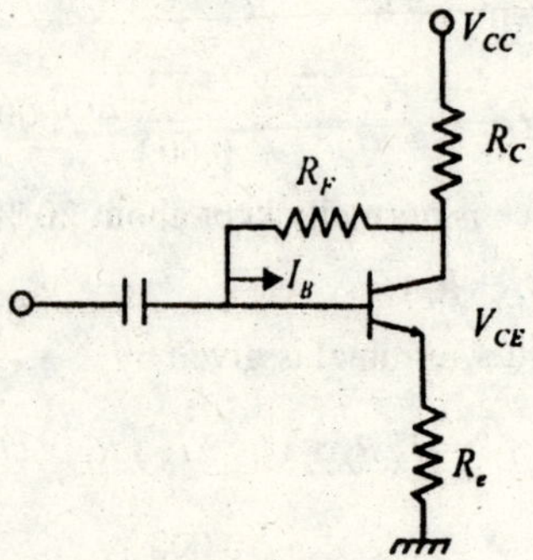

Fig. P. 3.35

Solution. By KVL of *CE* circuit.

$$-20 + \beta I_B \times 5000 + 5 + (\beta + 1) I_B R_e = 0$$

$$3 \times 10^5 \, I_B + 101 \times 100 \, I_B = 15$$

$$I_B = \frac{15}{5101 \times 100} = 0.029 \text{ mA}$$

$$I_B R_F + 0.7 = V_{CE} = 5 \text{ V}, R_F = \frac{5 - 0.7}{0.029 \times 10^{-3}} = 148 \text{ K}$$

Problem 3.36. Calculate the *Q* point

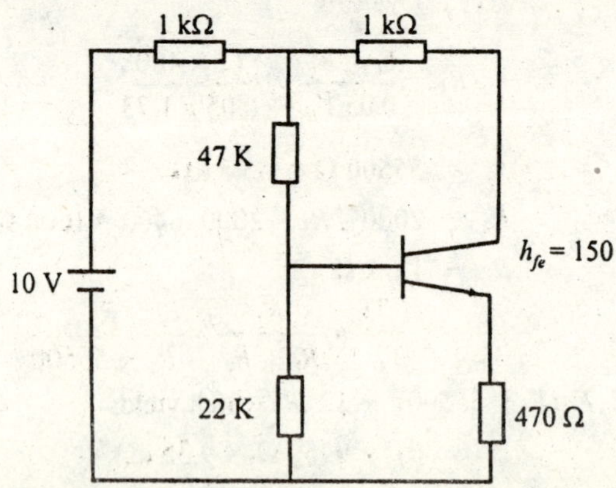

Fig. P. 3.36

Solution.

$$10 = (I_C + I_1)\,1\text{ K} + I_1\,47\text{ K} + 0.7 + 470\,(I_C + I_B) \quad \ldots (1)$$

$$9.3 = 1.47\text{ K}\,I_C + 48\text{ K}\,I_1 + 0.47\text{ K}\,I_B \quad \ldots (2)$$

$$10 = 1\text{ K}\,I_C + 48\text{ K}\,I_1 + 22\text{ K}\,I_2$$

$$9.3 = I_C\left(1.47\text{ K} + \frac{48\text{ K}}{150} + \frac{0.47\text{ K}}{150}\right) + 48\text{ K}\,I_2$$

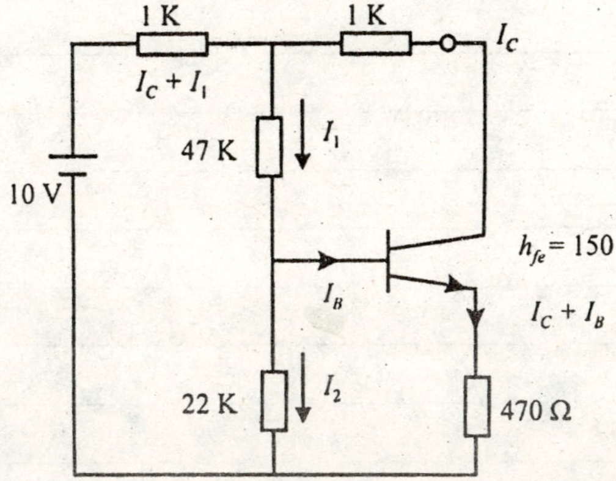

Fig. 3.36 (a)

$$10 = I_C\left(1\text{ K} + \frac{48\text{K}}{150}\right) + 70\text{ K}\;I_2$$

$$651 = 125.51\text{ K}\,I_C + 3360\text{ K}\,I_2$$

$$480 = 63.36\text{ K}\,I_C + 3360\text{ K}\,I_2$$

$$I_C = \frac{171}{62.15\text{ K}} = 2.75\text{ mA}$$

$$I_B = \frac{2.75}{150} = 0.018\text{ mA}\,,\; I_{EQ} = 2.768\text{ mA}$$

$$V_{CEQ} = 10 - (2.768 \times 10^{-3} \times 1 \times 10^3 + 2.765 \times 10^{-3} \times 1 \times 10^3 + 2.768 \times 10^{-3} \times 0.47 \times 1 \times 10^3)$$

$$= 10 - 6.834 = 3.166\text{ V}$$

Reader's Notes

CHAPTER 4

BJT-Small Signal Analysis

Introduction

The r_e model for a BJT in the a.c. domain is sensitive to the actual d.c. operating conditions of the network. This parameter is normally not provided on a specification sheet, although h_{ie} of the normally provided hybrid parameters is equal to $\beta\, r_e$ but only under specific operation conditions.

(*i*) Common emitter fixed bias configuration

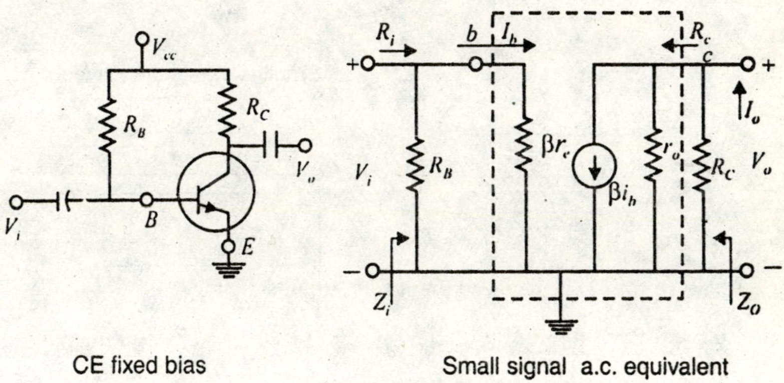

Fig. 4.1

The input impedance, $\boxed{Z_i \cong \beta r_e}$

The output impedance, $\boxed{Z_O \cong R_C}$

The voltage gain, $\boxed{A_v = \dfrac{-R_C}{r_e}}$

The current gain, $\boxed{A_i = -A_v \dfrac{Z_i}{R_C} \cong \beta}$

Phase relationship: The negative sign in the resulting equation for A_v reveals that a 180° phase shift occurs between the input and output signals.

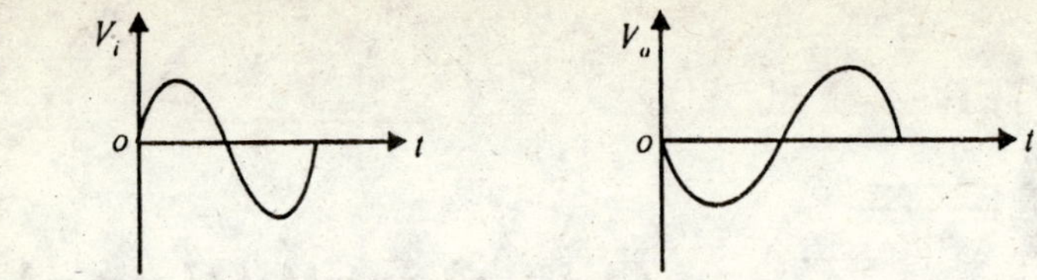

Fig. 4.2

(*ii*) Voltage divider bias configuration

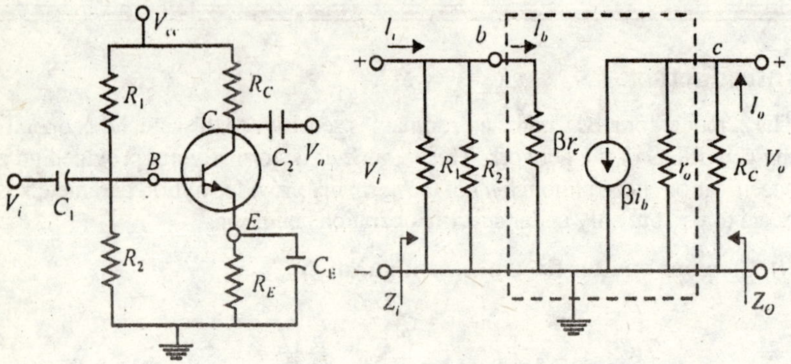

Circuit　　　Small signal a.c. equivalent

Fig. 4.3

$$Z_i = R_1 \| R_2 \| \beta r_e$$

$$Z_o \cong R_C$$

$$A_v = \frac{-R_C}{r_e}$$

$$A_i = -A_v \frac{Z_i}{R_C} \cong \beta$$

Phase shift of 180° between input and output signal.

(*iii*) CE emitter bias configuration

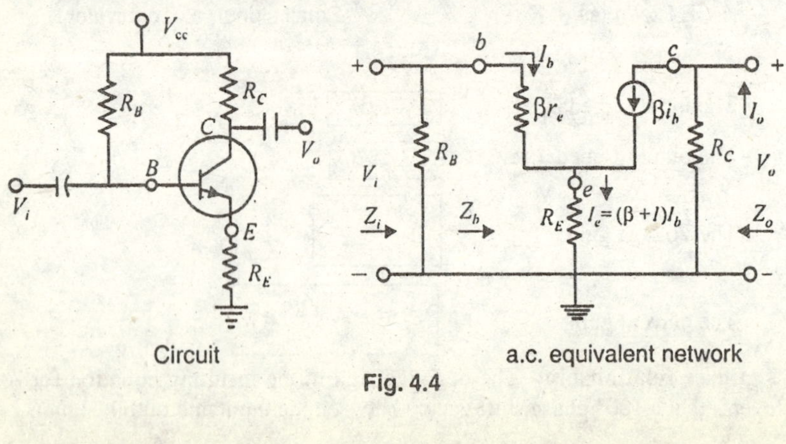

Circuit　　　a.c. equivalent network

Fig. 4.4

$$Z_i = R_B \parallel \beta R_E$$

$$A_v = \frac{-R_C}{R_E}$$

$$Z_o \cong R_C$$

$$A_i = \frac{\beta R_B}{R_B + \beta R_E}$$

(iv) Emitter follower configuration

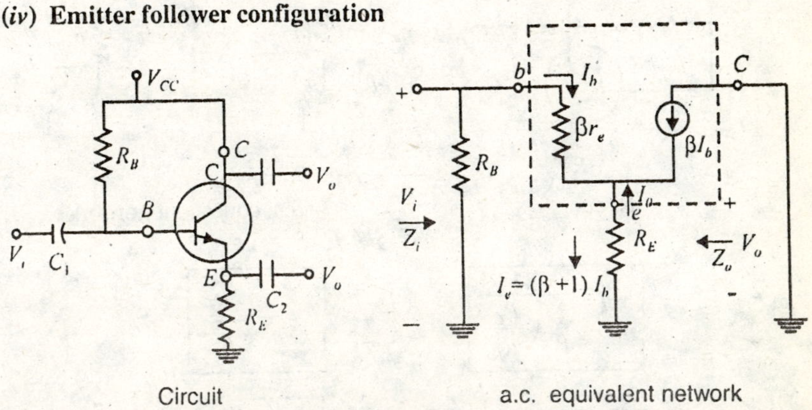

Fig. 4.5

$$Z_i \cong R_B \parallel \beta R_E$$

$$A_v \cong 1$$

$$Z_o \cong r_e$$

$$A_i = -A_v \frac{Z_i}{R_E}$$

(v) Common base configuration

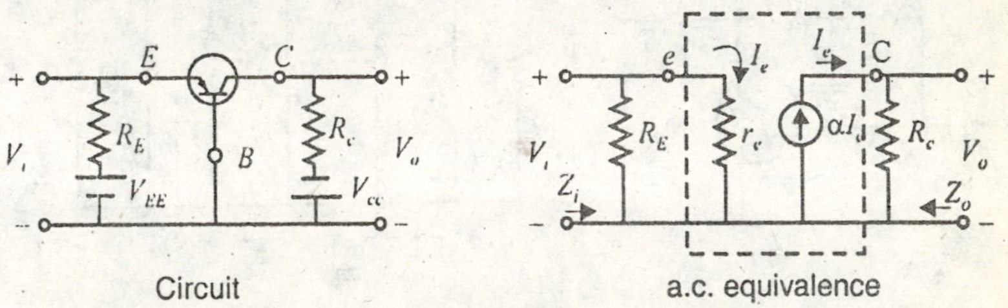

Fig. 4.6

$$Z_i \cong R_E \parallel r_e$$

$$A_v \cong \frac{R_C}{r_e}$$

$$Z_o \cong R_C$$

$$A_i \cong -1$$

No phase shift between the input and output signal as A_v in the equation is positive.

(vi) Collector feedback configuration

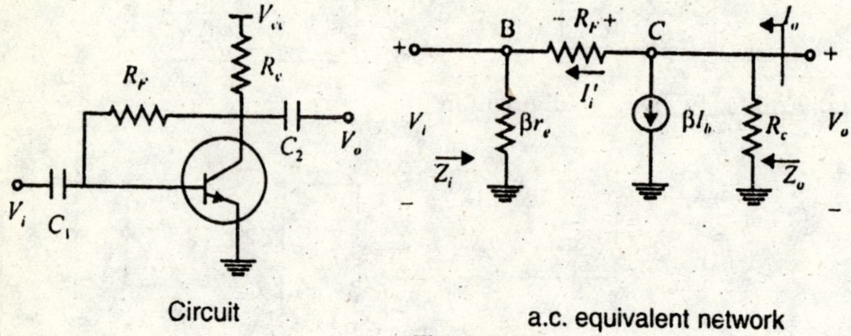

Fig. 4.7

$$Z_i \cong \frac{r_e}{\frac{1}{\beta} + \frac{R_C}{R_F}}$$

$$A_v = \frac{-R_C}{r_e}$$

$$Z_o \cong R_C \parallel R_F$$

$$A_i \cong \frac{R_F}{R_C}$$

(vii) Collector d.c. feedback configuration

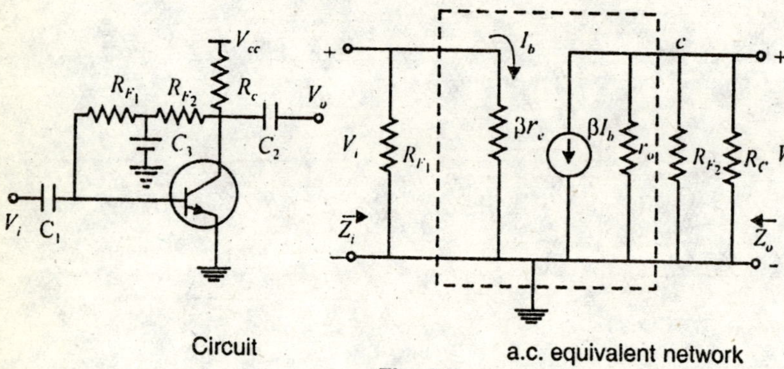

Fig. 4.8

$$Z_i \cong R_{F_1} \parallel \beta r_e$$

$$A_v = \frac{-R_{F_2} \parallel R_C}{r_e}$$

$$Z_o \cong R_C \parallel R_{F_2}$$

$$A_i = -A_v \frac{Z_i}{R_C}$$

Problem 4.1. For the network given below:

(*a*) Determine Z_i and Z_o (with $r_o = \infty\,\Omega$).

(*b*) Find A_v and A_i.

(*c*) Repeat part (*a*) with $r_o = 20\text{ k}\Omega$.

(*d*) Repeat part (*b*) with $r_o = 20\text{ k}\Omega$.

Solution.

(*a*) Given $\quad r_0 = \infty\,\Omega$ and $\beta = 60$

$$Z_i = R_B \parallel \beta\, r_e$$

To calculate r_e :

$$I_B = \frac{V_{CC} - V_{BE}}{R_B} = \frac{12\text{ V} - 0.7\text{ V}}{220\text{ k}\Omega} = 0.0514\text{ mA}$$

$$I_E = (\beta + 1)\, I_B = 3.133\text{ mA}$$

$$r_e = \frac{26\text{ mV}}{I_E} = \frac{26\text{ mV}}{3.133\text{ mA}} = 8.3\,\Omega$$

$$\beta\, r_e = 60 \times 8.3 = 498\,\Omega$$

$$Z_i = \frac{R_B \times \beta r_e}{R_B + \beta r_e} = \frac{2.20 \times 0.498}{220.498} = 0.497\text{ k}\Omega = 497\,\Omega.$$

Since $\quad r_o = \infty\,\Omega$, i.e. open circuit

$\therefore \quad Z_o = R_C = 2.2\text{ k}\Omega$

(*b*) Neglecting the effect of r_o

$$A_v = -\frac{R_C}{r_e} = -265.06$$

For A_i, we check whether $R_B \geq 10\,\beta\, r_e$

i.e. $\quad 220\text{ k}\Omega > 4.98\text{ k}\Omega$

Hence $\quad A_i \approx \beta = 60$

(*c*) Now given, $r_o = 20\text{ k}\Omega$

$$Z_i = R_B \parallel \beta\, r_e = 497\,\Omega$$

$$Z_o = r_o \parallel R_C = \frac{20\text{ k}\Omega \times 2.2\text{ k}\Omega}{22.2\text{ k}\Omega} = 1.98\text{ k}\Omega$$

(*d*)
$$A_v = -\frac{r_o \parallel R_C}{r_e} = \frac{-1.98 \times 1000}{8.3} = -238.55$$

Taking r_o into account

$$A_i = \frac{\beta R_B r_o}{(r_o + R_C)(R_B + \beta r_e)} = \frac{(60) \times (220\,\text{k}\Omega) \times (20\,\text{k}\Omega)}{(22.2\,\text{k}\Omega)\,(220\,\text{k}\Omega + 0.498\,\text{k}\Omega)}$$

$$= 53.93$$

We see that after including the effect of r_o, values of A_v and A_i get reduced.

Problem 4.2. For the network shown, determine V_{CC} for a voltage gain of $A_v = -200$.

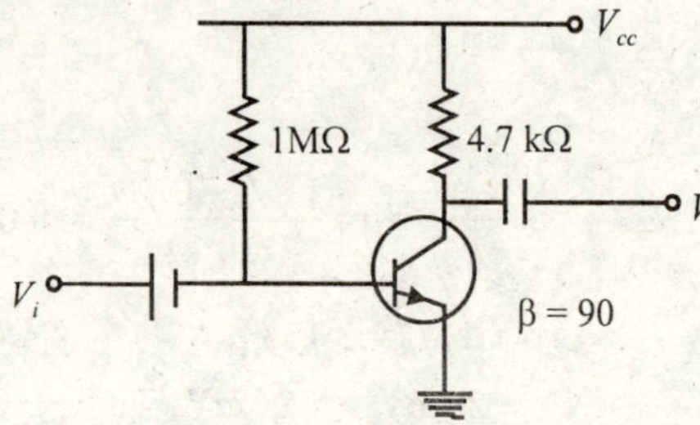

Fig. P. 4.2

Solution. Given $A_v = -200$

Since nothing is said about r_o so, we will neglect its effect.

Now $A_v = \frac{-R_C}{r_e}$ (for fixed-bias configuration)

$$\Rightarrow \quad -200 = -\frac{4.7\,\text{k}\Omega}{r_e}$$

$$\Rightarrow \quad r_e = \frac{4.7\,\text{k}\Omega}{200} = \frac{4700}{200} = 23.5\,\Omega$$

Since $r_e = \frac{26\text{ mV}}{I_E}$

$$\Rightarrow \quad 23.5 = \frac{26\,\text{mV}}{I_E} \quad \Rightarrow \quad I_E = \frac{26}{23.5}\,\text{mA} = 1.106\text{ mA}$$

Now $I_E = (\beta + 1)\, I_B$

or $I_B = \frac{I_E}{(\beta + 1)} = \frac{1.106\text{ mA}}{91} = 0.012\text{ mA} = 12.16\,\mu\text{A}$

From d.c. analysis

$$I_B = \frac{V_{CC} - V_{BE}}{R_B}$$

$$\Rightarrow \quad 12.16 \times 10^{-6} = \frac{V_{CC} - 0.7\text{ V}}{1 \times 10^6}$$

$$\Rightarrow \quad (V_{CC} - 0.7) = 12.16 \times 10^{-6} \times 1 \times 10^6 = 12.16$$

or $$V_{CC} = 12.16 + 0.7 = 12.86\text{ V}$$

Problem 4.3. For the network given below:

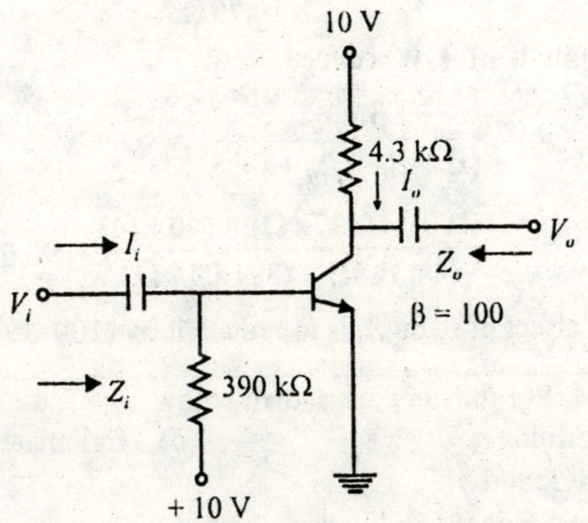

Fig. P. 4.3

(*a*) Calculate I_B, I_C and r_e. (*b*) Determine Z_i and Z_o.

(*c*) Calculate A_v and A_i.

(*d*) Determine the effect of $r_o = 40\text{ k}\Omega$ on A_v and A_i.

Solution. (*a*) Applying KVL to the loop containing $R_B = 390\text{ k}\Omega$

$$I_B = \frac{10\text{ V} - 0.7\text{ V}}{390\text{ k}\Omega} = 23.85\,\mu\text{A}$$

Since $$I_C = \beta I_B$$

$$\Rightarrow \quad I_C = 100 \times 23.85 = 2.38\text{ mA}$$

Also $$I_E = (\beta + 1) I_B = 101 \times 0.02385\text{ mA} = 2.409\text{ mA}$$

Hence $$r_e = \frac{26\text{ mV}}{I_E}$$

$$\Rightarrow \quad r_e = \frac{26\text{ mV}}{2.409\text{ mA}} = 10.79\ \Omega$$

(*b*) Now $$Z_i = \beta r_e$$

or $$Z_i = 100 \times 10.79\ \Omega = 1.079\text{ k}\Omega \approx 1.08\text{ k}\Omega$$

and $$Z_o = R_C = 4.3\text{ k}\Omega$$

(c) $$A_v = -\frac{R_C}{r_e} = -\frac{4.3\ \text{k}\Omega}{10.79\ \Omega} = -398.52$$

and current gain $A_i \approx \beta$ since the condition $R_B > 10\,\beta r_e$, i.e. 390 kΩ > 10.79 kΩ is fulfilled.

Hence $A_i = \beta = 100$

(d) Including the effect of $r_o = 40$ kΩ

$$A_v = -\frac{r_o \parallel R_C}{r_e} = -\frac{3.883\ \text{k}\Omega}{10.79\ \Omega} = -359.83$$

Thus, magnitude of A_v is reduced.

Now $$A_i = \frac{\beta R_B r_o}{(r_o + R_C)(R_B + \beta r_e)}$$

$$= \frac{(100) \times (390\ \text{k}\Omega) \times (40\ \text{k}\Omega)}{(44.3\ \text{k}\Omega) \times (391.08\ \text{k}\Omega)} = 90.044.$$

Hence, the effect of r_o on A_i is to reduce it by (100 – 90) = 10.

Problem 4.4. For the network shown below:

(a) Determine r_e (b) Calculate Z_i and Z_o

(c) Find A_v and A_i

(d) Repeat part (b) and with r_o = 50 kΩ

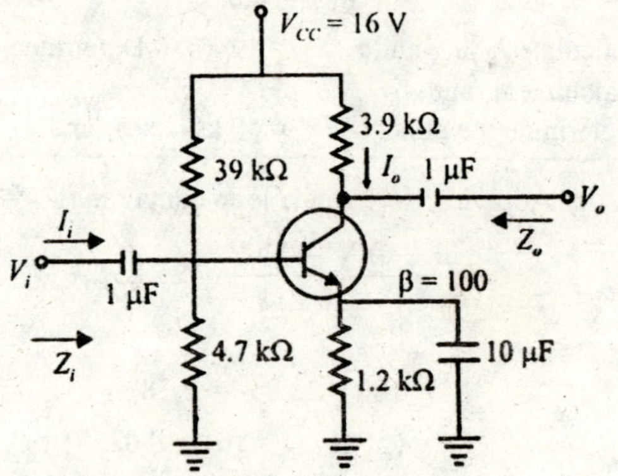

Fig. P. 4.4

Solution. The above configuration can be equivalently drawn as shown in the figure.

$$R' = R_1 \parallel R_2 = \frac{R_1 R_2}{R_1 + R_2} = \frac{(39\ \text{k}\Omega) \times (4.7\ \text{k}\Omega)}{43.7\ \text{k}\Omega}$$

$\Rightarrow$ $$R' = 4.194\ \text{k}\Omega$$

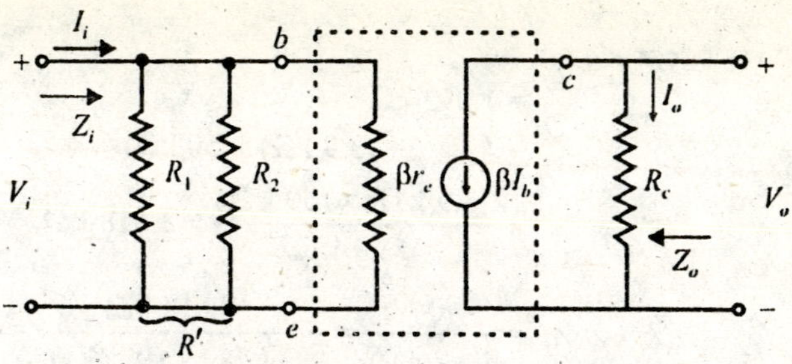

Fig. 4.4 (a)

(*a*) d.c. testing

$$\beta R_E > 10\, R_2$$

$$\Rightarrow \quad (100) \times (1.2\text{ k}\Omega) > 10\,(4.7\text{ k}\Omega)$$

$$120\text{ k}\Omega > 47\text{ k}\Omega \text{ satisfied}$$

Using the approximate approach

$$V_B = V_{CC}\cdot\frac{R_2}{R_1 + R_2} = (16\ \text{V})\cdot\frac{(4.7\,\text{k}\Omega)}{39\,\text{k}\Omega + 4.7\,\text{k}\Omega} = 1.72\ \text{V}$$

$$V_E = V_B - V_{BE} = 1.72\ \text{V} - 0.7\ \text{V} = 1.02\ \text{V}$$

$$I_E = \frac{V_E}{R_E} = \frac{1.02\text{V}}{1.2\ \text{k}\Omega} = 0.850\ \text{mA}$$

$$\therefore \quad r_e = \frac{26\,\text{mV}}{I_E} = \frac{26\ \text{mV}}{0.850\ \text{mA}} = 30.56\ \Omega$$

(*b*) $Z_i = R' \,\|\, \beta\, r_e$

Now $\beta\, r_e = 100 \times 14.789\ \Omega = 1.479\text{ k}\Omega$

$$\therefore \quad Z_i = (4.194\ \text{k}\Omega) \,\|\, (1.479\ \text{k}\Omega) = \frac{(4.194\ \text{k}\Omega)\times(1.479\ \text{k}\Omega)}{5.673\ \text{k}\Omega}$$

$$\Rightarrow \quad Z_i = 1.093\text{ k}\Omega,\ \ Z_o = R_C = 3.9\text{ k}\Omega$$

(*c*) Voltage gain $A_v = -\dfrac{R_C}{r_e} = -\dfrac{3.9\,\text{k}\Omega}{14.789\times10^{-3}\,\text{k}\Omega}$

$$\Rightarrow \quad A_v = -263.71$$

For voltage divider bias

$$A_i = \frac{R'\beta}{R' + \beta r_e} = \frac{(4.194\ \text{k}\Omega)\times(100)}{4.194\ \text{k}\Omega + 1.479\ \text{k}\Omega} = 73.93$$

(*d*) Now given $r_o = 50\text{ k}\Omega$

Z_i will not depend on r_o and

Hence $\quad Z_i = 1.093\ \text{k}\Omega$

$$Z_o = R_e \parallel r_o = (3.9\ \text{k}\Omega) \parallel (50\ \text{k}\Omega)$$

$$= \frac{(3.9\ \text{k}\Omega)\times(50\ \text{k}\Omega)}{53.9\ \text{k}\Omega} = 3.618\ \text{k}\Omega$$

$$A_v = -\frac{R_C \parallel r_o}{r_e} = -\frac{3.618\ \text{k}\Omega}{14.789\times 10^{-3}\ \text{k}\Omega} = -244.64$$

Current gain $\quad A_i = \dfrac{R'}{R'+\beta r_e}\cdot\dfrac{r_o\beta}{r_o + R_C}$

$\Rightarrow \quad A_i = \dfrac{4.194\ \text{k}\Omega}{5.673\ \text{k}\Omega}\times\dfrac{(50\ \text{k}\Omega)\times(100)}{53.9\ \text{k}\Omega} = 68.58.$

Problem 4.5. Determine V_{CC} for the network given below, if $A_v = -160$ and $r_o = \infty\ \Omega$.

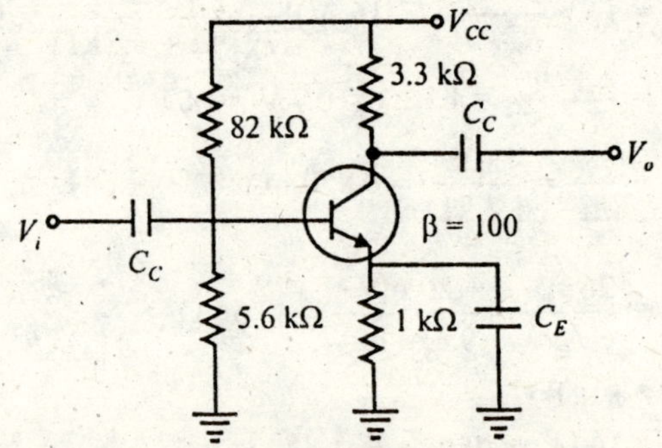

Fig. P. 4.5

Solution. Given: $A_v = -160$ and $r_o = \infty\ \Omega$

i.e. $\quad -\dfrac{R_c}{r_e} = -160$

$\Rightarrow \quad r_e = \dfrac{3.3\text{k}\Omega}{160} = 20.625\Omega$

Now $\quad \beta R_E > 10\ R_2$

$\Rightarrow \quad 100\times 1\ \text{k}\Omega > 10\times 5.6\ \text{k}\Omega$

$\Rightarrow \quad 100\ \text{k}\Omega > 56\ \text{k}\Omega$ satisfied.

$\therefore \quad V_B = V_{CC}\cdot\dfrac{R_2}{R_1+R_2} = V_{CC}\cdot\dfrac{5.6\ \text{k}\Omega}{82\ \text{k}\Omega + 5.6\ \text{k}\Omega}$

$\Rightarrow \quad V_B = V_{CC}\times 0.0639 \Rightarrow V_{CC} = 15.643\cdot V_B \qquad \ldots(1)$

Now, since $r_e = \frac{26 \text{ mV}}{I_E}$

$$\Rightarrow \quad 20.625 = \frac{26 \text{ mV}}{I_E} \quad \Rightarrow \quad I_E = \frac{26 \text{ mV}}{20.625} = 1.26 \text{ mA}$$

But $I_E = \frac{V_E}{R_E}$, $\quad \therefore \; V_E = I_E \times R_E = 1.26 \text{ mA} \times 1 \text{ k}\Omega = 1.26 \text{ V}$

Now $\quad V_{BE} = V_B - V_E$

$\Rightarrow \quad 0.7 \text{ V} = V_B - 1.26 \text{ V}$

Which yields $\quad V_B = 0.7 + 1.26 = 1.96 \text{ V}$

Hence, from equation $\quad V_{CC} = 15.643 \times V_B$

$$= 15.643 \times 1.96 \text{ V}$$

$$= 30.67 \text{ V}$$

Problem 4.6. For the network shown below:

(*a*) Determine r_e (*b*) Calculate V_B and V_C

(*c*). Determine Z_i and $A_v = \frac{V_o}{V_i}$

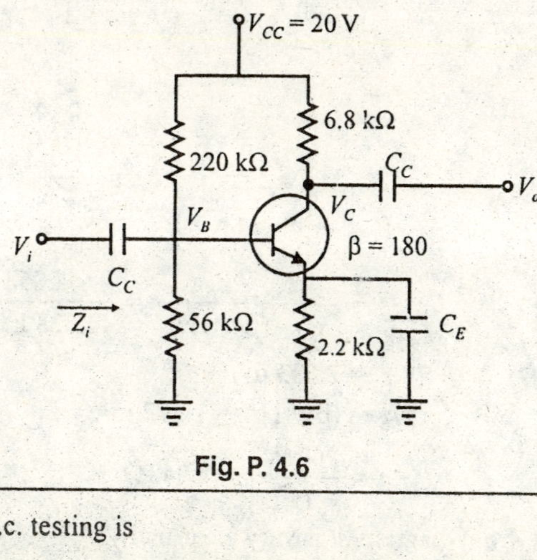

Fig. P. 4.6

Solution. d.c. testing is

$$\beta R_E > 10 \; R_2$$

$\Rightarrow \quad (180) \times (2.2 \text{ k}\Omega) > 10 \times 56 \text{ k}\Omega$

⇒ 396 kΩ > 560 kΩ, which is not true

Hence, we will proceed in a general way.

Drawing the base-emitter portion of the network.

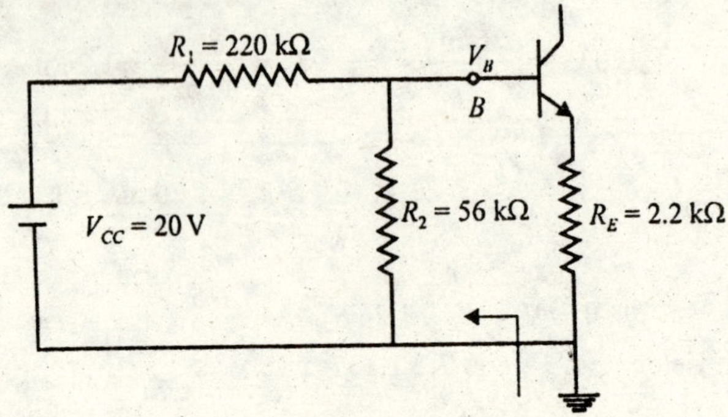

Fig. 4.6 (a)

$$R_{th} = R_1 \parallel R_2 = 220\text{ k}\Omega \parallel 50\text{ k}\Omega = 44.638\text{ k}\Omega$$

$$E_{th} = V_{R_2} = \frac{R_2 \cdot V_{CC}}{R_1 + R_2} = 4.058\text{ V}$$

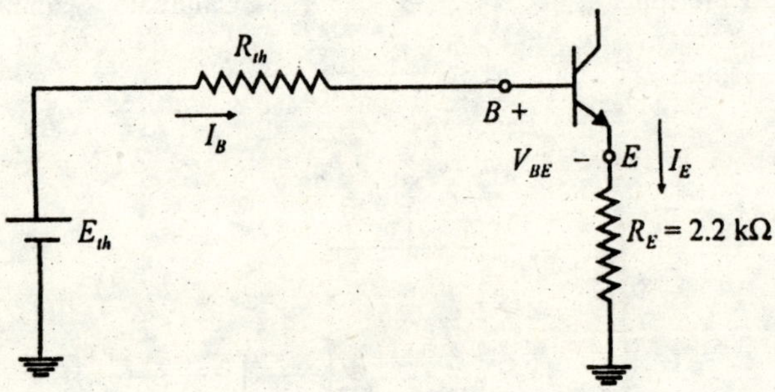

Fig. 4.6 (b)

Hence $$I_B = \frac{E_{th} - V_{BE}}{R_{th} + (\beta + 1) R_E} = \frac{4.058\text{ V} - 0.7\text{ V}}{44.638\text{ k}\Omega + 398.2\text{ k}\Omega}$$

$$= 7.583\ \mu\text{A}$$

Hence $$I_E = (\beta + 1) I_B = 1.372\text{ mA}$$

(*a*) Thus $$r_e = \frac{26\text{ mV}}{I_E} = 18.943\Omega$$

(*b*) Now, from diagram, it is very clear that

$$V_B = E_{th} - I_B \times R_{th}$$

⇒ $$V_B = 4.058\text{ V} - 7.583 \times 10^{-6} \times 44.638 \times 10^3\text{ V} = 3.719\text{ V}.$$

Also ∵ $I_C = \beta I_B$

⇒ $I_C = 180 \times 7.583 \times 10^{-6}\text{ A} = 1.365\text{ mA}$

Now, from the original diagram, we find

$$V_C = V_{CC} - I_C \times R_C$$

$$= 20 - 1.365 \times 10^{-3} \times 6.8 \times 10^3 = 10.718\text{ V}$$

(c) $Z_i = R' \parallel \beta r_e$ where $R' = R_{th}$

or $$Z_i = \frac{R_{th} \times \beta r_e}{R_{th} + \beta r_e} = \frac{(44.638\text{ k}\Omega) \times (3.41\text{ k}\Omega)}{48.048\text{ k}\Omega}$$

or $Z_i = 3.168\text{ k}\Omega$

Voltage gain $$A_v = \frac{V_o}{V_i}$$

But $$A_v = -\frac{R_C}{r_e} = -\frac{6.8 \times 10^3}{18.943} = -358.97$$

Problem 4.7. For the network shown:

(*a*) Determine r_e (*b*) Find Z_i and Z_o

(*c*) Calculate A_v and A_i

(*d*) Repeat parts (*b*) and with $r_o = 50\text{ k}\Omega$

(Assume $r_o = \infty\ \Omega$ for parts (*b*) and (*c*) unless otherwise stated)

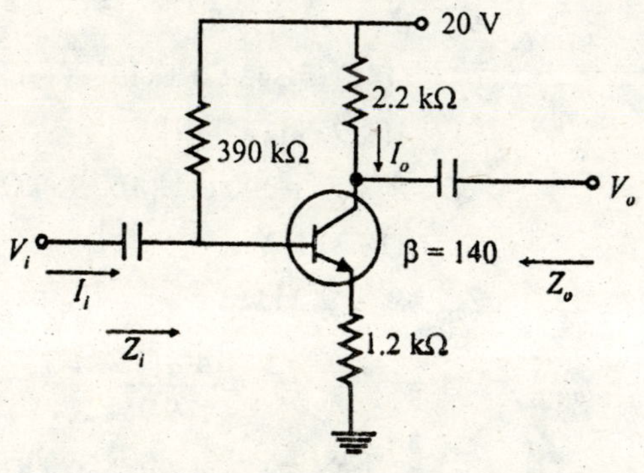

Fig. P. 4.7

Solution. Substituting the r_e equivalent circuit into the a.c. equivalent network of above figure, we get

(*a*) From d.c. analysis, we get

$$I_B = \frac{V_{CC} - V_{BE}}{R_B + (\beta + 1) R_E} = \frac{20\text{ V} - 0.7\text{ V}}{390\text{ k}\Omega + (141)1.2\text{ k}\Omega}$$

$$\Rightarrow \qquad I_B = 34.51\ \mu\text{A}$$

$$I_E = (\beta + 1)\, I_B = 141 \times (34.51\ \mu\text{A}) = 4.866\ \text{mA}$$

and
$$r_e = \frac{26\,\text{mV}}{I_E} = \frac{26\,\text{mV}}{4.866\,\text{mA}} = 5.343\,\Omega$$

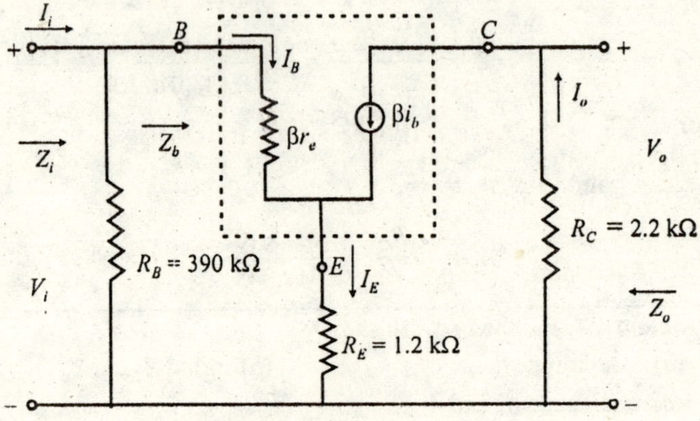

Fig. 4.7 (a)

(*b*) From equivalent diagram

$$Z_b = \beta r_e + (\beta + 1)\, R_E$$

$$= (140)\,(5.343\ \Omega) + (141)\,(1.2\ \text{k}\Omega)$$

$$= 169.95\ \text{k}\Omega \approx 170\ \text{k}\Omega$$

Now
$$Z_i = R_B \,\|\, Z_b = 390\ \text{k}\Omega \,\|\, 169.95\ \text{k}\Omega$$

$$\Rightarrow \qquad Z_i = 118.37\ \text{k}\Omega$$

$$Z_o = R_C = 2.2\ \text{k}\Omega$$

(*c*) Voltage gain
$$A_v = -\frac{\beta R_C}{Z_b} = -\frac{(140)(2.2\,\text{k}\Omega)}{169.95\,\text{k}\Omega} = -1.812$$

$$A_i = \frac{\beta R_B}{R_B + Z_b} = \frac{(140) \times 390\ \text{k}\Omega}{390\ \text{k}\Omega + 169.95\ \text{k}\Omega}$$

$$\Rightarrow \qquad A_i = 97.5$$

(*d*) The placement of r_o for this particular configuration is such that for typical parameter values, the effect of r_o on the output impedance and voltage gain can be ignored.

i.e. there will be no change by including $r_o = 50\ \text{k}\Omega$ and the results are all same as calculated before.

Problem 4.8. Repeat problem 4.7 with R_E bypassed. Compare results.

Solution. (*a*) The d.c. analysis is the same and $r_e = 5.343\ \Omega$. R_E is shorted out by C_E for the a.c. analysis.

Therefore $Z_i = R_B \parallel Z_b = R_B \parallel \beta r_e = (390\ \text{k}\Omega) \parallel \beta r_e$

$\Rightarrow$ $Z_i = (390\ \text{k}\Omega) \parallel (140)(5.343\ \Omega) = 0.7466\ \text{k}\Omega$

or $Z_i = 746.6\ \Omega$

$Z_o = R_C = 2.2\ \text{k}\Omega$

(*c*) $\because$ $Z_b = \beta r_e + (\beta + 1) R_E$

Here, R_E is shorted

$\therefore$ $Z_b = \beta r_e$

$\therefore$ Voltage gain $A_v = -\dfrac{\beta R_C}{Z_b} = -\dfrac{\beta R_e}{\beta r_e} = -\dfrac{R_C}{r_e}$

$\Rightarrow$ $A_v = -\dfrac{2.2\,\text{k}\Omega}{5.343\,\Omega} = -411.75$ (a significant increase)

$$A_i = \frac{\beta R_B}{R_B + Z_b} = \frac{\beta R_B}{R_B + \beta r_e} = \frac{(140)(390\,\text{k}\Omega)}{390\,\text{k}\Omega + (140)(5.343)}$$

$= 139.73$

Thus, comparing the results we find that values of Z_i, A_v and A_i increase, whereas value of Z_o remains same.

(*d*) Z_i is independent of r_o hence, its values will be unchanged.

i.e. $Z_i = 746.6\ \Omega$

$Z_o = R_e \parallel r_o = (2.2\ \text{k}\Omega) \parallel (50\ \text{k}\Omega) = 2.1073\ \text{k}\Omega$

$\Rightarrow$ $Z_o \approx 2.11\ \text{k}\Omega$

Now, without r_o $A_v = -\dfrac{R_C}{r_e}$. But with r_o

$$A_v = -\frac{R_C \parallel r_o}{r_e} = -\frac{2.1073 \times 1000}{5.343} = -394.4$$

Now $A_i = \dfrac{r_o \beta}{r_o + R_C} \cdot \dfrac{R_B}{R_B + \beta r_e}$

Or, we can use the simple relation

$$A_i = -A_v \times \frac{Z_i}{R_C} = \frac{-394.4 \times 746.6}{2.2 \times 1000} = 133.84$$

Problem 4.9. For the network configuration given below, determine R_E and R_B, if $A_v = -10$ and $r_e = 3.8\ \Omega$. Assume that $Z_b = \beta R_E$.

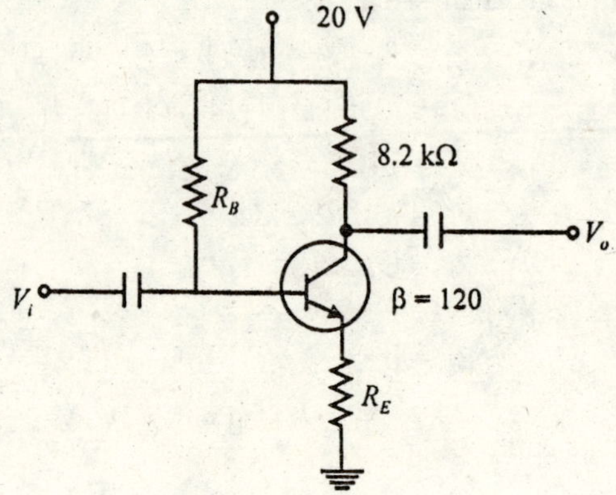

Fig. P. 4.9

Solution. Given $A_v = -10$

We know that $A_v = -\dfrac{\beta R_C}{Z_b}$

or $A_v = -\dfrac{\beta R_C}{\beta R_E} = -\dfrac{R_C}{R_E}$ (as given in the problem)

$\Rightarrow \quad -10 = -\dfrac{8.2\ \text{k}\Omega}{R_E}$

or $R_E = \dfrac{8.2\ \text{k}\Omega}{10} = 0.82\ \text{k}\Omega = 820\ \Omega$

Now, given $r_e = 3.8\ \Omega$

From $r_e = \dfrac{26\ \text{mV}}{I_E}$, we get $I_E = \dfrac{26\,\text{mV}}{r_e}$

$$I_E = \frac{26\ \text{mV}}{3.8\ \Omega} = 6.842\ \text{mA}$$

Also, since $I_E = (\beta + 1)\, I_B$

Therefore $I_B = \dfrac{I_E}{(\beta + 1)} = \dfrac{6.842\,\text{mA}}{121} = 56.546\,\mu\text{A}$

But, d.c. analysis suggests that

$$I_B = \frac{V_{CC} - V_{BE}}{R_B + (\beta + 1) R_E}$$

or $$56.546 \times 10^{-6} = \frac{20\,V - 0.7\,V}{R_B + (121)(820\,\Omega)}$$

$$\Rightarrow \quad R_B + 99.22\text{ k}\Omega = \frac{19.3 \times 10^6}{56.546} = 341.315\text{ k}\Omega$$

or $$R_B = 341.315\text{ k}\Omega - 99.22\text{ k}\Omega = 242.1\text{ k}\Omega.$$

Problem 4.10. For the network of following configuration:
(*a*) Determine r_e and βr_e (*b*) Find Z_i and Z_o (*c*) Calculate A_v and A_i

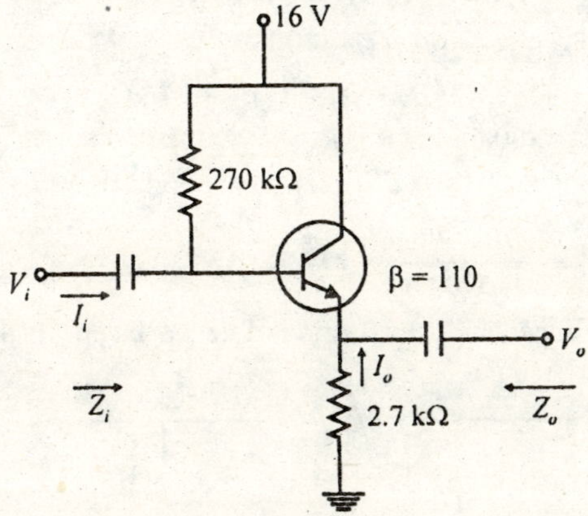

Fig. P. 4.10

Solution. (*a*) Substituting the r_e equivalent circuit into the a.c. equivalent network of given figure.

This emitter follower type configuration.

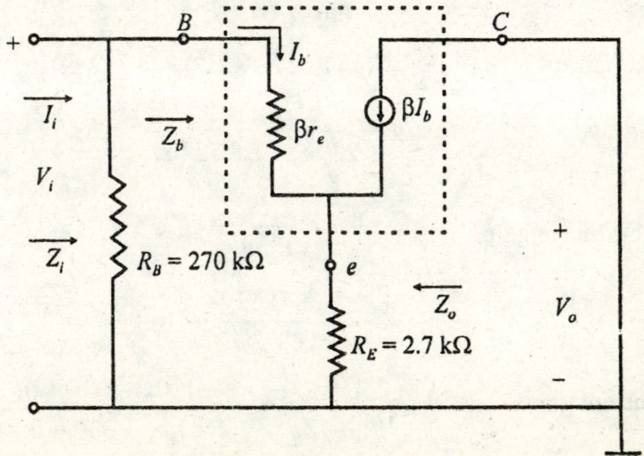

Fig. 4.10 (a)

$$\therefore \quad I_B = \frac{V_{CC} - V_{BE}}{R_B + (\beta + 1) R_E}$$

$$= \frac{16\text{V} - 0.7\text{V}}{270\,\text{k}\Omega + (111)(2.7\,\text{k}\Omega)} = 26.86\,\mu\text{A}$$

Thus $\quad I_E = (\beta + 1)\, I_B = (111)\,(26.86\ \mu\text{A}) = 2.981\ \text{mA}$

$$r_e = \frac{26\ \text{mV}}{I_E} = \frac{26\ \text{mV}}{2.981\ \text{mA}} = 8.722\ \Omega$$

and $\quad \beta\, r_e = (110)\,(8.722\ \Omega) = 959.42\ \Omega$

(*b*) $\quad Z_b = \beta\, r_e + (\beta + 1)\, R_E$

$$= 110 \times (8.722\ \Omega) + 111 \times (2.7\ \text{k}\Omega)$$

$$= 300.66\ \text{k}\Omega \approx \beta\, R_E$$

Thus $\quad Z_i = R_B \parallel Z_b = (270\ \text{k}\Omega) \parallel (300.66\ \text{k}\Omega)$

$$= \frac{270 \times 300.66}{570.66}\ \text{k}\Omega = 142.25\ \text{k}\Omega$$

For output impedance, the portion of network appears as

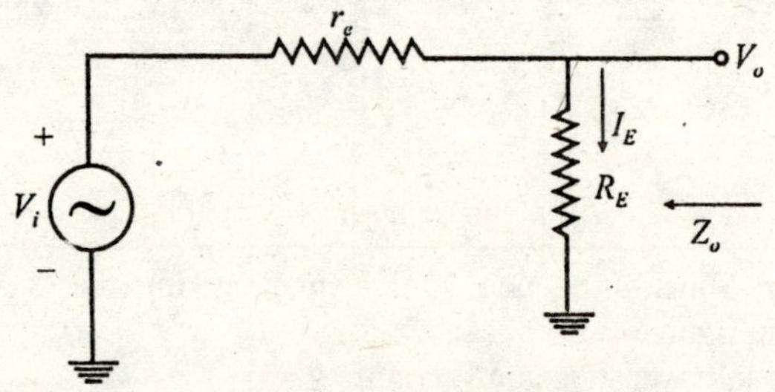

Fig. 4.10 (b)

Therefore $\quad Z_o = R_E \parallel r_e = (2.7\ \text{k}\Omega) \parallel (8.722\ \Omega)$

$$= \frac{(2.7\ \text{k}\Omega) \times (8.722\ \Omega)}{2.7\ \text{k}\Omega + 8.722\ \Omega} = 8.694 \approx r_e$$

(*c*) Voltage gain $\quad A_v = \dfrac{V_o}{V_i} = \dfrac{R_E}{R_E + r_e}$

$$\Rightarrow \quad A_v = \frac{2.7\,\text{k}\Omega}{2.7\,\text{k}\Omega + 8.722\,\Omega} = 0.997 \approx 1$$

Current gain $\quad A_i \ -\dfrac{\beta\, R_B}{R_B + Z_b} = -\dfrac{(110) \times 270\,\text{k}\Omega)}{270\,\text{k}\Omega + 300.66\,\text{k}\Omega}$

$$\Rightarrow \quad A_i = -\ 52.045$$

A_i can also be calculated with the help of following relation.

$$A_i = -A_v \frac{Z_i}{R_E} = -(0.997) \times \frac{142.25\,\text{k}\Omega}{2.7\,\text{k}\Omega} = 52.52$$

Problem 4.11. For the network of given figure:

(*a*) Determine Z_i and Z_o (*b*) Find A_v

(*c*) Calculate V_o if $V_i = 1$ mV

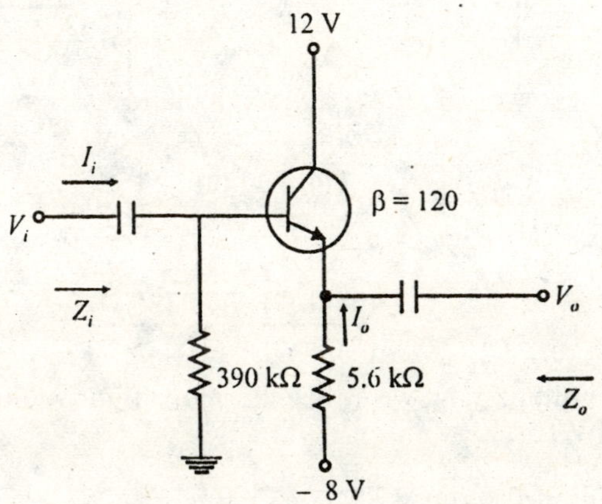

Fig. P. 4.11

Solution. This is the configuration of emitter following type.

(*a*) From KVL applying in the loop containing R_B and R_E

We get $$I_B = \frac{8\,\text{V} - 0.7\,\text{V}}{390\,\text{k}\Omega + (120+1)5.6\,\text{k}\Omega}$$

$$= \frac{7.3\,\text{V}}{1067.6\,\text{k}\Omega} = 6.84\,\mu\text{A}$$

Hence $$I_E = (\beta + 1)\, I_B = 0.83\ \text{mA}$$

Therefore $$r_e = \frac{26\ \text{mV}}{I_E} = \frac{26\ \text{mV}}{0.83\ \text{mA}} = 31.325\,\Omega$$

and $$\beta r_e = 120 \times 31.325\,\Omega = 3759.04\,\Omega = 3.76\,\text{k}\Omega.$$

Now, as we shown in the r_e equivalent circuit of last problem

$$Z_b = \beta r_e + (\beta + 1) R_E = 3.76\,\text{k}\Omega + 121 \times 5.6\,\text{k}\Omega$$
$$= 681.36\,\text{k}\Omega$$

Thus $$Z_i = R_B \parallel Z_b = (681.36\,\text{k}\Omega) \parallel (390\,\text{k}\Omega)$$

$$= \frac{681.36 \times 390}{681.36 + 390}\,\text{k}\Omega = 248\,\text{k}\Omega$$

Calculate Z_o, drawing the portion of the network is below:

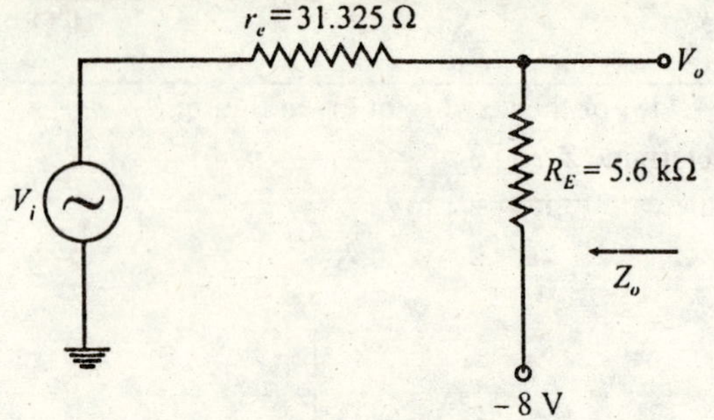

Fig. 4.11 (a)

Therefore $$Z_o = R_E \parallel r_e$$

or $$Z_o = \frac{(5.6\text{ k}\Omega) \times (31.325\ \Omega)}{(5.6\text{ k}\Omega) + (31.325\ \Omega)} = 31.15\ \Omega \approx r_e$$

(*b*) For voltage gain, we utilize the above figure voltage drop across $R_E = V_o + 8$ V.

But that drop also equals $\left(\dfrac{V_i + 8}{r_e + R_E}\right) \times R_E$

or $$\frac{V_i + 8}{5.631} \times 5.6 = V_o + 8$$

or $$(V_i + 8) \times 0.994 = (V_o + 8)$$

or $$A_v = 0.994$$

Alternatively we can also calculate by simple voltage division rule

$$V_o = \frac{R_E V_i}{r_e + R_E}$$

or $$A_v = \frac{V_o}{V_i} = \frac{5.6\text{ k}\Omega}{31.325\ \Omega + 5.6\text{ k}\Omega} = 0.994$$

(*c*) We have $A_v = 0.994$ or $\dfrac{V_o}{V_i} = 0.994$

Given $V_i = 1$ mV

Hence $V_o = 0.994 \times V_i$

or, $V_o = 0.994 \times 1\text{ mV} = 0.994$ mV.

Problem 4.12. For the network shown in Fig. P. 4.12:

(*a*) Calculate I_B and I_C (*b*) Determine r_e

(*c*) Determine Z_i and Z_o (*d*) Find A_v and A_i

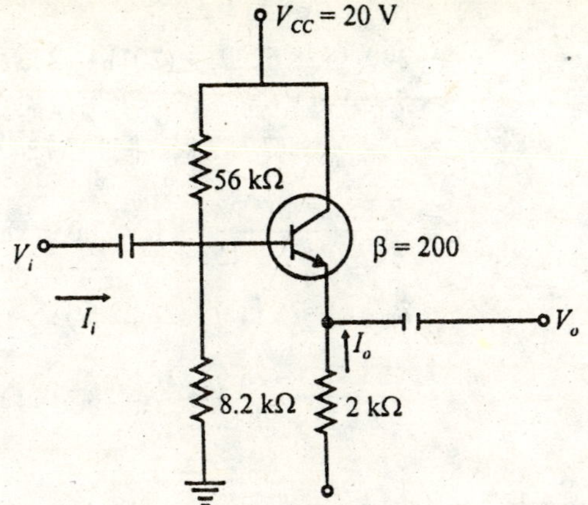

Fig. P. 4.12

Solution. (*a*) For voltage divider bias

$$R_{th} = R_1 \parallel R_2$$

or $$R_{th} = (56 \text{ k}\Omega) \parallel (8.2 \text{ k}\Omega) = \frac{(56 \text{ k}\Omega) \times (8.2 \text{ k}\Omega)}{64.2 \text{ k}\Omega}$$

$$= 7.153 \text{ k}\Omega$$

Also, since the condition $\beta R_E > 10\ R_2$

i.e. $200 \times (2 \text{ k}\Omega) > 10 \times (8.2 \text{ k}\Omega)$

⇒ $(400 \text{ k}\Omega) > (82 \text{ k}\Omega)$ is satisfied so, we can apply the approximate analysis and E_{th} can be found as

$$E_{th} = V_{CC} \cdot \frac{R_2}{R_1 + R_2} = (20 \text{ V}) \times \frac{(8.2 \text{ k}\Omega)}{(64.2 \text{ k}\Omega)} = 2.55 \text{ V}$$

Now $$I_B = \frac{E_{th} - V_{BE}}{R_{th} + (\beta + 1) R_E}$$

⇒ $$I_B = \frac{2.55 - 0.7}{7.153 \text{ k}\Omega + 201 \times 2 \text{ k}\Omega} = \frac{1.85 \text{ V}}{409.153 \text{ k}\Omega} = 4.52\ \mu\text{A}$$

and $$I_C = \beta\, I_B = (200 \times 4.52)\ \mu\text{A}$$

or $$I_C = 0.904 \text{ mA}$$

(*b*) $$I_E = (\beta + 1)\, I_B = 201 \times 4.52\ \mu\text{A} = 0.908 \text{ mA}$$

Now $$r_e = \frac{26 \text{ mV}}{I_E} = \frac{26 \text{ mV}}{0.908 \text{ mA}} = 28.62\ \Omega$$

(*c*) For determination of Z_i and Z_o simply replace R_B by R_{th}.

$$Z_b = \beta r_e + (\beta + 1) R_E$$
$$= 200 \times (28.62\ \Omega) + (201) \times (2\ k\Omega)$$
$$= 5.724\ k\Omega + 402\ k\Omega = 407.724\ k\Omega$$
$$Z_i = R_{th} \parallel Z_b = (7.153\ k\Omega) \parallel (407.724\ k\Omega)$$
$$= \frac{7.153 \times 407.724}{407.724 + 7.153}\ k\Omega = 7.03\ k\Omega$$
$$Z_o = R_E \parallel r_e = (2\ k\Omega) \parallel (28.62\ \Omega)$$
$$= \frac{(2\ k\Omega) \times (28.62\ \Omega)}{2\ k\Omega + 28.62\ \Omega} = 28.216\ \Omega \approx r_e$$

(*d*) Voltage gain $A_v = \frac{V_o}{V_i} = \frac{R_E}{R_E + r_e} = \frac{2\ k\Omega}{2\ k\Omega + 28.62\ \Omega}$

or $A_v = 0.986 \approx 1$

Currentgain $A_i = -\frac{\beta R_{th}}{R_{th} + Z_b} = -\frac{200(7.153\ k\Omega)}{7.153\ k\Omega + 407.724\ k\Omega} = -3.45$

It can also be found with the relation of

$$A_i = -A_v \frac{Z_i}{R_E} = -0.986 \times \frac{7.03\ k\Omega}{2\ k\Omega} = -3.46$$

Problem 4.13. For the common base configuration of given figure:

(*a*) Determine r_e (*b*) Find Z_i and Z_o (*c*) Calculate A_v and A_i

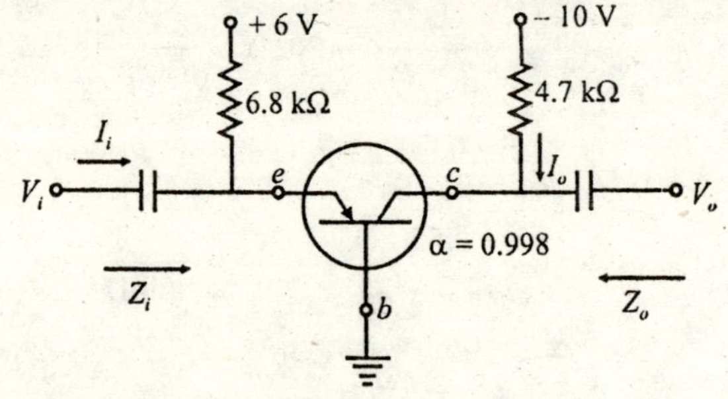

Fig. P. 4.13

Solution. This is a common base type configuration. Figure can be redrawn as the following:

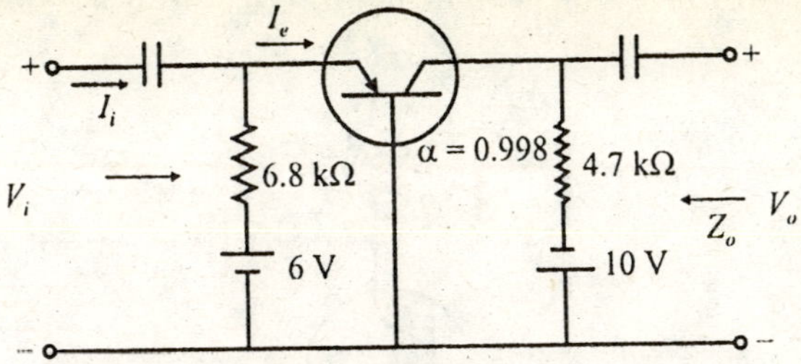

Fig. 4.13 (a)

Given $V_{EE} = 6\text{ V}, V_{CC} = -10\text{ V}, R_C = 4.7\text{ k}\Omega$ and $R_E = 6.8\text{ k}\Omega$

$$I_E = \frac{V_{EE} - V_{BE}}{R_E} = \frac{6\text{V} - 0.7\text{V}}{6.8\text{ k}\Omega} = \frac{5.3\text{V}}{6.8\text{ k}\Omega}$$

or $I_E = 0.78$ mA

From $$r_e = \frac{26\text{ mV}}{I_E}$$

We get $$r_e = \frac{26\text{ mV}}{0.78\text{ mA}} = 33.36\ \Omega$$

(*b*) $$Z_i = R_E \parallel r_e = (6.8\text{ k}\Omega) \parallel (33.36\ \Omega)$$

$$= \frac{(6.8\text{ k}\Omega) \times (33.36\ \Omega)}{6.8\text{ k}\Omega + 33.36\ \Omega} = \frac{226.85\ (\Omega)^2}{6.833\ \Omega}$$

or $Z_i = 33.2\ \Omega \approx r_e$

Also, output impedance $Z_o = R_C = 4.7\text{ k}\Omega$

(*c*) To calculate A_v and A_i :

Voltage gain for *CB*-configuration is given by

$$A_v = \frac{R_C}{r_e} = \frac{4.7\text{ k}\Omega}{33.36\ \Omega} = 140.887$$

and current gain $$A_i = \frac{I_o}{I_i} = -\alpha$$

or $A_i = -0.998 \approx -1$

The exact value of $$A_v = \frac{\alpha R_C}{r_e}$$

$$\Rightarrow \quad A_v = \frac{0.998 \times 4.7\text{ k}\Omega}{33.36\ \Omega}$$

or $$A_v = 140.605 \approx \frac{R_C}{r_e}$$

Problem 4.14. For the network shown in Fig. P. 4.14, determine A_v and A_i.

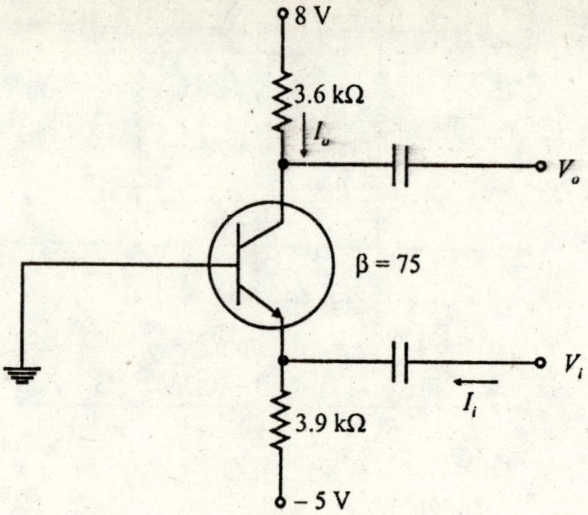

Fig. P. 4.14

Solution. Even though in the given problem β is specified, it is a common base type configuration.

Here given $V_{EE} = -5$ V, $V_{CC} = 8$ V, $R_C = 3.6$ kΩ and $R_E = 3.9$ kΩ

From expression $$I_E = \frac{V_{EE} - V_{BE}}{R_E} = \frac{-5\text{V} - 0.7\text{V}}{3.9\text{ k}\Omega}$$

or $$I_E = \frac{-5.7\text{ V}}{3.9\text{ k}\Omega} = -1.46\text{ mA}$$

– ve sign shows that the direction of I_E is opposite to that we assumed.

Therefore $$r_e = \frac{26\text{ mV}}{I_E} = \frac{26\text{ mV}}{1.46\text{ mA}} = 17.79\ \Omega$$

Given β = 75, from relation $\beta = \frac{\alpha}{1-\alpha}$, we get

$$\alpha = \frac{\beta}{\beta+1} = \frac{75}{76} = 0.9868$$

Therefore, voltage gain $$A_v = \frac{V_o}{V_i} = \frac{\alpha R_C}{r_e}$$

$$A_v = \frac{0.9868 \times 3.6\text{ k}\Omega}{17.79\ \Omega} = 199.689$$

$$A_i = -\alpha = -0.9868 \approx -1.$$

Problem 4.15. For the collector *FB* configuration of figure given below:

(*a*) Determine r_e (*b*) Find Z_i and Z_o

(*c*) Calculate A_v and A_i

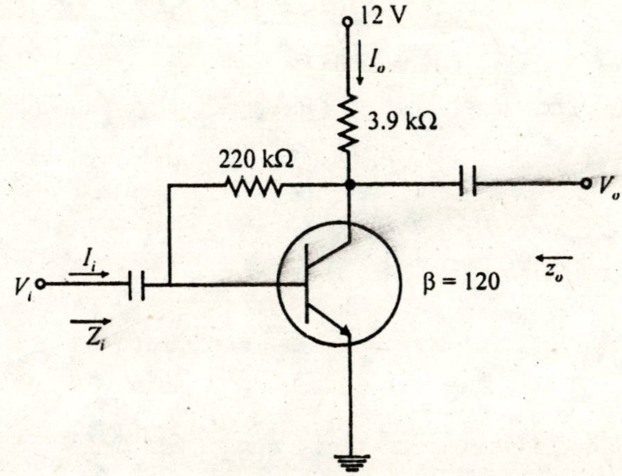

Fig. P. 4.15

Solution. Substituting the r_e equivalent of the given figure:

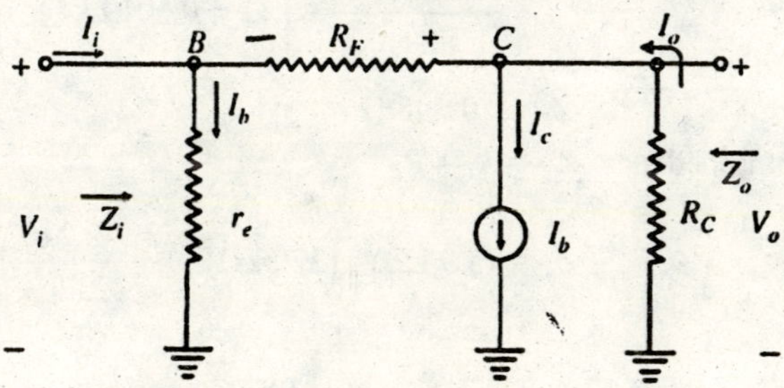

Fig. 4.15 (a)

Given $R_F = 220\text{ k}\Omega,\ \beta = 120,\ R_C = 3.9\text{ k}\Omega,\ V_{CC} = 12\text{ V}$

From KVL in base collector loop, we get I_B as

(*a*) $$I_B = \frac{V_{CC} - V_{BE}}{R_F + \beta R_C} = \frac{12\text{ V} - 0.7\text{ V}}{220\text{ k}\Omega + 120 \times (3.9\text{ k}\Omega)}$$

or $$I_B = \frac{11.3\text{ V}}{688\text{ k}\Omega} = 16.42\ \mu\text{A}$$

Thus from $I_E = (\beta + 1)\, I_B$, we get value of I_E as

$$I_E = (120 + 1) \times 16.42\ \mu\text{A} = 121 \times 16.42\ \mu\text{A}$$

or $$I_E = 1.987 \text{ mA}$$

Therefore $$r_e = \frac{26 \text{ mV}}{I_E} = \frac{26 \text{ mV}}{1.987 \text{ mA}}$$

or $$r_e = 13.08 \ \Omega$$

(*b*) We know that for collector feedback configuration input inpedance:

$$Z_i = \beta r_e \parallel \frac{R_F}{|A_v|}$$

A_v is given by $$A_v = -\frac{R_C}{r_e}$$

or $$A_v = -\frac{3.9 \text{ k}\Omega}{13.08 \ \Omega} = -298.165$$

Therefore $$Z_i = (120 \times 13.08 \ \Omega) \parallel \frac{220 \text{ k}\Omega}{298.165}$$

or $$Z_i = (1.57 \text{ k}\Omega) \parallel (0.738 \text{ k}\Omega)$$

$$= \frac{1.57 \times 0.738}{1.57 + 0.738} \text{ k}\Omega = 0.5019 \text{ k}\Omega$$

or $$Z_i = 501.95 \ \Omega$$

Output impedance $$Z_o = R_C \parallel R_F$$

or $$Z_o = (3.9 \text{ k}\Omega) \parallel (220 \text{ k}\Omega) = \frac{3.9 \times 220}{223.9} \text{ k}\Omega$$

$$Z_o = \frac{858}{223.9} \text{ k}\Omega = 3.832 \text{ k}\Omega$$

(*c*) To calculate A_v and A_i :

Calculated in part (*b*) that

$$A_v = -\frac{R_C}{r_e} = -298.165$$

Current gain A_i is given by

$$A_i = \frac{\beta R_F}{R_F + \beta R_C} = \frac{120 \times (220 \text{ k}\Omega)}{220 \text{ k}\Omega + 120 \times 3.9 \text{ k}\Omega}$$

or $$A_i = \frac{26400}{688} = 38.372$$

Problem 4.16. Given $r_e = 10\ \Omega$, $\beta = 200$, $A_v = -160$, and $A_i = 19$ for the network of figure given below, determine R_C, R_F and V_{CC}.

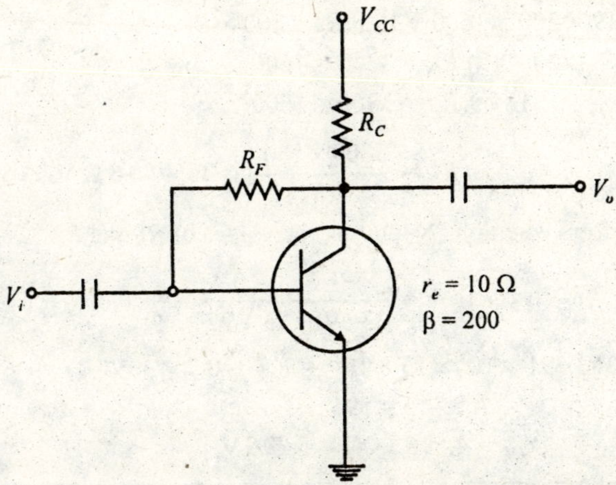

Fig. P. 4.16

Solution. Given $\quad r_e = 10\ \Omega$

But $\quad r_e = \dfrac{26\text{ mV}}{I_E} \Rightarrow 10\ \Omega = \dfrac{26\text{ mV}}{I_E}$

or $\quad I_E = \dfrac{26\text{ mV}}{10\ \Omega} = 2.6\text{ mA}$

But $\quad I_E = (\beta + 1)\, I_B$

or $\quad I_B = \dfrac{I_E}{\beta + 1} = \dfrac{2.6\text{ mA}}{200 + 1} = \dfrac{2.6\text{ mA}}{200}$

or $\quad I_B = 13\ \mu\text{A}$

From $\quad I_B = \dfrac{V_{CC} - V_{BE}}{R_F + \beta R_C}$

$$13 \times 10^{-6} = \frac{V_{CC} - 0.7}{R_F + 200 \times R_C} \qquad \dots (1)$$

Given $\quad A_v = -160$

or $\quad \dfrac{-R_C}{r_e} = -160 \quad \Rightarrow \quad R_C = 160 \times r_e$

or $\quad R_C = 160 \times 10\ \Omega = 1.6\text{ k}\Omega$

From $\quad A_i = \dfrac{\beta R_F}{R_F + \beta R_C}$, given $A_i = 19$

$$\therefore \qquad 19 = \frac{200 \times R_F}{R_F + 200 \times 1600\,\Omega}$$

or $\quad 19 \times 200 \times 1600 + 19 \times R_F = 200 \times R_F$

or $\quad (200 - 19)\, R_F = 3800 \times 1600$

or $\quad 181 \times R_F = 3800 \times 1600$

$$\Rightarrow \qquad R_F = \frac{3800 \times 1600}{181} = 33591.16\ \Omega = 33.591\ \text{k}\Omega$$

Now, from equation (1), putting the values of R_F and R_C, we get

$$13 \times 10^{-6} = \frac{V_{CC} - 0.7}{33.591\ \text{k}\Omega + 320\ \text{k}\Omega}$$

or $\quad 353.591 \times 10^3 \times 13 \times 10^{-6} = V_{CC} - 0.7$

or $\quad V_{CC} - 0.7 = 4.596$

or $\quad V_{CC} = 4.596 + 0.7 = 5.296$ V

Problem 4.17. For the network of figure given below:

(*a*) Determine Z_i and Z_o (*b*) Find A_v and A_i

(*c*) Repeat parts (*a*) and (*b*) if $r_o = 45$ kΩ

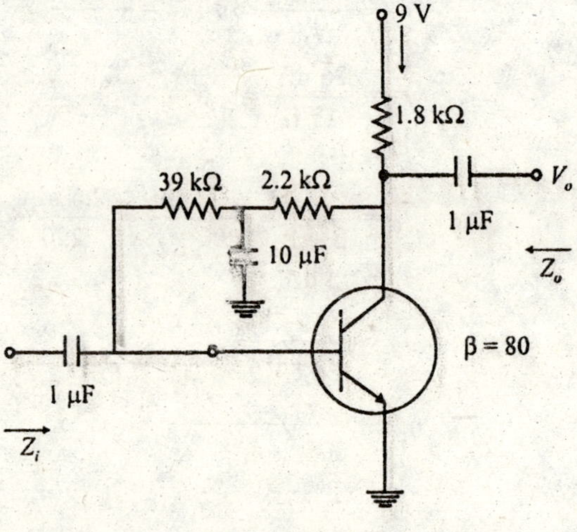

Fig. P. 4.17

Solution. This is a type of collector d.c. feedback configuration substituting the r_e equivalent circuit into the a.c. equivalent network given above.

Given $R_{F_1} = 39\,\text{k}\Omega, R_{F_2} = 22\ \text{k}\Omega$, $R_C = 1.8$ kΩ, β = 80, and $V_{CC} = 9$ V.

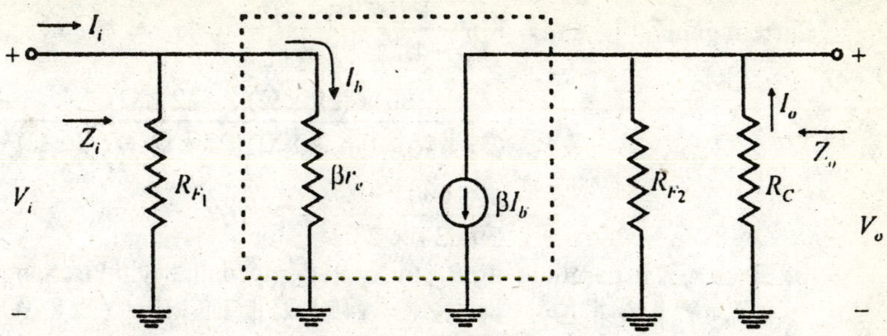

Fig. 4.17 (a)

From d.c. analysis

$$I_B = \frac{V_{CC} - V_{BE}}{(R_{F_1} + R_{F_2}) + \beta R_C}$$

$$\Rightarrow \quad I_B = \frac{9\,\text{V} - 0.7\,\text{V}}{(39\,\text{k}\Omega + 22\,\text{k}\Omega) + 80 \times (1.8\,\text{k}\Omega)} = \frac{8.3\,\text{V}}{61\,\text{k}\Omega + 144\,\text{k}\Omega}$$

or $$I_B = \frac{8.3\,\text{V}}{205\,\text{k}\Omega} = 40.487\,\mu\text{A}$$

$$I_E = (\beta + 1)\,I_B = 81 \times 40.487\,\mu\text{A} = 3.279\,\text{mA}$$

Hence $$r_e = \frac{26\,\text{mV}}{I_E} = \frac{26\,\text{mV}}{3.279\,\text{mA}} = 7.928\,\Omega$$

and $$\beta\, r_e = 80 \times 7.928\,\Omega = 0.634\,\text{k}\Omega$$

(*a*) $$Z_i = R_{F_1} \parallel \beta r_e$$ (from a.c. equivalent network)

or $$Z_i = (39\,\text{k}\Omega) \parallel (0.634\,\text{k}\Omega)$$

$$= \frac{39\,\text{k}\Omega \times 0.634\,\text{k}\Omega}{39.634\,\text{k}\Omega} = 0.6241\,\text{k}\Omega = 624.1\,\Omega$$

$$Z_o = R_{F_2} \parallel R_C = (22\,\text{k}\Omega) \parallel (1.8\,\text{k}\Omega) = \frac{22 \times 1.8}{23.8}\,\text{k}\Omega$$

or $$Z_o = \frac{39.6}{23.8}\,\text{k}\Omega = 1.664\,\text{k}\Omega$$

(*b*) Voltage gain $$A_v = -\frac{R_C \parallel R_{F_2}}{r_e}$$

$$R_C \parallel R_{F_2} = (22\,\text{k}\Omega \parallel (1.8\,\text{k}\Omega) = 1.664\,\text{k}\Omega$$

$$\therefore \quad A_v = -\frac{1.664\,\text{k}\Omega}{7.928\,\Omega} = -\,209.87$$

Now, for the type of configuration

Current gain $$A_i = \frac{\beta R_{F_1} R_{F_2}}{(R_{F_1} + \beta_{r_e})(R_{r_2} + R_C)}$$

or $$A_i = \frac{80 \times (39\ \text{k}\Omega) \times (22\ \text{k}\Omega)}{(39\ \text{k}\Omega + 0.634\ \text{k}\Omega)(22\ \text{k}\Omega + 1.8\ \text{k}\Omega)}$$

$$= \frac{68640}{39.634 \times 23.8} = 72.767$$

(*c*) Since r_o is in parallel with R_C, the output impedance will be changed to

$$Z_o = r_o \parallel R_C \parallel R_{F_2} \quad \text{or} \quad Z_o = (45\ \text{k}\Omega) \parallel (1.8\ \text{k}\Omega) \parallel (22\ \text{k}\Omega)$$

Solving the parallel combination of resistances, we get

$$\frac{1}{Z_o} = \frac{1}{45\ \text{k}\Omega} + \frac{1}{1.8\ \text{k}\Omega} + \frac{1}{22\ \text{k}\Omega} = \frac{0.623}{\text{k}\Omega} \quad \text{or} \quad Z_o = 1.6045\ \text{k}\Omega$$

and the voltage gain A_v is modified as

$$A_v = -\frac{r_o \parallel R_C \parallel R_{F_2}}{r_e} \quad \text{or} \quad A_v = -\frac{(45\ \text{k}\Omega) \parallel (1.8\ \text{k}\Omega) \parallel (22\ \text{k}\Omega)}{7.928\ \Omega}$$

or $$A_v = -\frac{1.6045\ \text{k}\Omega}{7.928\ \Omega} = -202.389$$

Other things are independent of r_o and hence, their values remain unchanged.

Problem 4.18. For the network of figure given below:

(*a*) Determine Z_i and Z_o (*b*) Calculate A_v and A_i

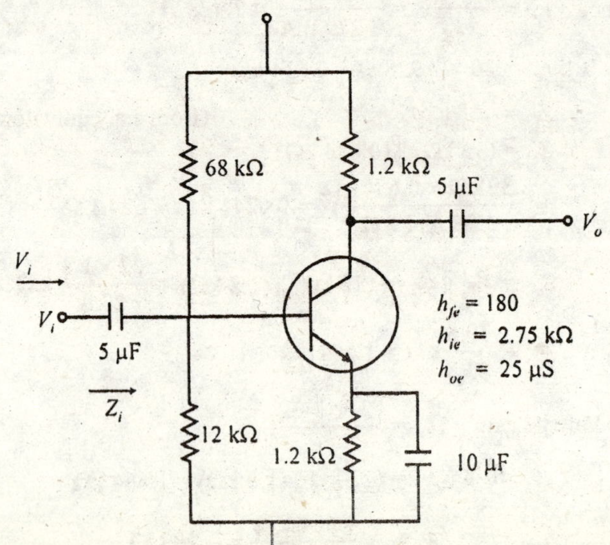

Fig. P. 4.18

Solution.

(*a*) $$R' = R_1 \parallel R_2 = (68\ \text{k}\Omega) \parallel (12\ \text{k}\Omega)$$

$$= \frac{(68\ \text{k}\Omega) \times (12\ \text{k}\Omega)}{68 + 12\ \text{k}\Omega} = 10.2\ \text{k}\Omega$$

Now $$Z_i = R' \parallel h_{ie} = (10.2\ \text{k}\Omega) \parallel (2.75\ \text{k}\Omega)$$

or $$Z_i = \frac{(10.2\ \text{k}\Omega) \times (2.75\ \text{k}\Omega)}{12.95\ \text{k}\Omega} = 2.166\ \text{k}\Omega$$

$$Z_o = R_C = 2.2\ \text{k}\Omega$$

(*b*) Voltage gain $$A_v = -\frac{h_{fe} R_C}{h_{ie}} = -\frac{180 \times 2.2\ \text{k}\Omega}{2.75\ \text{k}\Omega} = -144$$

Current gain $$A_i = -\frac{h_{fe} R'}{R' + h_{ie}}$$

or $$A_i = -\frac{180 \times (10.2\ \text{k}\Omega)}{10.2\ \text{k}\Omega + 2.75\ \text{k}\Omega}$$

⇒ $$A_i = -\frac{1836\ \text{k}\Omega}{12.95\ \text{k}\Omega} = -141.776$$

Problem 4.19. For the common base network of figure given below:

(*a*) Determine Z_i and Z_o

(*b*) Calculate A_v and A_i

(*c*) Determine α and β

$h_{fb} = -0.992$

$h_{ib} = 9.45\ \Omega$

$h_{ob} = 1\ \mu\text{A/V}$

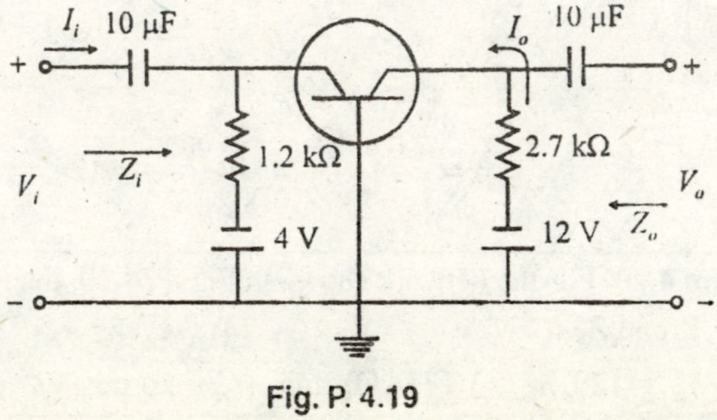

Fig. P. 4.19

Solution. Substituting the approximate hybrid equivalent circuit:

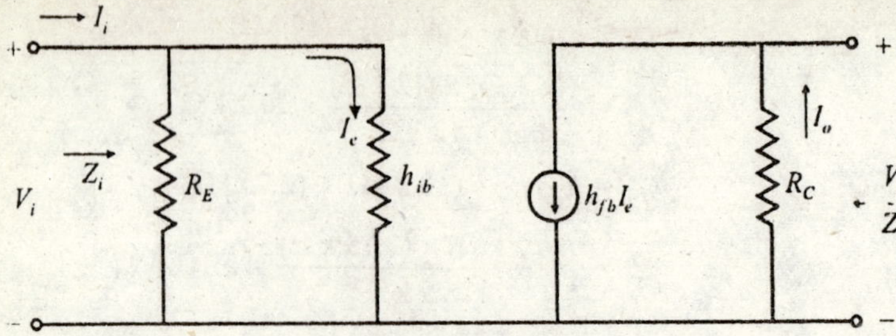

Fig. 4.19 (a)

(*a*) $Z_i = R_E \parallel h_{ib}$

Given $V_{EE} = 4$ V, $V_{CC} = 12$ V, $R_C = 2.7$ kΩ, $R_E = 1.2$ kΩ

$\therefore$ $Z_i = R_E \parallel h_{ib} = (1.2 \text{ k}\Omega) \parallel (9.45\ \Omega)$

$\Rightarrow$ $$Z_i = \frac{(1.2 \text{ k}\Omega) \times (9.45\ \Omega)}{1.20945 \text{ k}\Omega} = 9.376\ \Omega \approx 9.38\ \Omega$$

Output impedance $Z_o = R_C$

or $A_o = 2.7$ kΩ

(*b*) Voltage gain $$A_v = \frac{V_o}{V_i} = -\frac{h_{fb} R_C}{h_{ib}}$$

or $$A_v = -\frac{(-0.992) \times 2.7 \text{k}\Omega}{9.45\Omega} = 283.43$$

and current gain $A_i = h_{fb} = -0.992 \approx -1$

(*c*) To determine α and β.

For common base configuration

Current gain is h_{fb} is also called α

$\therefore$ magnitude of α = 0.992

also since $$\beta = \frac{\alpha}{1 - \alpha}$$

or $$\beta = \frac{0.992}{1 - 0.992} = \frac{0.992}{8 \times 10^{-3}} = 124$$

Problem 4.20. For the network shown in Fig. P. 4.20, determine:

(*a*) Z_i and Z_o (*b*) A_v and A_i

Given $h_{fe} = 120$, $h_{ie} = 1.175$ kΩ and $h_{oe} = 20\ \mu$ A/V

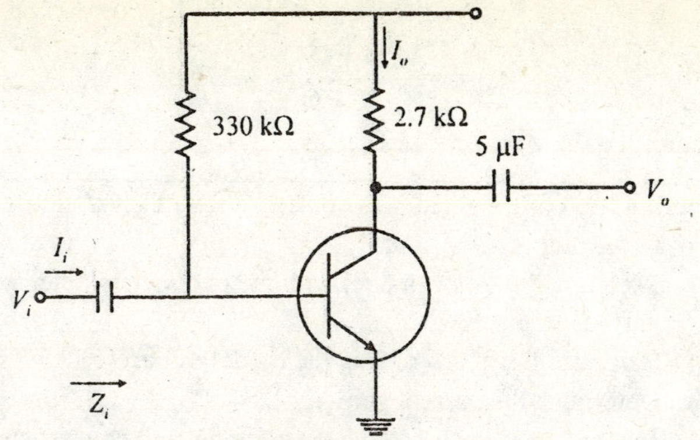

Fig. P. 4.20

Solution. This is a configuration of fixed bias type substituting the approximate hybrid model.

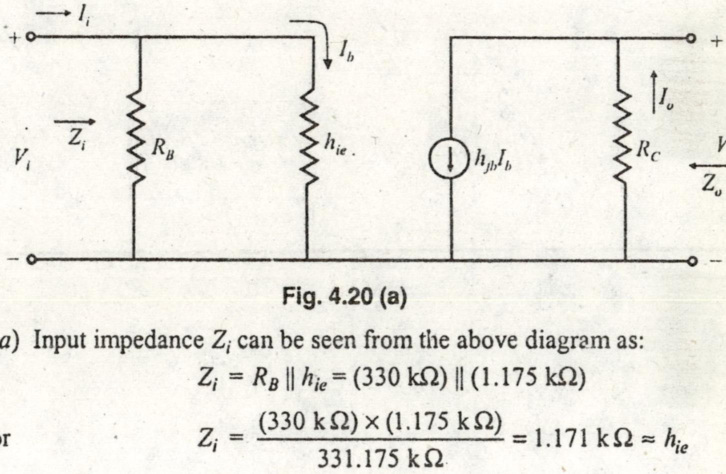

Fig. 4.20 (a)

(*a*) Input impedance Z_i can be seen from the above diagram as:

$$Z_i = R_B \parallel h_{ie} = (330\text{ k}\Omega) \parallel (1.175\text{ k}\Omega)$$

or
$$Z_i = \frac{(330\text{ k}\Omega) \times (1.175\text{ k}\Omega)}{331.175\text{ k}\Omega} = 1.171\text{ k}\Omega \approx h_{ie}$$

$$Z_o = \frac{1}{h_{oe}} \parallel R_C,\ r_o = \frac{1}{h_{oe}} = \frac{1}{20\mu\text{A/V}} = 50\text{ k}\Omega$$

$\therefore$
$$Z_o = (50\text{ k}\Omega) \parallel (2.7\text{ k}\Omega)$$

$$= \frac{50\text{ k}\Omega \times 2.7\text{ k}\Omega}{52.7\text{ k}\Omega} = 2.56\text{ k}\Omega \approx R_C$$

or
$$Z_o = 2.56\text{ k}\Omega$$

(*b*) For calculating of A_v and A_i:

$$A_v = -\frac{h_{fe} R_C}{h_{ie}} = -\frac{(120)(2.7\text{ k}\Omega)}{1.171\text{ k}\Omega}$$

or $$A_v = -\frac{324\ \text{k}\Omega}{1.171\ \text{k}\Omega} = -276.69$$

and current gain $A_i \approx h_{fe} = 120$

Problem 4.21. For the fixed bias configuration of given figure:

(*a*) Determine A_{VNL}, Z_i and Z_o.

(*b*) Sketch the two-port model with the parameters determined in part (*a*) in place.

(*c*) Calculate the gain A_v using the model of part (*b*).

(*d*) Determine the current gain A_i.

(*e*) Determine A_v, and A_i using the r_e model and compare the results above.

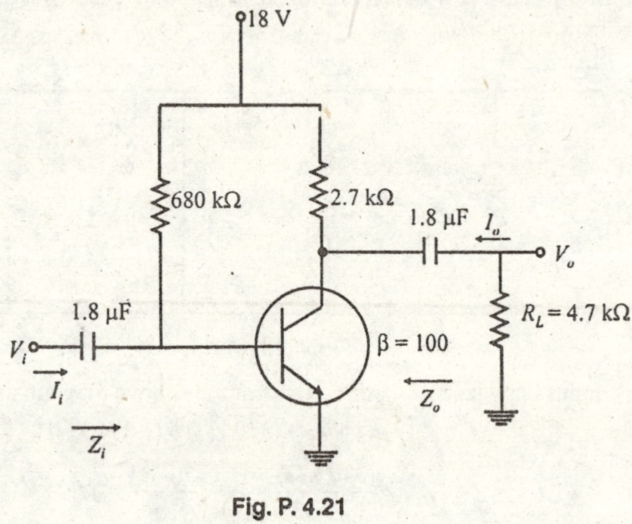

Fig. P. 4.21

Solution. (*a*) d.c. analysis:

$$I_B = \frac{V_{CC} - V_{BE}}{R_B} = \frac{18 - 0.7}{680\text{K}} = 25.44\ \mu\text{A}$$

$$I_E = (\beta + 1)I_B = 101 \times (25.44\ \mu\text{A}) = 2.569\ \text{mA}$$

$$r_e = \frac{26\ \text{mV}}{I_E} = \frac{26\ \text{mV}}{2.569\ \text{mA}} = 10.12\ \Omega$$

$$A_{VNL} = -\frac{R_C}{r_e} = -\frac{3.3\ \text{k}\Omega}{10.12\ \Omega} = -326.087$$

$$Z_i = R_B \parallel \beta r_e = 680\ \text{k}\Omega \parallel 1.012\ \text{k}\Omega = 1.0105\ \text{k}\Omega$$

$$Z_o = R_C = 3.3\ \text{k}\Omega.$$

(*b*) Using the information above, the two-port equivalent can be drawn as follow:

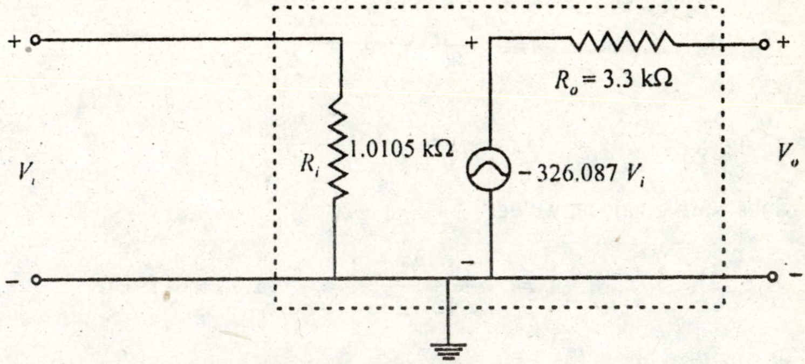

Fig. 4.21 (a)

(*c*) From the above model:

$$V_o = (A_{VNL} \times V_i) \times \frac{R_L}{R_L + R_o}$$

$$\Rightarrow \quad A_v = \frac{R_L}{R_L + R_o} A_{VNL} = \frac{4.7\text{ k}\Omega}{4.7\text{ k}\Omega + 3.3\text{ k}\Omega}(-326.087) = -191.6$$

(*d*) $$A_i = -A_v \frac{Z_i}{R_L}$$

$$A_i = -(-191.6) \times \frac{1.0105\text{ k}\Omega}{4.7\text{ k}\Omega} = 41.19$$

(*e*) Substituting the r_e model will result in the network shown:
Note here that $R_L \parallel R_C$

i.e. $R_L' = R_L \parallel R_C = 4.7\text{ k}\Omega \parallel 3.3\text{ k}\Omega$

or $$R_L' = \frac{(4.7\text{ k}\Omega) \times (3.3\text{ k}\Omega)}{8.0\text{ k}\Omega} = 1.9387\text{ k}\Omega$$

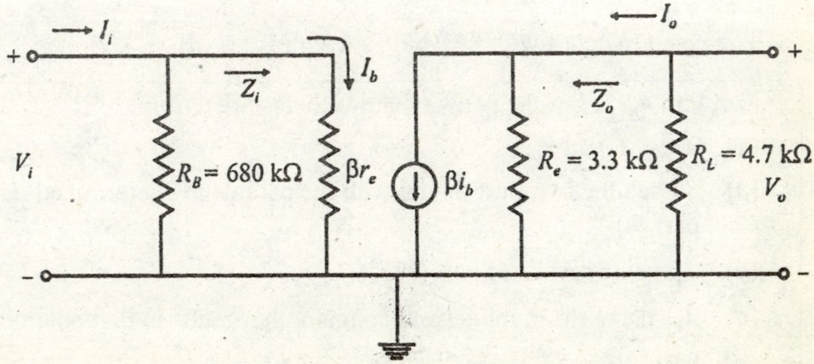

Fig. 4.21 (b)

The output voltage

$$V_o = -\beta I_B R'_L$$

with $I_b = \frac{V_i}{\beta r_e}$, and $V_o = -\beta \frac{V_i}{\beta r_e} \times R'_L$

So that $A_v = \frac{V_o}{V_i} = -\frac{R'_L}{r_e} = -\frac{R_C \| R_L}{r_e}$

Substituting values, we get

$$A_v = -\frac{1.9387 \text{ k}\Omega}{10.12 \ \Omega} = -191.58 \approx -191.6$$

as obtained above.

For the current gain, by the current division rule

$$I_b = \frac{(680 \text{ k}\Omega) I_i}{(680 \text{ k}\Omega) + 1.012 \text{ k}\Omega} = \frac{(680 \text{ k}\Omega) \times I_i}{(681.012 \text{ k}\Omega)}$$

or $I_b = 0.9985 \, I_i \cong I_i$ ... (1)

and $I_o = \frac{3.3 \text{ k}\Omega \, (\beta I_b)}{3.3 \text{ k}\Omega + 4.7 \text{ k}\Omega} = \frac{(3.3 \text{ k}\Omega)\beta I_b}{8.0 \text{ k}\Omega}$

or $I_o = \frac{3.3}{8} \beta I_b = 0.4125 \beta I_b$

Now $A_i = \frac{I_o}{I_i} = \frac{0.4125 \beta I_b}{I_i} = \frac{0.4125 \beta I_i}{I_i}$ (from 1)

or $A_i = 0.4125 \, (100)$

$= 41.25$

The result matches with the value solved in part (*d*).

Problem 4.22. For the network shown below, determine:

(*a*) A_{VNL}, Z_i and Z_o.

(*b*) Sketch the two-port model with the parameters determined in part (*a*).

(*c*) A_{vs} using results of part (*b*).

(*d*) A_{vs} using the r_e model and compare the results to that obtained in part (*c*).

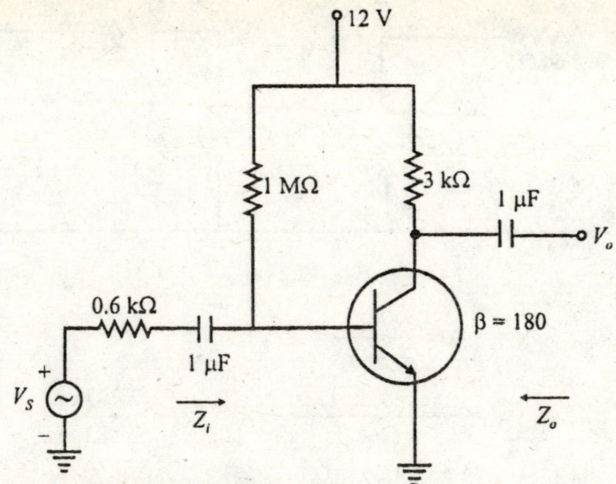

Fig. P. 4.22

Solution. (*a*) For calculation of A_{VNL}, Z_i and Z_o, from d.c. analysis.

$$I_B = \frac{V_{CC} - V_{BE}}{R_B} = \frac{12\text{V} - 0.7\text{V}}{1\text{M}\Omega} = 11.3\mu\text{A}$$

and $$I_E = (\beta + 1)\, I_B = (181)\,(11.3\ \mu\text{A}) = 2.0453\ \text{mA}$$

Thus $$r_e = \frac{26\ \text{mV}}{I_E} = \frac{26\ \text{mV}}{2.0453\ \text{mA}} = 12.712\ \Omega$$

$$A_{VNL} = -\frac{R_C}{r_e} = -\frac{3\ \text{k}\Omega}{12.712\ \Omega} = -235.997$$

or $$A_{VNL} \cong -236$$

Input impedance $Z_i = R_B \,\|\, \beta r_e = (1\ \text{M}\Omega) \,\|\, (180 \times 12.712\ \Omega)$

or $$Z_i = \frac{(1000\ \text{k}\Omega) \times (2.288\ \text{k}\Omega)}{1002.288\ \text{k}\Omega} = 2.283\ \text{k}\Omega \approx \beta r_e$$

Output impedance $Z_o = R_C = 3\ \text{k}\Omega$

(*b*) The two-port equivalent network with the relevant parameters is shown below:

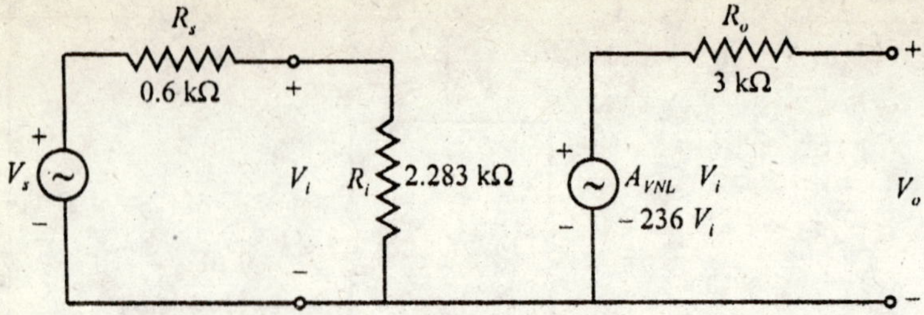

Fig. 4.22 (a)

(c) $$A_{vs} = \frac{V_o}{V_S} = \frac{R_i}{R_i + R_s} \times A_{VNL}$$

or $$A_{vs} = \frac{2.283\,\text{k}\Omega}{2.283\,\text{k}\Omega + 0.6\,\text{k}\Omega} \times (-236) = \frac{2.283\,\text{k}\Omega}{2.883\,\text{k}\Omega} \times (-236)$$

or $$A_{vs} = (0.792)\,(-236) = -186.88 \approx -187.$$

(*d*) Substituting the r_e model will result in the equivalent circuit shown below:

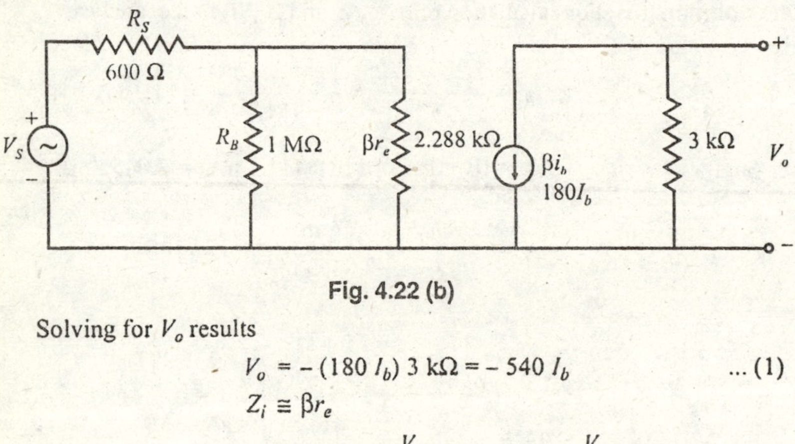

Fig. 4.22 (b)

Solving for V_o results

$$V_o = -(180\ I_b)\ 3\ \text{k}\Omega = -540\ I_b \qquad \ldots (1)$$

$$Z_i \cong \beta r_e$$

and $$I_b \cong I_i = \frac{V_s}{R_s + \beta r_e} = \frac{V_s}{0.6\ \text{k}\Omega + 2.288\ \text{k}\Omega}$$

Therefore, putting the value of I_b in equation (1) yields

$$V_o = -\frac{540 V_S}{2.888}$$

or $$\frac{V_o}{V_S} = -\frac{540}{2.888} = -186.98$$

So that $$A_{vs} = \frac{V_o}{V_S} \cong -187$$

Problem 4.23. For the network of figure given below:

(*a*) Determine A_{VNL}, Z_i and Z_o.

(*b*) Sketch the two-port model with the parameters determined in part (*a*) in place.

(*c*) Determine A_v and A_{vs}.

(*d*) Calculate A_i.

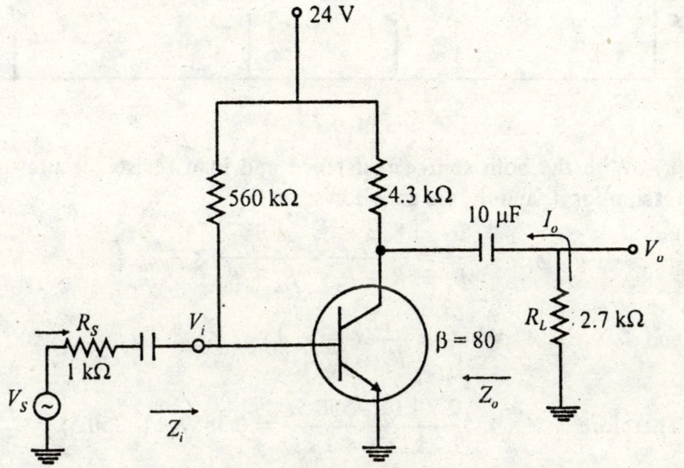

Fig. P. 4.23

Solution: (*a*) From d.c. analysis:

$$I_B = \frac{V_{CC} - V_{BE}}{R_B} = \frac{24\text{ V} - 0.7\text{ V}}{560\text{ k}\Omega} = \frac{23.3\text{ V}}{560\text{ k}\Omega} = 41.6\ \mu\text{A}$$

$\therefore$
$$I_E = (\beta + 1)\, I_B = 81 \times 41.6\ \mu\text{A} = 3.328\text{ mA}$$

Thus
$$r_e = \frac{26\text{ mV}}{I_E} = \frac{26\text{ mV}}{3.328\text{ mA}} = 7.811\ \Omega$$

Now
$$A_{VNL} = -\frac{R_C}{r_e} = -\frac{4.3\text{k}\Omega}{7.811} = -550.5$$

Input impedance

$$Z_i = R_B \parallel \beta r_e = (560\text{ k}\Omega) \parallel (80)\,(7.811\ \Omega)$$
$$= (560\text{ k}\Omega) \parallel (0.6249\text{ k}\Omega) = 0.6242\text{ k}\Omega$$

and
$$Z_o = R_C = 4.3\text{ k}\Omega$$

(*b*) The two-port model for the given circuit can be sketched as below:

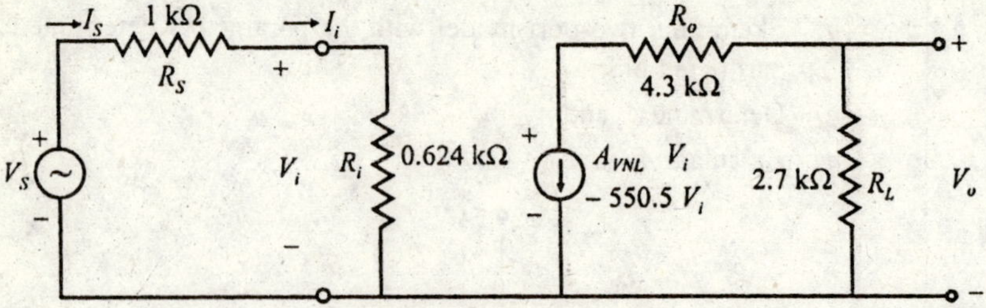

Fig. 4.23 (a)

(*c*) With the both source resistance and load resistance present, the expression for A_v and A_{vs} are as follows:

$$A_{vs} = \frac{V_o}{V_s} = \frac{R_i}{R_i + R_s} \times \frac{R_L}{R_L + R_o} \times A_{VNL}$$

and $$A_v = \frac{V_o}{V_i} = \frac{R_L}{R_L + R_o} \times A_{VNL}$$

Therefore $$A_v = \frac{2.7\ \text{k}\Omega(-550.5)}{2.7\ \text{k}\Omega + 4.3\ \text{k}\Omega} = 0.3857 \times (-550.5)$$

or $$A_v = -212.33$$

and $$A_{vs} = \frac{0.624\ \text{k}\Omega}{0.624\ \text{k}\Omega + 1\ \text{k}\Omega} \times \frac{2.7\ \text{k}\Omega}{2.7\ \text{k}\Omega + 4.3\ \text{k}\Omega}(-550.5)$$

$$= (0.3842)\,(0.3857)\,(-550.5) = -81.58$$

(*d*) $$A_i = -A_v \cdot \frac{R_i}{R_L} = -(-212.33)\left(\frac{0.624\ \text{k}\Omega}{2.7\ \text{k}\Omega}\right)$$

or $$A_i = 49.072$$

also $$A_{is} = -A_{vs}\frac{R_s + R_i}{R_L} = 49.072$$

Problem 4.24. For the voltage-divider configuration of Fig. P. 4.24:

(*a*) Determine A_{VNL}, Z_i and Z_o.

(*b*) Sketch the two-port model with the parameters determined in part (*a*) in place.

(*c*) Calculate the gain A_v using the model of part (*b*).

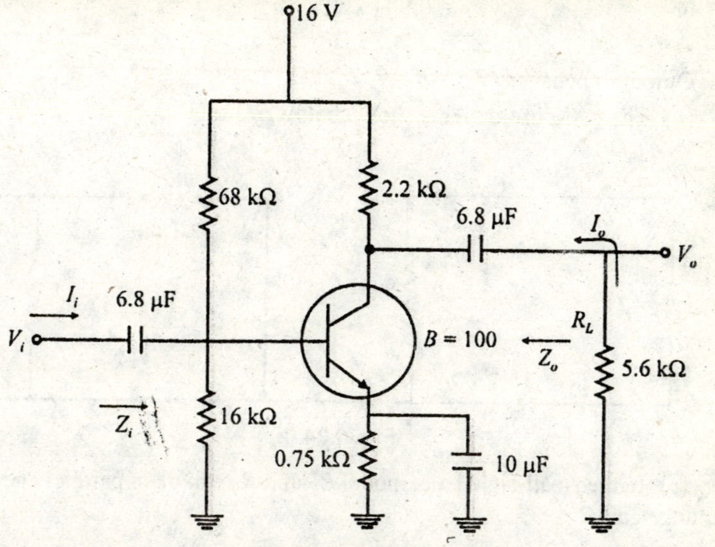

Fig. P. 2.24

Solution.

$$R' = R_1 \| R_2 = (68\text{ k}\Omega) \| (16\text{ k}\Omega) = \frac{(68\text{ k}\Omega)\times(16\text{ k}\Omega)}{84\text{ k}\Omega}$$

or, $R' = 12.95\text{ k}\Omega$ or $R_{th} = R' = 12.95\text{ k}\Omega$

$$A_{VNL} = \frac{V_o}{V_i} = -\frac{R_C}{r_e}$$

To calculate r_e

$$E_{th} = \frac{R_2 V_{CC}}{R_1 + R_2} = \frac{16\text{ k}\Omega \times V_{CC}}{68\text{ k}\Omega + 16\text{ k}\Omega} = \frac{16\text{ k}\Omega}{84\text{ k}\Omega} \times 16\text{ V} = 3.048\text{ V}$$

Now $$I_B = \frac{E_{th} - V_{BE}}{R_{th} + (\beta+1)R_E} = \frac{3.048\text{ V} - 0.7\text{ V}}{12.95\text{ k}\Omega + 101 \times 0.75\text{ k}\Omega}$$

or $I_B = 26.47\ \mu\text{A}$

Therefore $I_E = (b+1)I_B = 101(26.47\ \mu\text{A}) = 2.67\text{ mA}$

Now $$r_e = \frac{26\text{ mV}}{I_E} = \frac{26\text{ mV}}{2.67\text{ mA}} = 9.725\ \Omega$$

Now $$A_{VNL} = -\frac{R_C}{r_e} = -\frac{2.2\text{ k}\Omega}{9.725\ \Omega} = -226.23$$

Input impedance $Z_i = R' \parallel \beta r_e = (12.95\ \text{k}\Omega) \parallel (0.9725\ \text{k}\Omega)$

or $Z_i = 0.904\ \text{k}\Omega \cong \beta r_e$

∞ $Z_i = 904.55\ \Omega$

Output impedance $Z_o = R_C = 2.2\ \text{k}\Omega$

(*b*) The two-port model is shown below:

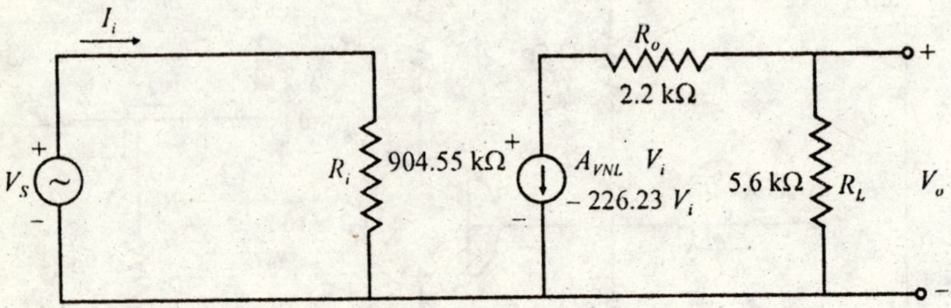

Fig. 4.24 (a)

(*c*) In the small-signal a.c. model R_C and R_L will be in parallel and thus, voltage gain:

$$A_v = -\frac{R_C \parallel R_L}{r_e}$$

or $$A_v = -\frac{R_C\, R_L}{R_C + R_L} \times \frac{1}{r_e} = \frac{R_L}{R_C + R_L}\left(-\frac{R_C}{r_e}\right)$$

or $$A_v = \frac{R_L}{R_C + R_L} A_{VNL} = \frac{5.6\ \text{k}\Omega}{2.2\ \text{k}\Omega + 5.6\ \text{k}\Omega}(-226.23)$$

or $$A_v = (0.718)(-226.23) = -162.42$$

As was solved in the last problem, the values can be obtained by substituting the r_e model as well.

This is an example of the circuit having only load resistance with voltage divider bias.

Problem 4.25. The transistor in the amplifier circuit shown in Fig. P. 4.25, is biased at $I_C = 1$ mA. Use $V_T \left(= \frac{KT}{q}\right) = 26$ mV,

$$b_o = 200,\ r_b = 0, \text{ and } r_o \to \infty$$

(*a*) Determine the a.c. small signal midband voltage gain $\frac{V_o}{V_i}$ of the circuit.

(*b*) Determine the required value of C_E for the circuit to have a lower cut-off frequency of 10 Hz.

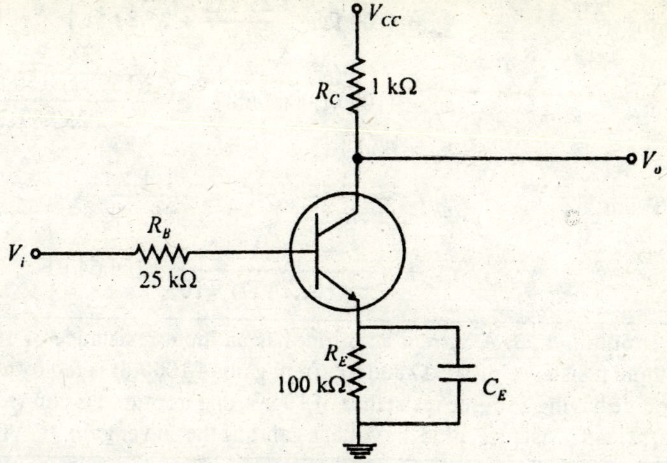

Fig. P. 4.25

(GATE 2001)

Solution. To determine r_e:

Given $I_C = 1$ mA and $\beta_o = 200$

Now, from the relation, $I_C = \beta I_B$ and $I_E = (\beta + 1) I_B$, we find $I_E = \dfrac{(\beta + 1)}{\beta} I_C$

or $$I_E = \frac{201}{200} \times (1\,\text{mA}) = 1.005\,\text{mA}$$

Hence $$r_e = \frac{V_T}{I_E} = \frac{26\text{ mV}}{I_E} = \frac{26\text{ mV}}{1.005\text{ mA}} = 25.87\ \Omega$$

(*a*) Midband gain A_v is given by:

$$A_v = \frac{V_o}{V_i} = \frac{-R_C}{r_e} = -\frac{(1\text{ k}\Omega)}{25.87\ \Omega}$$

or $$A_v = -\frac{1000\ \Omega}{25.87\ \Omega} = -38.65$$

(*b*) Given lower cut-off frequency = 10 Hz:

i.e. $f_{LE} = \dfrac{1}{2\pi R_e C_E} = 10\,\text{Hz}$

R_e is given by

$$R_e = R_E \,\|\, \left(\frac{R_B}{\beta} + r_e\right)$$

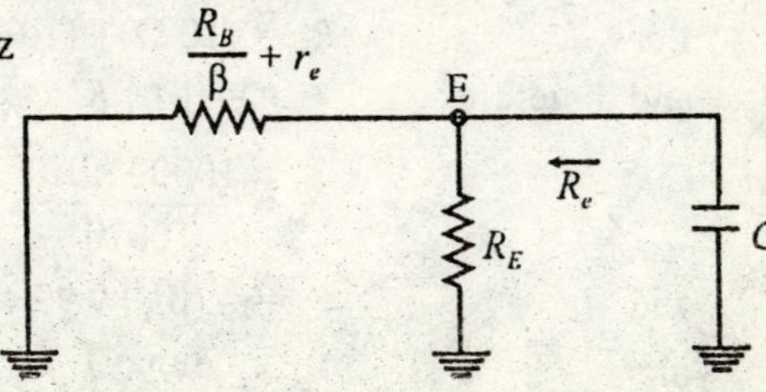

Fig. 4.25 (a)

or $$R_e = (100\ \Omega) \parallel \left(\frac{25\ \text{k}\Omega}{200} + 25.87\ \Omega\right)$$

or $$R_e = (100\ \Omega) \parallel (150.87\ \Omega) = \frac{(100\ \Omega)(150.87\ \Omega)}{250.87\ \Omega}$$

or $$R_e \simeq 60.14\ \Omega$$

From $$f_{LE} = \frac{1}{2\pi R_e C_E} \Rightarrow C_E = \frac{1}{2\pi R_e f_{LE}}$$

or $$C_E = \frac{1}{2\pi \times (60.14\ \Omega) \times 10} = 264.64\ \mu\text{F}$$

Problem 4.26. A current amplifier has an input resistance of 10 Ω, an output resistance of 10 kΩ and a current gain of 1000. It is fed by a current source having a source resistance of 10 kΩ, and its output is connected to a 10 Ω load resistance. Find the voltage gain and the power gain. (GATE 2000)

Solution. The problem can be solved with the help of two-port model as shown below:

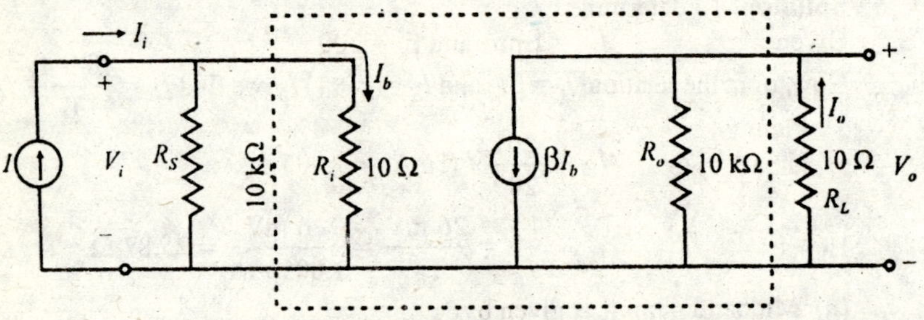

Fig. 4.26

The voltage gain and the current gain can be found simply by voltage and current division rule.

From the two-port model, it is clear that

$$V_i = I_b R_i = I_b(10\ \Omega) = 10\ I_b \qquad \text{... (1)}$$

For V_o

Let $$R' = (10\ \Omega) \parallel (10\ \text{k}\Omega)$$

and therefore $$V_o = -(\beta I_b)\ R'$$

$$R' = \frac{10000 \times 10}{10010}\ \Omega = 9.99\ \Omega$$

and $$V_o = -(\beta I_b)\ 9.99 = -I_b \times 1000 \times 9.99$$

or $$V_o = -9990\ I_b \qquad \text{... (2)}$$

From (2)/(1), we get the voltage gain

$$A_v = \frac{V_o}{V_i} = \frac{-9990\, I_b}{10\, I_b} = -999$$

or voltage gain $A_v = -999.$

For current gain $I_i = I$, also, $I_b = I \times \dfrac{10\text{ k}\Omega}{10\text{ k}\Omega + 10\text{ k}\Omega} = I \times 0.999$

or $I_i = \dfrac{I_b}{0.999} = 1.001\, I_b$ ... (3)

Output current $I_o = \dfrac{(\beta I_b)\, 10\text{ k}\Omega}{10\text{ k}\Omega + 10\ \Omega} = (\beta I_b)\, 0.999$

or $I_o = I_b \times 1000 \times 0.999 = 999\, I_b$ ... (4)

Applying (4) divided by (3), we get

$$\frac{I_o}{I_i} = \frac{I_o}{I} = \frac{999\, I_b}{1.001\, I_b} = 998$$

or Current gain $A_i = \dfrac{I_o}{I_i} = 998$

Now Power gain $= A_v \times A_i = (-999) \times 998 = 997\,002$

or Power gain $= 997.002 \times 10^3$

Problem 4.27. For the small signal BJT amplifier shown in Fig. P. 4.27, determine at 1 kHz the following. (*a*) quiescent collector current, I_{CQ}, (*b*) small signal voltage gain, $\left(\dfrac{V_o}{V_i}\right)$, (*c*) maximum possible swing of the collector current.

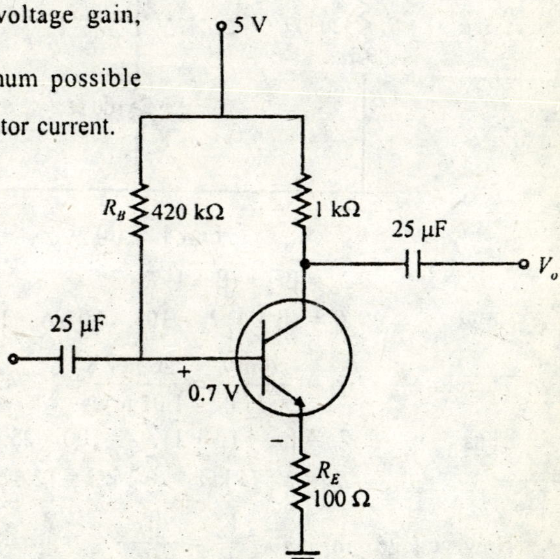

Fig. P. 4.27 (GATE' 1999)

Solution. Though the current gain b should be mentioned in the problem, it is not given here. Hence, we will assume some appropriate value between 50 to 200, this will not score-off the paper

Here, we assume $\beta = 100$.

Now (*a*) from d.c. analysis, we have

$$I_B = \frac{V_{CC} - V_{BE}}{R_B + (\beta+1)R_E} = \frac{5\text{ V} - 0.7\text{ V}}{420\text{ k}\Omega + 101 \times 100\ \Omega}$$

or $$I_B = \frac{4.3\text{ V}}{420\text{ k}\Omega + 10.1\text{ k}\Omega} = \frac{4.3\text{ V}}{430.1\text{ k}\Omega} = 9.998\ \mu\text{A}$$

or $$I_B \cong 10\ \mu\text{A}$$

Hence $$I_{CQ} = \beta I_B = 100 \times 10\ \mu\text{A} = 1\text{ mA}$$

(*b*) **To calculate small signal voltage gain:** Since, this is a CE emitter bias configuration, hence we will solve this by substituting the equivalent r_e model into the given network.

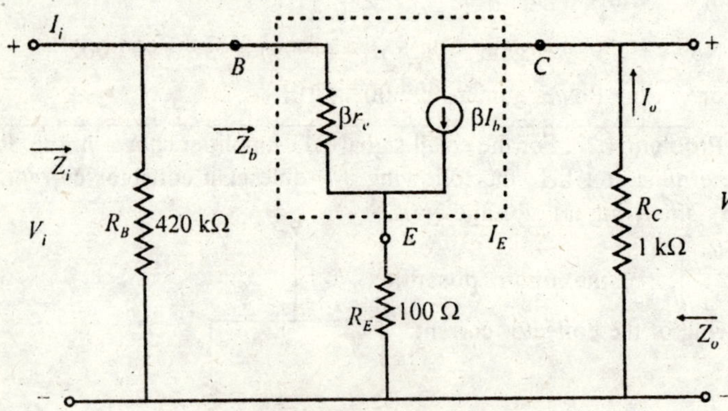

Fig. 4.27 (a)

$$Z_b = \beta r_e + (\beta + 1) R_E$$

Now $$I_E = (\beta + 1) I_B = 101 \times 10\ \mu\text{A} = 1.01\text{ mA}$$

Therefore $$r_e = \frac{26\text{ mV}}{I_E} = \frac{26\text{ mV}}{1.01\text{ mV}} = 25.742\ \Omega$$

Thus $$Z_b = \beta r_e + (\beta + 1) R_E = 100 \times 25.742\ \Omega + 101 \times 100\ \Omega$$

$\Rightarrow$ $$Z_b = 2.574\text{ k}\Omega + 10.1\text{ k}\Omega = 12.674\text{ k}\Omega.$$

Now, voltage gain $$A_v = \frac{V_o}{V_i} = -\frac{\beta R_c}{Z_b}$$

or $$A_v = -\frac{100 \times 1\text{ k}\Omega}{12.674\text{ k}\Omega} \cong -7.9$$

(*c*) Maximum possible swing of the collector current is given by:

$$I_{c\,max} = \frac{V_C}{R_C + R_E} = \frac{5}{1\text{ k}\Omega + 100\ \Omega} = \frac{5}{1.1\text{ k}\Omega}$$

or $$I_{c\,max} = 4.545\text{ mA}$$

Problem 4.28. The BJT in amplifier of given figure has $h_{fe} = 100$, $V_{BE} = 0.7$ V, $I_{co} = 0$. Calculate the values of R_1 and R_C, such that its $I_C = 1$ mA and $V_{CE} = 2.5$ V.

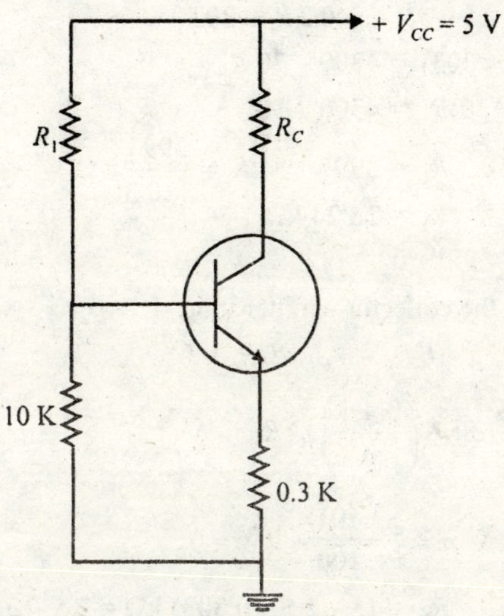

Fig. P. 4.28 (IES' 2001)

Solution. For voltage divider bias configuration

$$R_{th} = R_1 \parallel R_2 = R_1 \parallel (10\text{ K})$$

or $$R_{th} = \frac{10 R_1}{10 + R_1}$$

$$V_{th} = V_B = V_{CC} \times \frac{R_2}{R_1 + R_2} = (5\text{ V})\frac{10}{10 + R_1}$$

or $$V_{th} = \frac{50}{10 + R_1}$$

From d.c. analysis, we have

$$I_B = \frac{V_{th} - V_{BE}}{R_{th} + (h_{fe} + 1) R_E} = \frac{\dfrac{50}{10 + R_1} - 0.7}{\dfrac{10 R_1}{10 + R_1} + 101 \times 0.3}\text{ mA}$$

or $$I_B = \frac{43 - 0.7 R_1}{10 + R_1} \times \frac{(10 + R_1)}{40.3 R_1 + 303} \text{ mA}$$

or $$I_B = \frac{43 - 0.7 R_1}{40.3 R_1 + 303} \text{ mA}$$

Here $I_C = h_{fe} \times I_B = 100\, I_B$

Given, $I_C = 1$ mA so, we get the expression as

$$1 = \frac{100(43 - 0.7 R_1)}{40.3 R_1 + 303}$$

or $(40.3\, R_1 + 303) = 4300 - 70\, R_1$

or $(40.3 + 70)R_1 = 4300 - 303$

or $110.3\, R_1 = 3997 \Rightarrow R_1 = \dfrac{3997}{110.3} \text{ k}\Omega$

or $R_1 = 36.24 \text{ k}\Omega$

To calculate R_C

Apply KVL to the collector emitter loop

$$5 - I_C R_C - V_{CE} - I_E R_E = 0$$

or $$5 - I_C R_C - 2.5 - \left(\frac{\beta + 1}{\beta}\right) I_C\, R_E = 0$$

or $$5 - R_C - 2.5 - \frac{101}{100} \times 0.3 = 0$$

$\Rightarrow$ $R_C = (5 - 2.5 - 0.303) \text{ k}\Omega = 2.5 - 0.303$

or $R_C = 2.197 \text{ k}\Omega$

Problem 4.29. Show that the circuit shown below simulates an inductance.

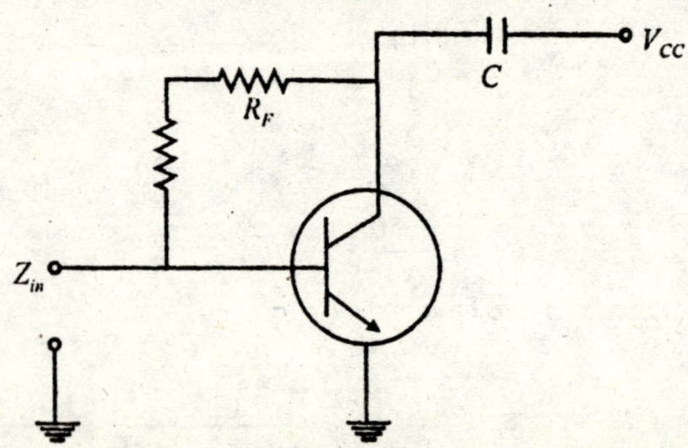

Fig. P. 4.29

Solution. $$Z_{in} = \frac{R_F}{|A_v|}$$

$$|A_v| = \frac{h_{fe} X_C}{r_\pi}$$

$$\therefore \quad Z_{in} = R_F \frac{r_\pi}{h_{fe} X_C} = \frac{R_F r_\pi}{h_{fe}}(j\omega C) = j\omega\left(\frac{R_F r_\pi}{h_{fe}}\right)C = j\omega L$$

Where $$L = \frac{R_F r_\pi C}{h_{fe}}$$

So a capacitance load appears as an inductance to the input terminals. A resistance will be connected across C for d.c. supply to transistor.

Problem 4.30. A transistor used in common emitter (CE) configuration has the following set of h-parameters, when the d.c. operating points is $V_{CE} = 10$ V and $I_C = 1$ mA.

$h_{ie} = 2000\ \Omega$; $h_{oe} = 10^{-4}$ S

$h_{re} = 10^{-3}$; $h_{fe} = 50$

If the a.c. load line seen by the transistor is $R_L = 600\ \Omega$, determi. e input impedance, current gain and voltage gain.

Solution. The equivalent circuit is:

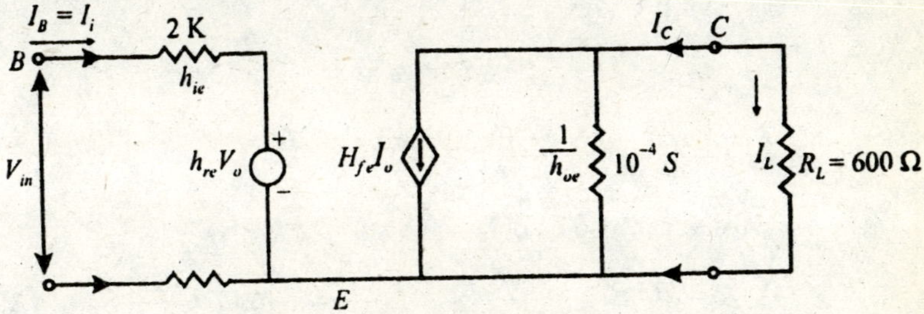

Fig. 4.30

Input impedance:

$$V_o = -I_C R_L = -1 \times 10^{-3} \times 600 = -0.6 \text{ V}$$

$$-V_{in} + I_B h_{ie} + h_{re} V_o = 0, \quad I_B = \frac{I_C}{h_{fe}} = \frac{1 \times 10^{-3}}{50} = 0.02 \times 10^{-3}$$

$$V_{in} = 0.02 \times 10^{-3} \times 2 \times 10^3 - 10^{-3} \times 0.6$$

$$= 0.04 - 0.0006 = 0.0394 \text{ V}$$

$$R_{in} = \frac{V_{in}}{I_B} = \frac{0.0394}{0.02 \times 10^{-3}} = 1.97 \text{ k}\Omega$$

Current gain:

By current division

$$I_o = -h_{fe} I_B \times \frac{\frac{l}{h_{oe}}}{R_L + \frac{l}{h_{oe}}} = \frac{-h_{fe} I_B}{1 + R_L h_{oe}}$$

$$I_{in} = I_B$$

$\therefore$ Current gain $A_I - \frac{I_o}{I_{in}} = \frac{-h_{fe}}{1 + R_L h_{oe}}$

$$= -\frac{50}{1 + 600 \times 10^{-4}} = -\frac{50}{1.06} = -47.17$$

Voltage gain:

$$V_o = -h_{fe} I_B \left(\frac{l}{h_{oe}} \parallel R_L\right) = -50 \times 0.02 \times 10^{-3} \times \left(\frac{10^4 \times 600}{10^4 + 600}\right)$$

$$= -0.5660 \text{ V}$$

$$V_{in} = h_{ie} I_B + h_{re} V_o = 0.0394 \text{ V}$$

Voltage gain $= \frac{V_o}{V_{in}} = -\frac{0.5660}{0.0394} = -14.37$

The same result can also be obtained by

$$A_v = -\frac{I_C R_L}{I_B R_{in}} = -A_i \frac{R_L}{R_{in}} = -46.17 \times \frac{600}{1970} = -14.37$$

For the Zener to regulate, its current should not go below 1 mA, i.e.

$$I_{Z\,\min} = I_{zk} = 1 \text{ mA}$$

This current through Zener should flow even when the load draws the maximum current, i.e. 30 mA + 50% = 45 mA

$\therefore$ $I_{S\,\min} = I_{Z\,\min} + I_{L\,\max}$

$$= 1 + 45 = 46 \text{ mA}$$

The value of R_S should be of a maximum value such that $I_{S\,min}$ is always allowed to flow for the Zener to regulate and load to draw maximum current at minimum supply voltage.

$$R_{S_{max}} \le \frac{V_{S\,min} - V_L}{I_{S\,min}}$$

$$V_{S\ min} = 20\text{ V} - 20\% = 16\text{ V}$$

$$\therefore \quad R_{S_{max}} = \frac{16 - 6.8}{46 \times 10^{-3}} = 200\ \Omega$$

The load resistance varies between minimum and maximum current.

$$\therefore \quad R_{L\ max} = \frac{6.8}{15 \times 10^{-3}} = 453\ \Omega$$

$$R_{L\ min} = \frac{6.8}{45 \times 10^{-3}} = 151\ \Omega$$

As long as the Zener draws the minimum current of 1 mA, it is on and so also the diode being across 6.8 V.

Now $\quad V_Z = 6.8 - 0.6 = 6.2\text{ V}$

At $\quad V_{S\,max}, I_{in} = \frac{24\text{ V} - 6.8\text{ V}}{200} = 0.086\text{ A} = 86\text{ mA}$

At $I_{L\ min}$, Zener has to bypass 86 – 15 = 71 mA.

So, Zener's minimum rating will be

$$= 6.2 \times 71 = 440\text{ mW}.$$

Reader's Notes

CHAPTER 5

FET and Its Biasing

FET: The field effect transistor is a three-terminal devices used for a variety of applications that match, to a large extent, those of the BJT transistor. The primary differences between the two are:

(*i*) BJT is a current controlled device whereas FET is a voltage controlled device.

(*ii*) BJT transistor is a bipolar device revealing that the conduction level is a function of two charge carriers, i.e. electrons and holes. While the FET transistor is unipolar device depending solely on either electrons (*n*-channel) or hole (*p*-channel) conduction.

One of the most important characteristics of the FET is its high input impedance. In general, FETs are more temperature stable than BJTs and FETs are usually smaller in construction than BJTs making them particularly useful in integrated circuit (IC) chips.

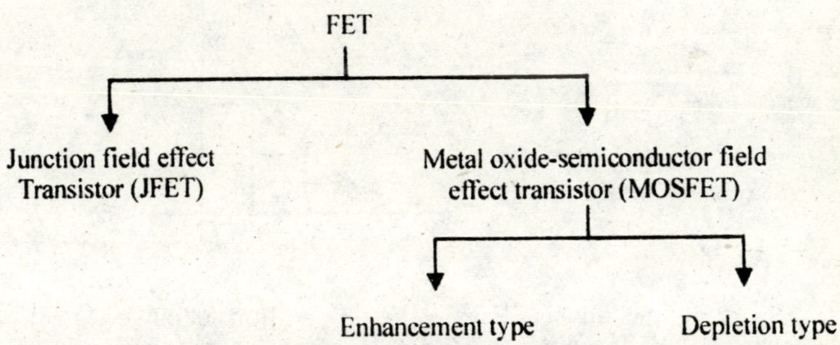

JFET : There are three terminals in JFET name, gate (G) source (S) and drain (D). In the absence of any applied potentials the JFET has two *p-n* junctions under no-bias conditions, thus results in depletion region. A depletion region is that region which is void of free carriers, and therefore unable to support conduction through the region.

Some Important Features of JFET

(*i*) The JFET can actually be used as a voltage controlled resistor, because of a unique sensitivity of the drain-to-source impedance to the gate-to-source voltage.

(*ii*) The maximum current for any JFET is labelled as I_{DSS} and occurs when $V_{GS} = 0$ V.

(*iii*) The minimum current for a JFET occurs at pinch off voltage defined by $V_{GS} = V_P$.

(*iv*) The relationship between the drain current and the gate-to-source voltage of a JFET is a non-linear one defined by Schokley's equation.

$$I_D = I_{DSS}\left(1 - \frac{V_{GS}}{V_P}\right)^2$$

$$I_D = I_{DSS} \quad \text{at} \quad V_{GS} = 0 \text{ V}$$
$$I_D = 0 \quad \text{at} \quad V_{GS} = V_P$$

(*v*) The resistance level in terms of applied voltage V_{GS} of JFET is

$$r_d = \frac{r_o}{\left(1 - \frac{V_{GS}}{V_P}\right)^2}$$ where, r_o = resistance at $V_{GS} = 0$ V

r_d = resistance at a particular level of V_{GS}.

The symbol of *n*-channel JFET is shown below:

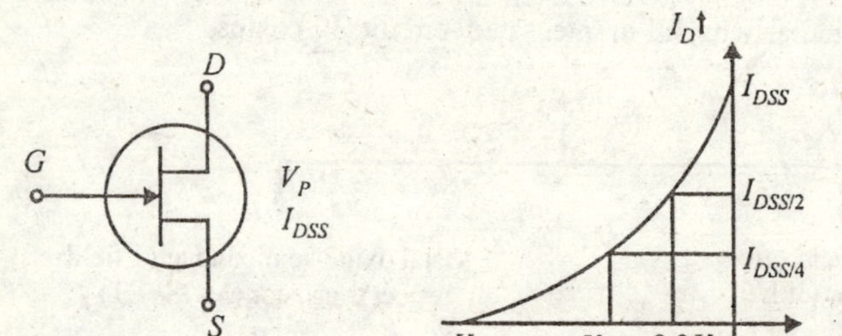

Symbol of *n*-channel JFET   Transfer curve

Fig. 5.1

MOSFET: MOSFETS is of two types:

(*i*) Depletion type MOSFET.

(*ii*) Enhancement type MOSFET.

In the *n*-channel depletion type MOSFET a slab of *p*-type material is formed from a silicon base and is referred to as the substrate. The source and drain terminals are connected through metallic contacts to *n*-doped regions linked by an *n*-channel. The gate is also connected to a metal contact surface, but remain insulated from the *n*-channel by a very thin *silicon dioxide* (SiO_2) *layer* SiO_2 is a particular type of insulator referred to as a *dielectric* that sets up opposing electric field within the dielectric when exposed to an externally applied field. Thus, there is no direct electrical connection between the gate terminal and the channel of a MOSFET.

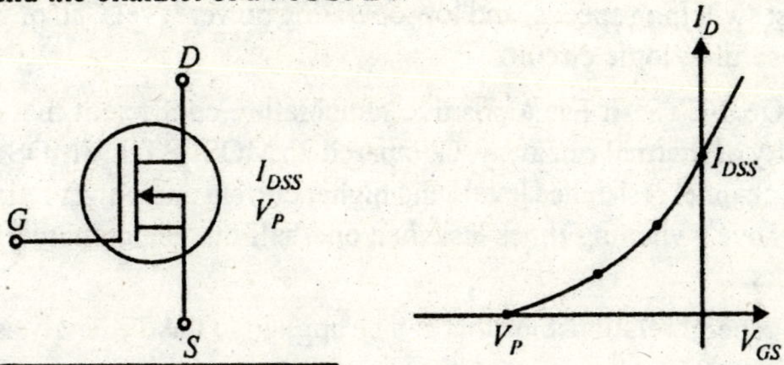

$$I_D = I_{DSS}\left(1 - \frac{V_{GS}}{V_P}\right)^2$$

Depletion type MOSFET (*n*-channel)

Transfer curve of depletion type MOSFET (*n*-channel)

Fig. 5.2

The relationship between the drain current and gate-to-source voltage for an enhancement type is not defined by Schokley' equation, but rather by a non-linear equation controlled by gate-to-source voltage, the threshold voltage and a constant K defined by the device employed. The drain current is given by

$$I_D = K(V_{GS} - V_T)^2$$

$V_T \rightarrow$ Threshold
$K \rightarrow$ Constant

$$K = \frac{I_{D(\text{on})}}{\left(V_{GS(\text{on})} - V_T\right)^2}$$

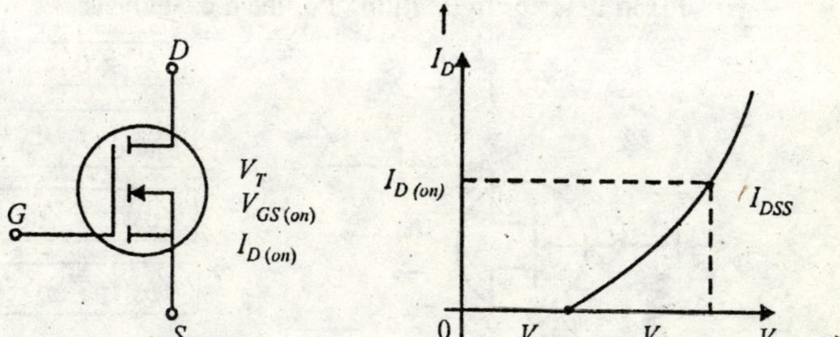

$$I_G = 0, I_D = I_S$$

Symbol of enhancement type MOSFET (*n*-channel)

Transfer curve C-MOSFET (*n*-channel)

Fig. 5.3

CMOS: A CMOS (i.e. complementary MOSFET) device is one that employ a unique combination of a *p*-channel and an *n*-channel MOSFET, with a single set of external leads. It has the advantages of very high input impedance, fast switching speeds, and low operating power levels, all of which make it very useful in logic circuit.

VMOS FETS: It has a positive temperature coefficient that combat the possibility of thermal run away. Compared to MOSFETS, VMOS FETS have reduced channel resistance levels and higher current and power ratings. VMOS devices have switching times less than one-half that encountered for the BJT transistors.

The general relationships that can be applied to the d.c. analysis of all FET amplifiers are.

$$I_G \cong 0\text{ A} \quad \text{and} \quad I_D = I_S$$

For JFET and depletion type MOSFET

$$I_D = I_{DSS}\left(1 - \frac{V_{GS}}{V_P}\right)^2$$

For enhancement type MOSFET

$$I_D = K(V_{GS} - V_T)^2 \quad \text{and} \quad K = \frac{I_{D(\text{on})}}{\left(V_{GS(\text{on})} - V_T\right)^2}$$

Types of Biasing in FET

(*a*) For JFET and depletion type MOSFETs there are three type of biasing:

(*i*) **Fixed bias configuration.** For the d.c. analysis:

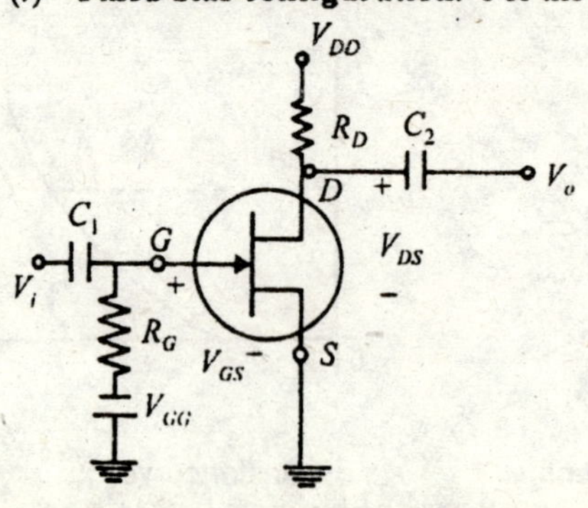

$$I_G \cong 0\text{ A}$$

$$V_{RG} = I_G R_G = 0\text{ V}$$

$$V_{GS} = -V_{GG}$$

$$V_{DS} = V_{DD} - I_D R_D$$

Fig. 5.4

(*ii*) **Self bias configuration**

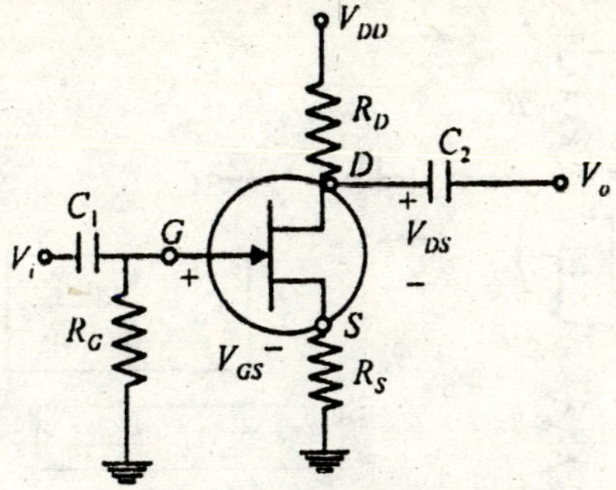

$$V_{GS} = -I_D R_S$$

$$V_{DS} = V_{DD} - I_D (R_S + R_D)$$

$$V_S = I_D R_S$$

$$I_D = I_{DSS}\left(1 - \frac{V_{GS}}{V_P}\right)^2$$

Fig. 5.5

(*iii*) **Voltage divider biasing configuration:** Using the Thevenin theorem:

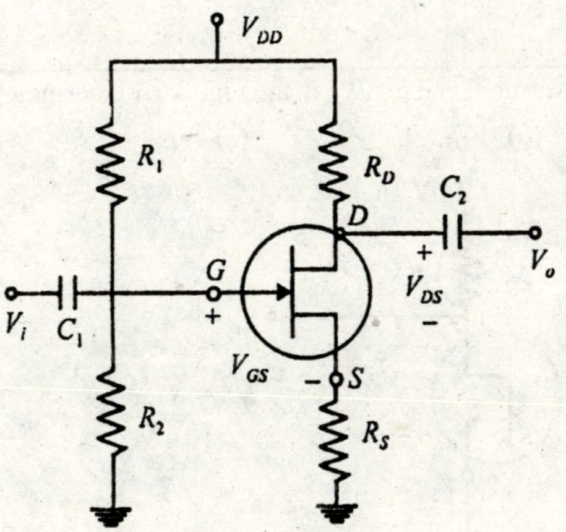

$$V_G = \frac{R_2 V_{DD}}{R_1 + R_2}$$

$$V_S = I_D R_S$$

$$V_{GS} = V_G - I_o R_S$$

$$V_{DS} = V_{DD} - I_D (R_D + R_S)$$

Fig. 5.6

(*b*) For enhancement type MOSFET, we have two types of biasing:

(*i*) **Feedback biasing**

$$V_{DS} = V_{GS}$$

$$V_{GS} = V_{DD} - I_o R_D$$

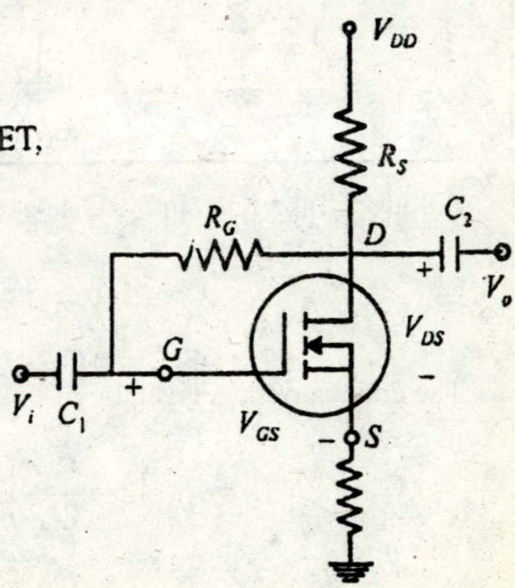

Fig. 5.7

(*ii*) Voltage divider biasing

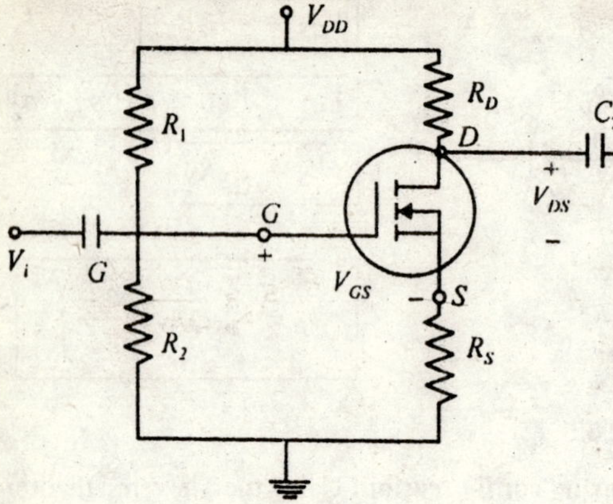

$$V_G = \frac{V_{DD} R_2}{R_1 + R_2}$$

$$V_{GS} = V_G - I_D R_S$$

Fig. 5.8

Problem 5.1. Given the measured value of V_D in the Fig. P. 5.1 determine:

(*a*) I_D (*b*) V_{DS} (*c*) V_{GG}

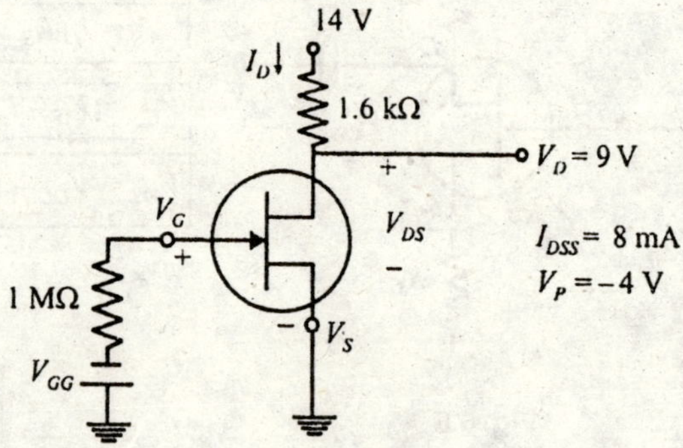

Fig. P. 5.1

Solution. Take KVL in the Gate-source loop

$$-V_{GG} - V_{GS} = 0 \qquad \{\because I_G = 0\}$$

$$\Rightarrow \quad V_{GS} = -V_{GG}$$

$$\Rightarrow \quad V_{GG} = -V_{GS} \qquad \dots (1)$$

The drain current is given by

$$I_D = I_{DSS}\left(1 - \frac{V_{GS}}{V_P}\right)^2, \qquad I_D = \frac{14-9}{1.6} = \frac{5}{1.6} = 3.125\,\text{mA}$$

$$\Rightarrow \quad 3.125 \times 10^{-3} = 8 \times 10^{-3}\left(1 + \frac{V_{GS}}{4}\right)^2$$

$$\Rightarrow \quad 0.625 = 1 + \frac{V_{GS}}{4}$$

$$\Rightarrow \quad V_{GS} = -1.5 \text{ V}$$

Hence $\quad V_{GG} = -V_{GS} = +1.5$ V

$$V_{DS} = V_D - V_S = 9 - 0 = 9 \text{ V}$$

Hence, we have

(*a*) $I_D = 3.125$ mA (*b*) $V_{DS} = 9$ V (*c*) $V_{GG} = +1.5$ V

Problem 5.2. For the fixed bias configuration shown in Fig. P. 5.2, calculate:

(*a*) V_D (*b*) V_{GS} (*c*) I_D

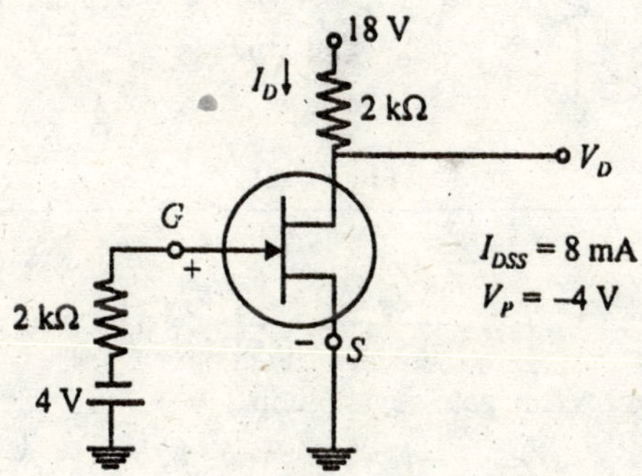

Fig. P. 5.2

Solution. Apply KVL in gate-source loop

$$-4 - V_{GS} = 0$$

$$V_{GS} = -4 \text{ V}$$

The drain current is given by

$$I_D = I_{DSS}\left(1 - \frac{V_{GS}}{V_P}\right)^2$$

$$= 8 \times 10^{-3}\left(1 - \frac{(-4)}{(-4)}\right)^2 = 0 \text{ A}$$

$$V_D = 18 - I_D \,.\, 2\text{ K} = 18 - 0 = 18\text{ V}$$

Therefore

(*a*) $V_D = 18$ V (*b*) $V_{GS} = -4$ V (*c*) $I_D = 0$

Problem 5.3. Given the measurement $V_S = 1.7$ V for the network shown in Fig. P. 5.3, determine:

(*a*) I_{DQ} (*b*) V_{GSQ} (*c*) I_{DSS}

(*d*) V_D (*e*) V_{DS}

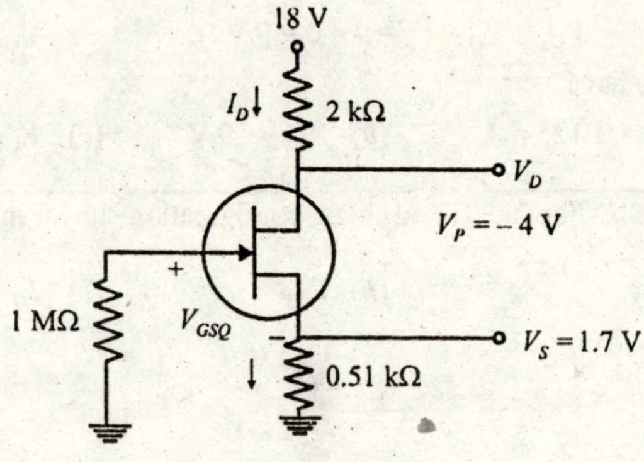

Fig. P. 5.3

Solution.

(*a*) $$I_{DQ} = \frac{V_S}{0.51\text{K}} = \frac{1.7}{0.51\text{K}} = 3.33\text{ mA}$$

(*b*) Apply KVL in gate-source loop

$$V_{GSQ} = -I_{DQ} \times 0.51 = -V_S = -1.7\text{ V}$$

(*c*) Now $$I_D = I_{DSS}\left(1 - \frac{V_{GS}}{V_D}\right)^2$$

$\Rightarrow$ $$I_D = I_{DSS}\left(1 - \frac{1.7}{4}\right)^2$$

$\Rightarrow$ $$I_{DSS} = \frac{3.33}{\left(1 - \frac{1.7}{4}\right)^2} = 10.082\text{ mA}$$

(*d*) $$V_D = 18 - 2\text{ K} \times I_{DQ} = 18 - 2\text{ K} \times 3.33 \times 10^{-3}$$
$$= 18 - 6.66 = 11.34\text{ V}$$

(*e*) $$V_{DS} = V_D - V_S = 11.34 - 1.7 = 9.64\text{ V}$$

Therefore

(a) $I_{DQ} = 3.33$ mA (b) $V_{GSQ} = -1.7$ V

(c) $I_{DSS} = 10.082$ mA (d) $V_D = 11.34$ V

(e) $V_{DS} = 9.64$ V

Problem 5.4. For the network shown in Fig. P. 5.4, find V_s and drain current I_D.

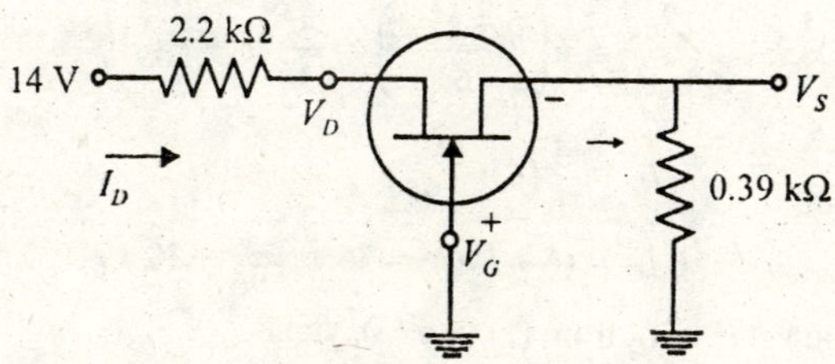

Fig. P. 5.4

$$I_{DSS} = 6 \text{ mA}, \quad V_P = -6 \text{ V}$$

Also, find the value of drain terminal voltage, i.e. V_D.

Solution. The circuit can be redrawn as follows:

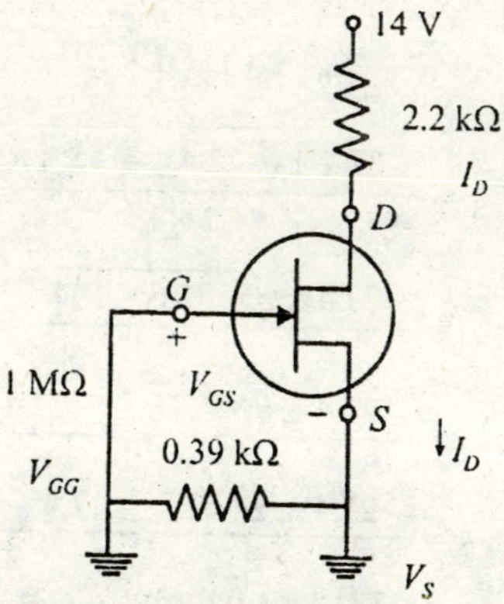

Fig. 5.4 (a)

Given $I_{DSS} = 6$ mA and $V_P = -6$ V

From the circuit diagram it is easily seen that

$$V_S = -V_{GS} \quad \text{... (1)}$$

From KVL to the loop containing gate and source, we get

$$V_{GS} = -0.39\, I_D \quad \text{... (2)}$$

I_D is in mA

The drain current is given by Shockley's equation as

$$I_D = I_{DSS}\left(1 - \frac{V_{GS}}{V_P}\right)^2$$

or
$$I_D = 6\left(1 - \frac{V_{GS}}{(-6)}\right)^2 = 6\left(1 + \frac{V_{GS}}{6}\right)^2$$

$$= 6\left(\frac{6 + V_{GS}}{6}\right)^2 = \frac{6}{36}(6 + V_{GS})^2$$

⇒
$$I_D = \frac{1}{6}(6 + V_{GS})^2$$

or
$$6\,I_D = (6 + V_{GS})^2 = 36 + V_{GS}^2 + 12\,V_{GS} \quad \text{... (3)}$$

Putting the value of I_D from (2) into (3), yields

$$-6 \times \frac{V_{GS}}{0.39} = 36 + V_{GS}^2 + 12\,V_{GS}$$

or
$$V_{GS}^2 + 12V_{GS} + 36 + 6 \times \frac{V_{GS}}{0.39} = 0$$

⇒
$$V_{GS}^2 + 12\,V_{GS} + 15.4\,V_{GS} + 36 = 0$$

⇒
$$V_{GS}^2 + 27.4\,V_{GS} + 36 = 0$$

or
$$V_{GS} = \frac{-27.4 \pm \sqrt{(27.4)^2 - 4 \times 1 \times 36}}{2 \times 1}$$

$$= \frac{-27.4 \pm \sqrt{750.76 - 144}}{2}$$

$$= \frac{-27.4 \pm \sqrt{606.76}}{2} = \frac{-27.4 \pm 24.632}{2}$$

or
$$V_{GS} = \frac{-27.4 + 24.632}{2}, \frac{-27.4 - 24.632}{2}$$

$$= -1.4\text{ V}, -26.02\text{ V (rejected)}$$

Hence $V_{GS} = -1.4$ V

Now, from (1) $V_S = -V_{GS} = 1.4$ V

I_{DQ} can also be calculated from equation (2)

$$I_D = -\frac{V_{GS}}{0.39} = -\frac{(-1.4)}{0.39}$$

or $$I_D = \frac{1.4}{0.39} = 3.59 \text{ mA}$$

and hence

$$V_D = V_{DD} - I_D R_D = 14 \text{ V} - (3.59 \text{ mA}) \times 2.2 \text{ k}\Omega$$
$$= 14 \text{ V} - 7.898 \text{ V} = 6.102 \text{ V}$$

Problem 5.5. For the network of Fig. P. 5.5, determine:

(*a*) I_{DQ} (*b*) V_{GSQ} (*c*) V_{DS} (*d*) V_S (*e*) V_D

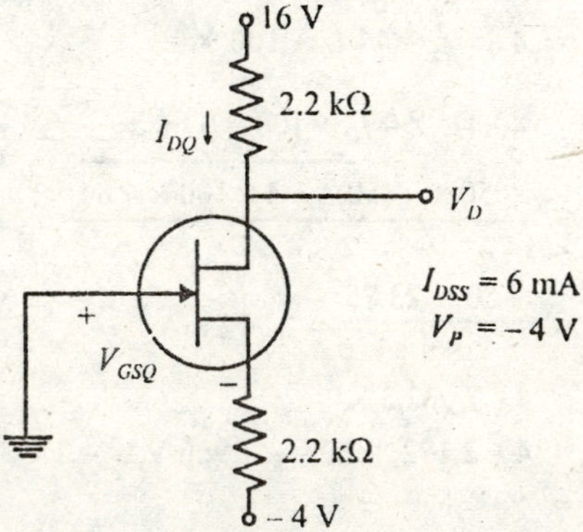

Fig. P. 5.5

Solution. Apply KVL in gate-source loop

$$-V_{GSQ} - I_{DQ} \,.\, 2.2 \text{ k}\Omega + 4 = 0$$

$$V_{GSQ} = (-I_{DQ}\, 2.2 \text{ K} + 4) \quad \text{... (1)}$$

The drain current is given by

$$I_D = I_{DSS}\left(1 - \frac{V_{GSQ}}{V_p}\right)^2$$

$$\Rightarrow \quad I_D = 6\left(1 + \frac{V_{GSQ}}{6}\right)^2$$

$$\Rightarrow \quad \sqrt{\frac{I_D}{6}} = 1 + \frac{V_{GSQ}}{6}$$

$$\Rightarrow \quad \frac{V_{GSQ}}{6} = \sqrt{\frac{I_D}{6}} - 1$$

$$\Rightarrow \quad V_{GSQ} = 6\left[\sqrt{\frac{I_D}{6}} - 1\right]$$

$$\Rightarrow \quad V_{GSQ} = \sqrt{6I_D} - 6$$

Put in equation (1)

$$V_{GSQ} = -I_D \times 2.2 + 4$$

$$\Rightarrow \quad \sqrt{6I_D} - 6 = -2.2I_D + 4$$

$$\Rightarrow \quad \sqrt{6I_D} = -2.2\, I_D + 10$$

$$\Rightarrow \quad 6\, I_D = (-2.2\, I_D + 10)^2$$

$$\Rightarrow \quad 6\, I_D = 4.84 I_D^2 - 44\, I_D + 100$$

$$\Rightarrow \quad 4.84 I_D^2\ 50\, I_D + 100 = 0$$

$$\Rightarrow \quad I_D = \frac{+50 \pm \sqrt{(50)^2 - 4 \times 100 \times 4.84}}{2 \times 4.84}$$

$$= \frac{+50 \pm 23.75}{2 \times 4.84} = 2.712 \text{ mA}$$

$$V_{GSQ} = 4 - I_{DQ}\, 2.2 \text{ K}$$

$$= 4 - 2.172 \times 2.2 = -1.966 \text{ V} \cong -2 \text{ V}$$

$$V_{GS} = V_G - V_S$$

(From the figure, we can calculate the drain current I_{DQ})

$$\Rightarrow \quad V_{GS} = 0 - V_S$$

$$\Rightarrow \quad V_S = -V_{GS}$$

$$\Rightarrow \quad V_S = +2 \text{ V}$$

$$I_{DQ} = \frac{16 - V_D}{2.2}$$

$$V_D = 16 - 2.2 \times 2.712$$

$$= 16 - 5.966 = 10.03 \text{ V}$$

$$V_{DS} = V_D - V_S$$

$$= 10.03 - 2 = 8.03 \text{ V}$$

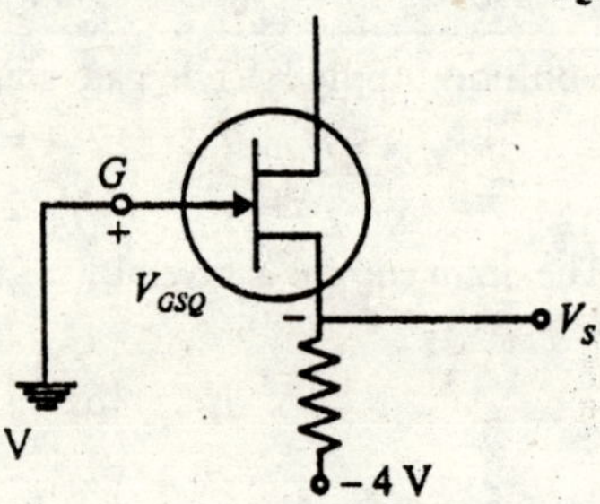

Fig. 5.5 (a)

For active anode of FET, drain is positive with respect to source, i.e. $V_{DS} \Rightarrow 0$.

Hence, we have

(a) $I_{DQ} = 2.712$ mA (b) $V_{GS}(Q) = -2$ V

(c) $V_{DS} = 8.03$ V (d) $V_S = 2$ V

(e) $V_D = 10.03$ V

Problem 5.6. For the self-bias configuration of depletion type MOSFET, determine:

(*a*) V_{GS} (*b*) I_D (*c*) V_{DS} (*d*) V_D

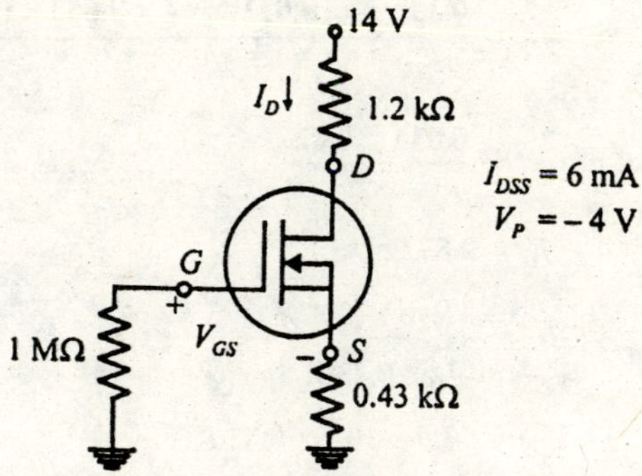

Fig. P. 5.6

Solution.

$$I_D = I_{DSS}\left(1 - \frac{V_{GS}}{V_P}\right)^2$$

$$\Rightarrow \quad \sqrt{\frac{I_D}{I_{SS}}} = 1 - \frac{V_{GS}}{V_P}$$

$$\Rightarrow \quad V_{GS} = V_P\left(1 - \sqrt{\frac{I_D}{I_{DSS}}}\right) \quad \text{... (1)}$$

From the KVL in source-gate loop

$$V_{GS} = -I_D \times 0.43\text{ K}, \text{ put this in equation (1)}$$

$$\Rightarrow \quad \sqrt{\frac{I_D}{I_{DSS}}} = 1 - \frac{V_{GS}}{V_P}$$

$$\Rightarrow \quad \sqrt{\frac{I_D}{6 \times 10^{-3}}} = 1 - \frac{0.43\text{K } I_D}{4}$$

$$\Rightarrow \quad \frac{I_D \times 10^3}{6} = \left(1 - \frac{430\, I_D}{4}\right)^2$$

$$\Rightarrow \quad \frac{1000\, I_D}{6} = \frac{16 + (430)^2 I_D - 4 \times 860\, I_D}{16}$$

$$\Rightarrow \quad 16000\, I_D = 96 + 1109400\, I_D^2 - 20640\, I_D.$$

$$\Rightarrow \quad 1109400\, I_D^2 - 36640\, I_D + 96 = 0$$

$$\Rightarrow \quad I_D^2 - 0.033 I_D + 8.653 \times 10^{-5} = 0$$

$$\Rightarrow \quad I_D = \frac{0.333 \pm \sqrt{(0.033)^2 - 4 \times 8.653 \times 10^{-5}}}{2}$$

$$\Rightarrow \quad I_D = \frac{0.033 \pm 0.0273}{2}$$

$$\Rightarrow \quad I_D = 2.872 \text{ mA}$$

$$\Rightarrow \quad I_D \cong 2.9 \text{ mA}$$

Now $\quad V_{GS} = -I_D \times 0.43 \text{ K}$

$$V_{GS} = -2.9 \times 10^{-3} \times 0.43 \text{ K}$$

$$V_{GS} = -1.247 \text{ V}$$

Apply KVL in the drain-source loop

$$14 = (0.43 + 1.2) \text{ K} \times I_D + V_{DS}$$

$$\Rightarrow \quad 14 = 1.63 \text{ K} \times 2.9 \times 10^{-3} + V_{DS}$$

$$\Rightarrow \quad 14 = 4.727 + V_{DS}$$

$$\Rightarrow \quad V_{DS} = 9.273 \text{ V}$$

Now $\quad V_{DS} = V_D - V_S$

$$\Rightarrow \quad V_D = V_{DS} + V_S$$

$$\Rightarrow \quad V_D = 9.273 + 0.43\text{K}\, I_D$$

$$\Rightarrow \quad V_D = 9.273 + 1.247$$

$$\Rightarrow \quad V_D = 10.52 \text{ V}$$

Hence

(*a*) $I_D = 2.9$ mA

(*b*) $V_{GS} = -1.247$

(*c*) $V_{DS} = 9.273$ V

(*d*) $V_D = 10.52$ V

Problem 5.7. For the network of the enhancement type MOSFET of the Fig. P. 5.7 determine:

(*a*) I_{DQ} (*b*) V_{GSQ}

(*c*) V_{DSQ} (*d*) V_D

(*e*) V_S (*f*) V_{DS}

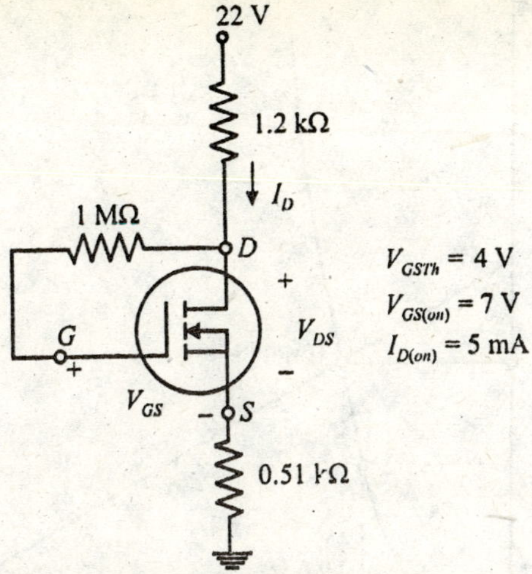

Fig. P. 5.7

Solution. For enhance type MOSFET the drain current is given as

$$I_D = K(V_{GS} - V_{GSth})^2$$

Where $$K = \frac{I_{D(\text{on})}}{(V_{GS(\text{on})} - V_{GSth})^2}$$

Hence, the proper given value, we get the value of the constant K.

$$\Rightarrow \quad K = \frac{5\times10^{-3}}{(7-4)^2} = \frac{5\times10^{-3}}{(3)^2} = \frac{5}{9}\times10^{-3}\ \text{A/V}^2$$

Now, draw a curve between I_D and V_{GS} as shown in Fig. 5.7 (a)

$$I_D = K\left(V_{GS} - V_{GSth}\right)^2$$

$$\Rightarrow \quad I_D = \frac{5}{9}\times10^{-3}\left(V_{GS}-4\right)^2$$

Let take a point between 7 and 4, (say = 6) (V_{GS} = 6 V)

$$\Rightarrow \quad I_D = \frac{5}{9}\times10^{-3}(6-4)^2$$

$$\Rightarrow \quad I_D = \frac{5}{9}\times10^{-3}\times4$$

$$\Rightarrow \quad I_D = 2.22\ \text{mA}$$

Now, take one point slightly greater than 7 say N_{GS} = 8.

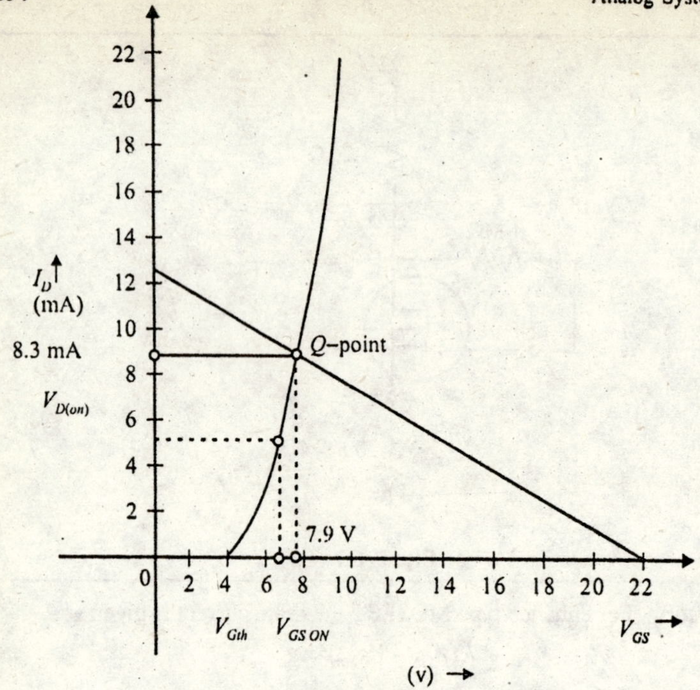

Fig. 5.7 (a)

$$I_D = \frac{5}{9} \times 10^{-3} (8-4)^2 = 8.8 \text{ mA}$$

Thus, we draw the graph using the three points.

Now, for the network bias line

$$22 = V_{GS} + I_D (1.2 + 0.51) \text{ K}$$

$$V_{GS} = 22 - I_D (1.71) \text{ K} \qquad \text{... (1)}$$

When $\quad I_D = 0, V_{GS} = 22 \text{ V}$

When $\quad V_{GS} = 0, I_D = \dfrac{22}{1.71 \text{ K}} = 12.86 \text{ mA}$

Now, draw a line join the two points. This line cut the curve and the point of intersection is called the Q-point, the (I_D) drain current correspond to it is given as

$$I_{DQ} = 8.3 \text{ mA}, V_{GSQ} = 7.9 \text{ V}$$

From figure, we have $\quad V_{DSQ} = V_{GSQ} = 7.9 \text{ V}$

Now $\quad \dfrac{22 - V_D}{1.2\text{K}} = I_D$

$$\Rightarrow \qquad V_D = 22 - I_D \times 1.2 \text{ K}$$

$$\Rightarrow \qquad V_D = 22 - 8.3 \times 1.2 = 12.04 \text{ V}$$

$$V_S = I_D \times 0.51 \text{ K} = 8.3 \times 0.51 = 4.233 \text{ V}$$

For active mode of FET the drain terminal point higher than source terminal, i.e. V_{DS} = (+)ve

Now $\qquad V_{DS} = V_D - V_S = 12.04 - 4.233 = 7.807 \text{ V}$

Hence, we have

(*a*) $I_{DQ} = 8.3$ mA (*b*) $V_{GSQ} = 7.9$ V

(*c*) $V_{DSQ} = 7.9$ V (*d*) $V_D = 12.04$ V

(*e*) $V_S = 4.233$ V (*f*) $V_{DS} = 7.807$ V

Problem 5.8. For the network of *p*-channel FET, determine:

(*a*) I_{DQ} (*b*) V_{GSQ}

(*c*) V_{DS} (*d*) V_D

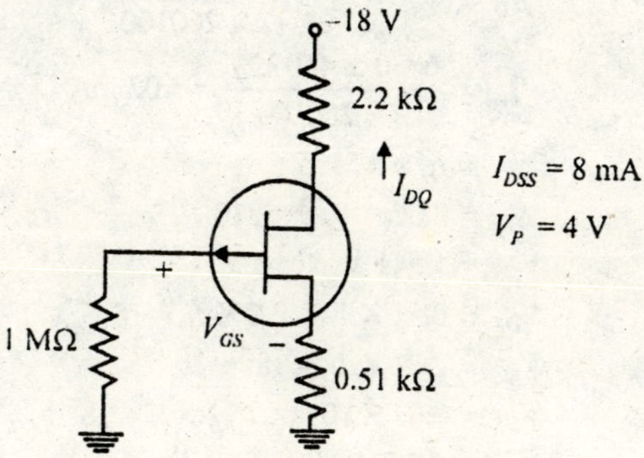

Fig. P. 5.8

Solution. Apply KVL in the gate-source loop

$$V_{GS} = + I_D \times 0.51 \text{ K}$$

The drain current of the *p*-channel FET is given by

$$I_D = I_{DSS}\left(1 - \frac{V_{GS}}{V_P}\right)^2$$

$$\Rightarrow \qquad I_D = 8 \times 10^{-3}\left(1 - \frac{V_{GS}}{4}\right)^2$$

$$\Rightarrow \quad \frac{I_D}{8 \times 10^{-3}} = \left(1 - \frac{V_{GS}}{4}\right)^2$$

$$\Rightarrow \quad \frac{V_{GS}}{4} = 1 - \sqrt{\frac{I_D}{8 \times 10^{-3}}}$$

$$\Rightarrow \quad \sqrt{\frac{I_D}{8 \times 10^{-3}}} = 1 - \frac{I_D \times 0.51\text{K}}{4}$$

$$\Rightarrow \quad \frac{1000\, I_D}{8} = \left(\frac{4 - 510\, I_D}{4}\right)^2$$

$$\Rightarrow \quad \frac{1000\, I_D}{8} = \frac{16 + 260100\, I_D{}^2 - 4080\, I_D}{16}$$

$$\Rightarrow \quad 260100\, I_D^2 - 6080\, I_D + 16 = 0$$

$$\Rightarrow \quad I_D = \frac{6080 \pm \sqrt{(6080)^2 - 4 \times 16 \times 260100}}{2 \times 260100}$$

$$\Rightarrow \quad I_D = \frac{6080 \pm 4507.77}{2 \times 260100} = 3.02 \text{ mA}$$

$$V_{GS} = I_D \times 0.51 \text{ K}$$

$$\Rightarrow \quad V_{GS} = 3.02 \times 10^{-3} \times 510$$

$$\Rightarrow \quad V_{GS} = 1.541 \text{ V}$$

$$V_{DS} = 0.51 \text{ K } I_D - 2.2 \text{ K } I_D - 18$$

$$\Rightarrow \quad V_{DS} = 2.71 \text{ K } I_D - 18$$

$$\Rightarrow \quad V_{DS} = 3.02 \times 10^{-3} \times 2.71 \times K - 18$$

$$\Rightarrow \quad V_{DS} = -9.815 \text{ V}$$

$$V_D = V_{DS} + V_S$$

$$= -9.815 - 1.541 = -11.356 \text{ V}$$

Hence, we have

(*a*) $I_{DQ} = 3.02$ mA (*b*) $V_{GS} = 1.541$ V

(*c*) $V_{DS} = -9.815$ V (*d*) $V_D = -11.356$ V

Problem 5.9. Given $I_{DSS} = 9$ mA and $V_P = -3.5$ V, determine I_D when

(*a*) $V_{GS} = 0$ V (*b*) $V_{GS} = -2$ V

(*c*) $V_{GS} = -3.5$ V (*d*) $V_{GS} = -5$ V

Solution. For the FET, the drain current is given by the equation:

$$I_D = I_{DSS}\left(1 - \frac{V_{GS}}{V_P}\right)^2$$

(*a*) $V_{GS} = 0$ V $\quad I_D = 9\,(1 - 0)^2$

$$I_D = 9 \text{ mA}$$

(*b*) $V_{GS} = -2$ V $\quad I_D = 9\left(1 - \frac{(-2)}{(-3.5)}\right)^2$

$$\Rightarrow \quad I_D = 9\left(1 - \frac{2}{3.5}\right)^2$$

$$\Rightarrow \quad I_D = 1.653 \text{ mA}$$

(*c*) $V_{GS} = -3.5$ V $\quad I_D = 9\left(1 - \frac{(-3.5)}{(-3.5)}\right)^2$

$$\Rightarrow \quad I_D = 9(1 - 1)^2 \Rightarrow I_D = 0 \text{ mA}$$

(*d*) $V_{GS} = -5$ V, $I_D = 0$, because the minimum value of V_{GS} to flow the drain current I_D is $V_{GS} = -3.5$ V for all values of V_{GS} less than this value, the drain current (I_D), will be zero.

Problem 5.10. For the fixed bias configuration, determine:

(*a*) I_{DQ} and V_{GSQ} using purely mathematical approach.

(*b*) Find V_{DS}, V_D, V_G and V_S using results of part (*a*).

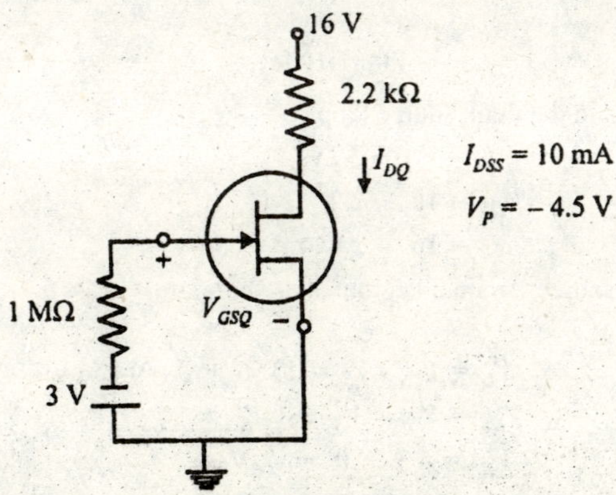

Fig. P. 5.10

Solution. Now, applying KVL in the gate-source loop

$$V_{GG} + V_{GSQ} = 0$$

$$\Rightarrow \quad V_{GSQ} = -V_{GG} = -3\text{ V}$$

The drain current is given as

$$I_D = I_{DSS}\left(1 - \frac{V_{GSQ}}{V_P}\right)^2$$

$$\Rightarrow \quad I_D = 10 \times 10^{-3}\left(1 - \frac{(-3)}{(-4.5)}\right)^2$$

$$\Rightarrow \quad I_D = 10 \times 10^{-3}\left(\frac{4.5 - 3}{4.5}\right)^2 = 10 \times 10^{-3}\left(\frac{1.5}{4.5}\right)^2$$

$$\Rightarrow \quad I_{DQ} = 1.11\text{ mA}$$

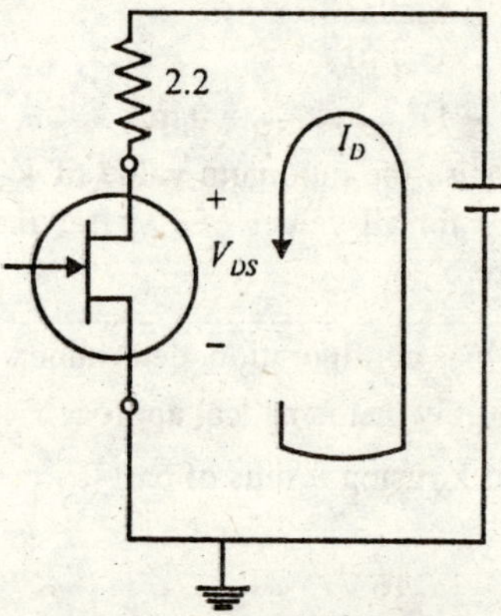

Fig. 5.10 (a)

Apply KVL in the drain source loop

$$16 = V_{DS} + 2.2\, I_D$$

$$\Rightarrow \quad V_{DS} = 16 - 2.2 \times 1.11$$

$$= 16 - 2.444 = 13.56\text{ V}$$

Since, the source terminal is grounded, therefore $V_S = 0$

Therefore

$$V_D = V_{DS} + V_S = 13.56 + 0 = 13.56\text{ V}$$

and $\quad V_{GS} = V_G - V_S \Rightarrow V_G = V_{GS} + V_S$

$$\Rightarrow \quad V_G = -3 + 0 = -3\text{V}$$

Thus, we get

(*a*) $I_{DQ} = 1.11$ mA

$V_{GSQ} = -3$ V

(*b*) $V_{DS} = 13.56$ V

$V_D = 13.56$ V

$V_G = -3$ V

$V_S = 0$ V

Problem 5.11. Determine V_D for the fixed bias configuration:

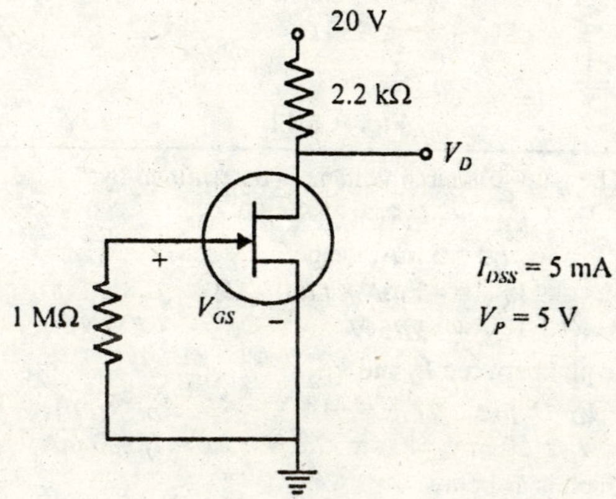

Fig. P. 5.11

Solution. Since, there is no d.c. voltage in the gate terminal, i.e. $V_{GG} = 0$.

Therefore $V_{GS} = -V_{GG} = 0$ V

So, the drain current is

$$I_D = I_{DSS}\left(1 - \frac{V_{GS}}{V_P}\right)^2 = I_{DSS}\left(1 - \frac{0}{(-5)}\right)^2$$

$\Rightarrow$ $I_D = I_{DSS} = 5$ mA

Now, apply KVL in the source-drain loop,

$$20 = 2.2\text{ k}\Omega \times I_D + V_D$$

$\Rightarrow$ $V_D = 20 - 2.2\text{ k}\Omega \times 5$ mA

$\Rightarrow$ $V_D = 20 - 11.0$

$\Rightarrow$ $V_D = 9$ V

Problem 5.12. For the self-bias configuration determine:

(*a*) I_{DQ} and V_{GSQ} (*b*) V_{DS}, V_D, V_G and V_S

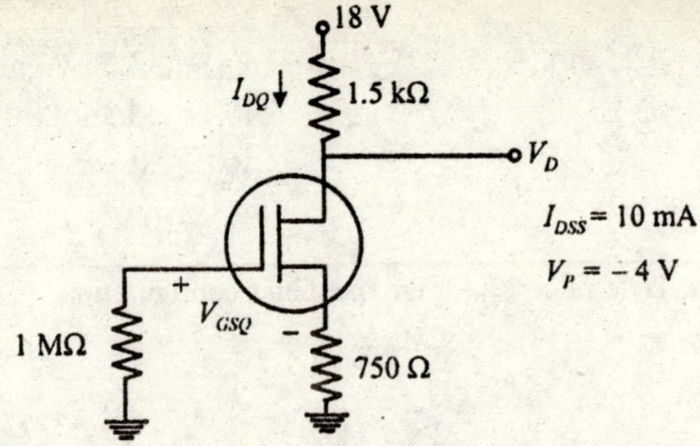

Fig. P. 5.12

Solution. The gate-to-source voltage is determined by

$$V_{GS} = -I_D R_S$$

Choosing $I_D = 5$ mA, we get

$$V_{GS} = -5 \text{ mA} \times (750 - \Omega)$$

$$V_{GS} = -3.75 \text{ V}$$

We draw the plat between I_D and V_{GS}.

When $I_D = 6$ mA, $V_{GS} = -4.5$ V $\quad I_D = 8$ mA, $V_{GS} = -6$ V

$I_D = 10$ mA, $V_{GS} = --7.5$ V $\quad I_D = 0$ mA, $V_{GS} = 0$ V

Thus, the graph is as below

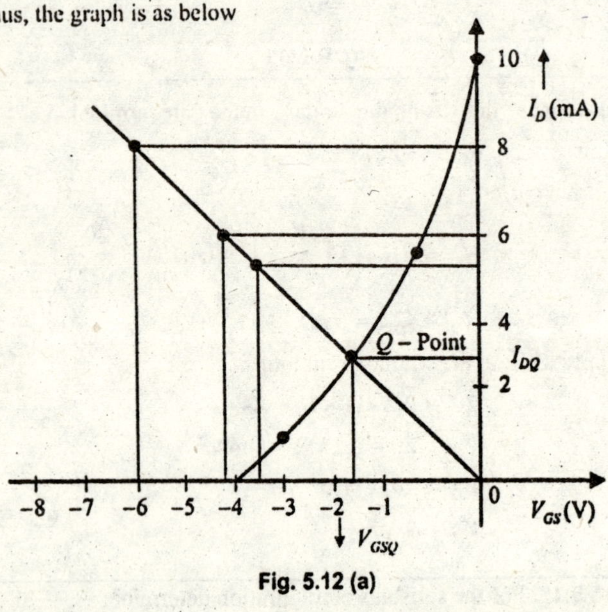

Fig. 5.12 (a)

And from Schokley's equation

$$I_D = I_{DSS}\left(1 - \frac{V_{GS}}{V_P}\right)^2$$

$$\Rightarrow \qquad I_D = 10 \times 10^{-3}\left(1 + \frac{V_{GS}}{4}\right)^2$$

For $V_{GS} = -4$ V, $I_D = 0$ mA

$V_{GS} = -3$ V, $I_D = 0.625$ mA

$V_{GS} = -2$ V, $I_D = 2.5$ mA

$V_{GS} = -1$ V, $I_D = 5.625$ mA

$V_{GS} = 0$ V, $I_D = 10$ mA

From the point of intersection, we get the quiscent point Q-point. The corresponding drain current and gate-to-source voltage is found to be.

$$I_{DQ} = 2.6 \text{ mA}$$

and $$V_{GSQ} = -1.75 \text{ V}$$

Now, apply KVL in the drain to source loop

$$18 = V_{DS} + I_{DQ}(1.5 + 0.75) \text{ k}\Omega$$

$$\Rightarrow \quad V_{DS} = 18 - 2.6 \times 2.25 = 12.15 \text{ V}$$

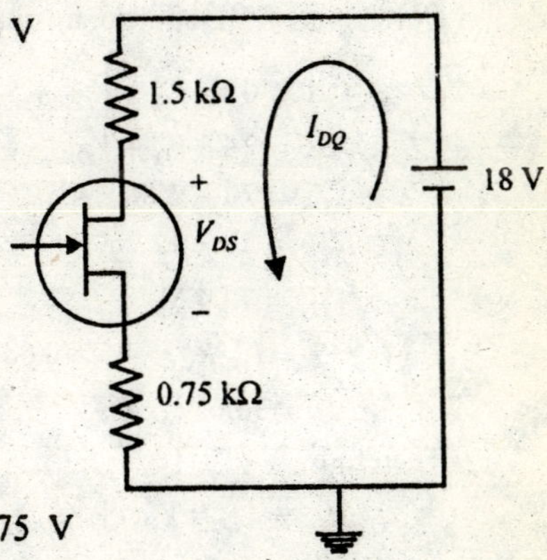

Fig. 5.12 (b)

Therefore

$$V_{GS} = -I_D R_S$$
$$= -2.6 \times 0.75 = -1.95 \text{ V}$$
$$V_S = I_D R_S = 1.95 \text{ V}$$

Now $V_{GS} = V_G - V_S$

$$\Rightarrow \quad V_G = V_{GS} + V_S$$
$$= -1.95 \text{ V} + 1.95 \text{ V} = 0 \text{ V}$$

Thus, we get

(*a*) $I_{DQ} = 2.6$ mA and $V_{GSQ} = -1.75$ V

(*b*) $V_{DS} = 12.15$ V, $V_G = 0$ V

$V_S = 1.95$ V, $V_{GS} = -1.95$ V

$V_D = V_{DS} + V_S$

$= 12.15 + 1.95 \text{ V} = 14.1$ V

Problem 5.13. For the network shown in Fig. P. 5.13, determine:

(*a*) I_D (*b*) V_{DS} (*c*) V_D (*d*) V_S

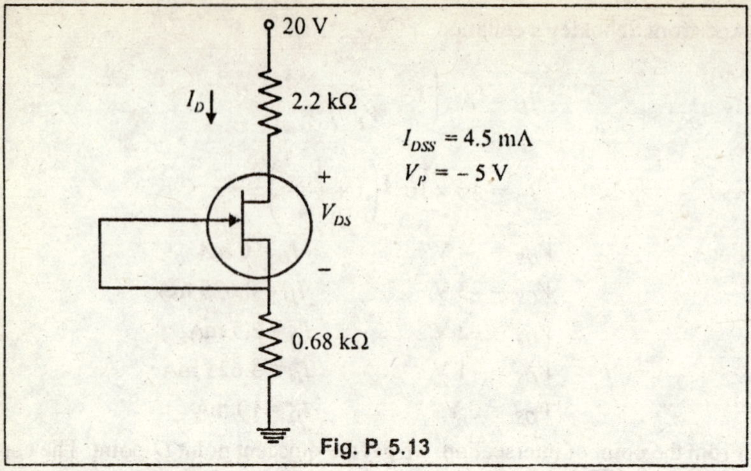

Fig. P. 5.13

Solution. The gate-to-source voltage is $V_{GS} = 0$, since it is short-circuited. Therefore, the drain current at $V_{GS} = 0$ is $I_D = I_{DSS} = 4.5$ mA, by putting $V_{GS} = 0$ in the Shockley's equation:

$$I_D = I_{DSS}\left(1 - \frac{V_{GS}}{V_P}\right)^2$$

Now, apply KVL in the drain-source loop

$$20 = 0.68\ I_D + V_{DS} + 2.2\ I_D$$

$$20 = 2.88\ I_D + V_{DS}$$

$$V_{DS} = 20 - 2.88 \times 4.5$$

$$V_{DS} = 20 - 12.96$$

$$V_{DS} = 7.04 \text{ V}$$

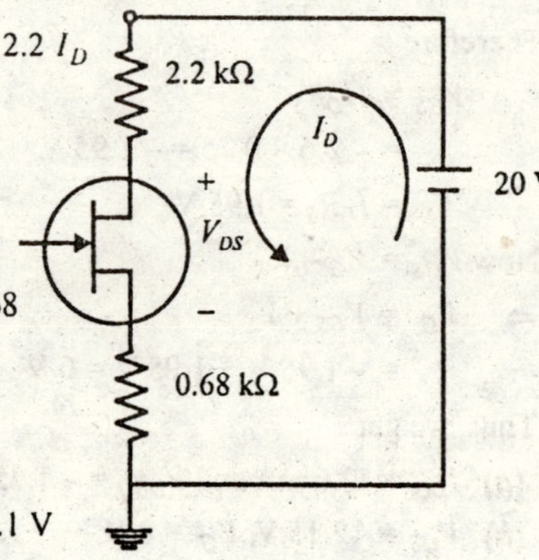

Now $V_S = I_D\ R_S = 4.5 \times 0.68$

$= 3.06$ V

Therefore

$$V_D = V_D = V_{DS} + V_S$$

$$= 7.04 + 3.06 = 10.1 \text{ V}$$

and $V_G = V_S = 3.06$ V

Fig. 5.13 (a)

Therefore

(*a*) $I_D = 4.5$ mA (*b*) $V_{DS} = 7.04$ V

(*c*) $V_D = 10.1$ V (*d*) $V_S = 3.06$ V.

Problem 5.14. For the network of voltag … T bias. Find:

(*a*) I_D … and V_{DS}

(*c*) V_G and V_{GS} … V_P

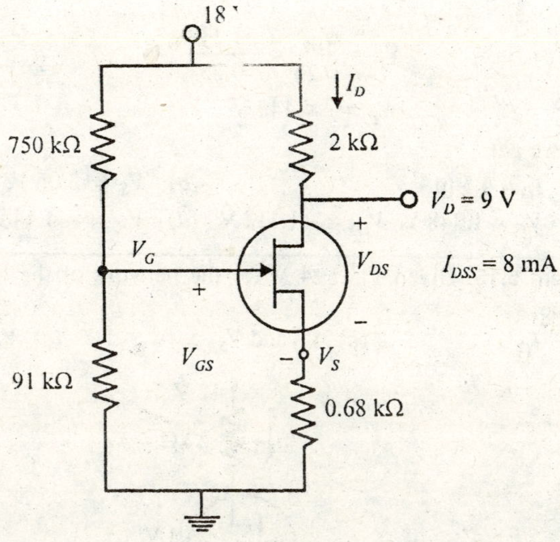

Fig. P. 5.14

Solution. The drain current I_D is

$$I_D = \frac{18 - 9}{2\text{K}} = \frac{9}{2\text{K}} = 4.5 \text{ mA}$$

Therefore $\quad V_S = I_D R_S = 4.5 \times 0.68 = 3.06 \text{ V}$

Therefore $\quad V_{DS} = V_D - V_S = 9 - 3.06 = 5.94 \text{ V}$

Now $\quad V_G = \dfrac{R_2}{R_1 + R_2} V_{DD} = \dfrac{91}{750 + 91} \times 18 = 1.948 \text{ V}$

Therefore $\quad V_{GS} = V_G - V_S = (1.948 - 3.06) \text{ V} = -1.112 \text{ V}$

To calculate V_p

$$I_D = I_{DSS}\left(1 - \frac{V_{GS}}{V_P}\right)^2$$

$$1 - \frac{V_{GS}}{V_P} = \sqrt{\frac{I_D}{I_{DSS}}}$$

$$\frac{V_{GS}}{V_P} = 1 - \sqrt{\frac{I_D}{I_{DSS}}} = 1 - \sqrt{\frac{4.5}{8}}$$

$$\Rightarrow \qquad \frac{V_{GS}}{V_P} = 0.25$$

$$\Rightarrow \qquad V_P = \frac{V_{GS}}{0.25} = \frac{-1.112}{0.25}$$

$$\Rightarrow \qquad V_P = -4.448 \text{ V}$$

Thus, we get

(*a*) $I_D = 4.5$ mA (*b*) $V_S = 3.06$ V, $V_{DS} = 5.94$ V

(*c*) $V_G = 1.948$ V, $V_{GS} = -1.112$ V (*d*) $V_P = -4.448$ V

Problem 5.15. Given $V_{DS} = 4$ V for the network of the Fig. P. 5.15, determine:

(*a*) I_D (*b*) V_D and V_S (*c*) V_{GS}

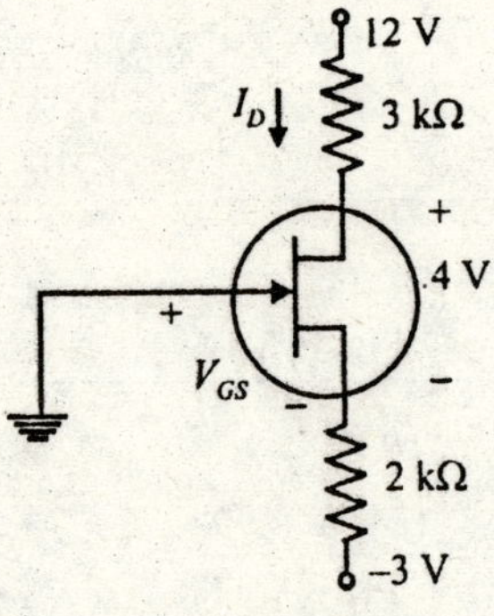

Fig. P. 5.15

Solution. Apply KVL in the gate-source loop

$$-V_{GS} - I_D R_S + 3 = 0$$

$$\Rightarrow \qquad V_{GS} = -I_D R_S + 3 = 0 \qquad \text{... (1)}$$

The drain current I_D is

$$I_D = I_{DSS}\left(1 - \frac{V_{GS}}{V_P}\right)^2$$

$$\Rightarrow \qquad \sqrt{\frac{I_D}{I_{DSS}}} = 1 - \frac{V_{GS}}{V_P}$$

$$\Rightarrow \qquad \frac{V_{GS}}{V_P} = 1 - \sqrt{\frac{I_D}{I_{DSS}}} \qquad \text{... (2)}$$

Apply KVL in the drain-source loop

$$12 + 3 = 2I_D + 4 + 3I_D$$

$$\Rightarrow \quad 5I_D = 15 - 4 = 11 \text{ V}$$

$$\Rightarrow \quad I_D = \frac{11}{5 \text{ K}} = 2.2 \text{ mA}$$

Put the value of I_D in equation (1)

$$V_{GS} = -2.2 \times 2 + 3$$

$$\Rightarrow \quad V_{GS} = -4.4 + 3 = -1.4 \text{ V}$$

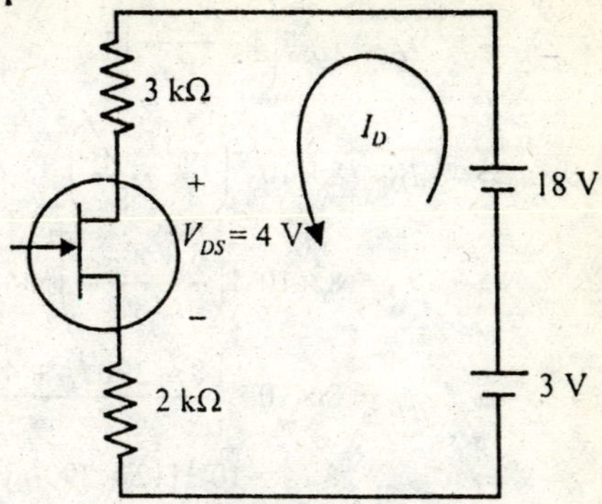

Fig. 5.15 (a)

Now, the source terminal voltage

$$V_S = V_G - V_{GS}$$

$$= 0 + 1.4 = +1.4 \text{ V}$$

So $\quad V_{DS} = V_D - V_S$

$$\Rightarrow \quad V_D = V_{DS} + V_S = 4 + 1.4$$

$$\Rightarrow \quad V_D = 5.4 \text{ V}$$

So, we have

(*a*) $I_D = 2.2$ mA (*b*) $V_D = 5.4$ V, $V_S = 1.4$ V (*c*) $V_{GS} = -1.4$ V

Problem 5.16. For the depletion type MOSFET shown in Fig. P. 5.16, determine:

(*a*) I_{DQ} and V_{GSQ} (*b*) V_{DS} and V_S

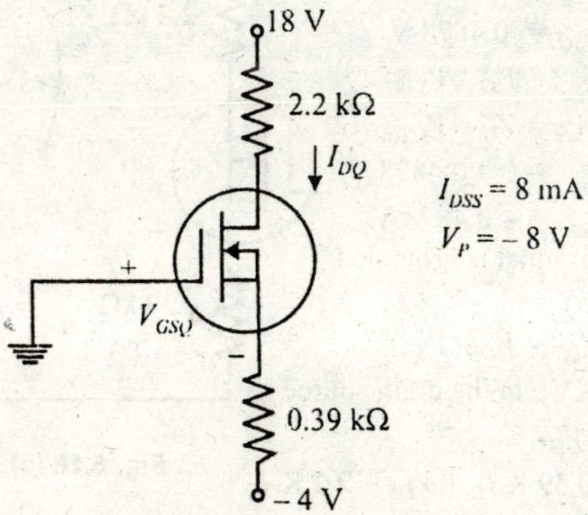

Fig. P. 5.16

Solution. Applying KVL in the gate-source loop

$$-V_{GSQ} - 0.39 \text{ K } I_D + 4 = 0$$

$$\Rightarrow \quad V_{GSQ} = -390 I_D + 4 \qquad \text{... (1)}$$

Frcm Schockley's equation, the drain current is

$$I_D = I_{DSS}\left(1 - \frac{V_{GSQ}}{V_P}\right)^2$$

$$\Rightarrow \quad I_D = 8\times10^{-3}\left(1 + \frac{V_{GSQ}}{8}\right)^2$$

$$\Rightarrow \quad I_D = 8\times10^{-3}\left(\frac{8 + V_{GSQ}}{8}\right)^2$$

$$\Rightarrow \quad I_D = 8\times10^{-3}\left[\frac{8 - 390I_D + 4}{8}\right]^2$$ Put the value of V_{GSQ} from Eq. (1)

$$\Rightarrow \quad 8\,I_D = 10^{-3}[12 - 390I_D]^2$$

$$\Rightarrow \quad 8000\,I_D = 144 - 9360\,I_D + 152100\,I_D^2$$

$$\Rightarrow \quad 152100\,I_D^2 - 17360\,I_D + 144 = 0$$

$$I_D^2 - 0.11413\,I_D + 9.467\times10^{-4} = 0$$

$$\Rightarrow \quad I_D = \frac{0.11413 \pm \sqrt{(0.11413)^2 - 4\times9.467\times10^{-4}}}{2}$$

$$\Rightarrow \quad I_{DQ} = \frac{0.11413 \pm 0.09612}{2} \cong 9\,\text{mA}$$

Therefore $V_{GSQ} = -390\,I_{DQ} + 4$

$\Rightarrow \quad V_{GSQ} = -390\times9\times10^{-3} + 4$

$\Rightarrow \quad V_{GSQ} = -3.512 + 4$

$\Rightarrow \quad V_{GSQ} = 0.4878$ V

Now, $\quad V_{GSQ} = V_G - V_S$

$\Rightarrow \quad V_S = V_G - V_{GSQ}$

$= 0 - 0.4878$ V

$= -0.4878$ V

[$\because$ the gate terminal is grounded $\therefore V_G = 0$]

Therefore $\quad V_{DS} = V_D - V_S$

Now, apply KVL in the drain source loop to calculate V_{DS}

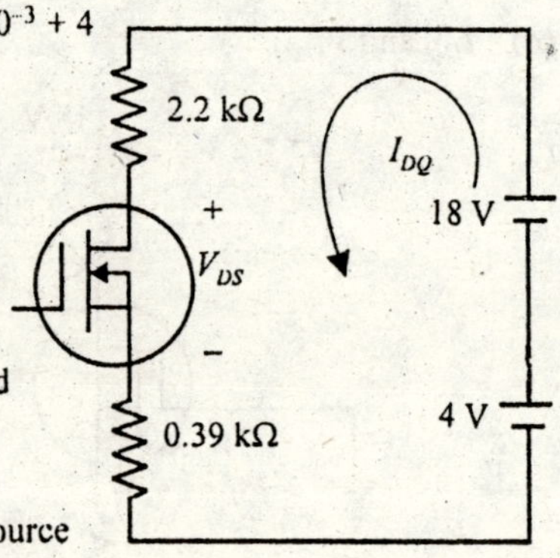

Fig. 5.16 (a)

$18 + 4 = 0.39\text{ K } I_D + V_{DS} + 2.2\text{ K } I_D$

$\Rightarrow \quad V_{DS} = 22 - 2.59\text{ K } I_D$

$\Rightarrow \quad V_{DS} = 22 - 2.59\text{ K} \times 9$ mA

$\Rightarrow \quad V_{DS} = 22 - 23.31$

$\Rightarrow \quad V_{DS} = -1.31$ V

So $V_D = V_S + V_{DS}$

$= -0.4878 - 1.31 = -1.798$ V

Therefore, we have

(*a*) $I_{DQ} = 9$ mA and $V_{GSQ} = 0.4878$ V

(*b*) $V_{DS} = -1.31$ V and $V_D = -1.798$ V and $V_S = -0.4878$ V

Problem 5.17. For the voltage divider configuration of enhancement type MOSFET, determine:

(*a*) I_{DQ} and V_{GSQ}

(*b*) V_D and V_S

$V_{GS\,(th)} = 3$ V

$I_{D\,(on)} = 5$ mA

$V_{GS\,(on)} = 6$ V

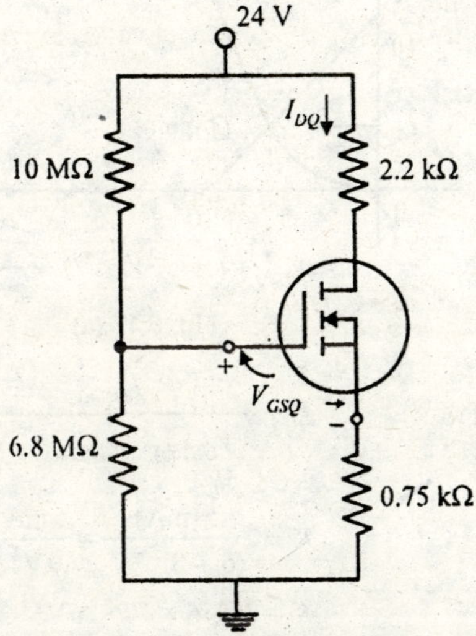

Fig. P. 5.17

Solution. Now, we get V_G as

$$V_G = \frac{R_2 V_{DD}}{R_1 + R_2} = \frac{6.8(\text{M}\Omega) \times 24\,\text{V}}{10\,\text{M}\Omega + 6.8\,\text{M}\Omega}$$

$\Rightarrow \quad V_G = 9.714$ V

Now $\quad V_{GS} = V_G - I_D R_S = 9.714\ \text{V} - I_D\ (0.75\ \text{k}\Omega)$

When $\quad I_D = 0$ mA, $V_{GS} = 9.714$ V

and when $\quad V_{GS} = 0$ V, $I_D = \dfrac{9.714\,\text{V}}{0.75\,\text{k}\Omega} = 12.952$ mA

which is plotted in the graph below.

For the device

$$V_{GS\,(th)} = 3\ \text{V},\ I_{D\,(on)} = 5\ \text{mA and } V_{GS\,(on)} = 6\ \text{V}.$$

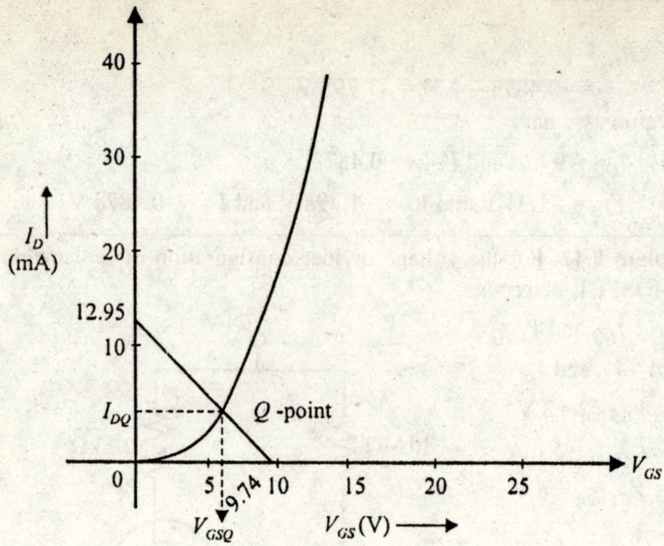

Fig. 5.17 (a)

Therefore
$$K = \frac{I_{D(\text{on})}}{V_{GS(\text{on})} - V_{GS(th)}{}^2 \text{ V}^2}$$

$$\Rightarrow \quad K = \frac{5(\text{mA})}{(6-3)^2 \text{ V}^2} = \frac{5 \text{ mA}}{9 \text{ V}^2}$$

$$\Rightarrow \quad K = 0.556 \times 10^{-3} \text{ A/V}^2$$

Thus, the drain current is

$$I_D = K\,[V_{GS} - V_{G(\text{th})}]^2$$

$$\Rightarrow \quad I_D = 0.556 \times 10^{-3}\,(V_{GS} - 3)^2$$

Substituting different values of V_{GS} and I_D, we plot the graph.

when $V_{GS} = 4$, $I_D = 0.556$ mA

$V_{GS} = 5$, $I_D = 2.224$ mA

$V_{GS} = 10$, $I_D = 27.25$ mA

$V_{GS} = 15$, $I_D = 80$ mA.

The point of intersection of straight line and the curve gives the quiecent operating point. The corresponding values are

$$V_{GSQ} \cong 6.3 \text{ V}$$

and $$I_{DQ} \cong 6 \text{ mA}$$

Thus, to calculate V_{DS}, we apply KVL in the drain-source loop

$$V_{DD} = V_{DS} + I_D(R_S + R_D)$$

$$\Rightarrow \quad V_{DS} = 24 - 6 \text{ mA } (2.2 + 0.75) \text{ k}\Omega$$

$$\Rightarrow \quad V_{DS} = 24 - 17.7 = 6.3 \text{ V}$$

Now $\quad V_S = I_D R_S = (0.75 \text{ k}\Omega) \times (6 \text{ mA}) = 4.5 \text{ V}$

Therefore $\quad V_D = V_{DS} + V_S = 6.3 \text{ V} + 4.5 \text{ V}$

and $\quad = 10.8 \text{ V}$

Thus, we get

(*a*) $I_{DQ} = 6$ mA and $V_{GSQ} = 6.3$ V

(*b*) $V_D = 10.8$ V and $V_S = 4.5$ V

Problem 5.18. Design a self-bias network using a JFET transistor with $I_{DSS} = 8$ mA and $V_P = -6$ V to have a *Q*-point at $I_{DQ} = 4$ mA using a supply of 14 V. Assume that $R_D = 3\,R_S$ and use standard values.

Solution. The network of a self-bias configuration is drawn below.

Apply KVL in the drain-source loop

$$14 = I_D(R_D + R_S) + V_{DS}$$

$$\Rightarrow \quad V_{DS} = 14 - I_D(4R_S)$$

$$\Rightarrow \quad V_{DS} = 14 - (4 \text{ mA} \times 4R_S)$$

$$= 14 - 16\,R_S$$

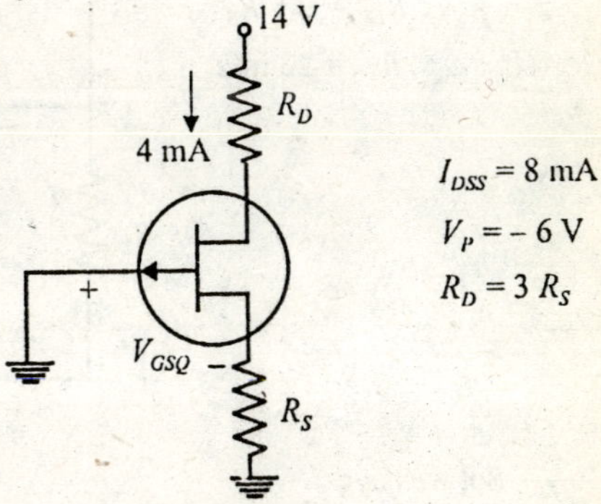

Fig. 5.18

To get V_{GSQ}, we have

$$I_{DQ} = I_{DSS}\left(1 - \frac{V_{GSQ}}{V_P}\right)^2$$

$$\Rightarrow \quad 4 \text{ mA} = 8 \text{ mA}\left(1 - \frac{V_{GSQ}}{(-6)}\right)^2$$

$$\Rightarrow \quad \sqrt{\frac{1}{2}} = 1 + \frac{V_{GSQ}}{6}$$

$$\Rightarrow \qquad V_{GSQ} = 6\;\; 1 - \frac{1}{\sqrt{2}} \;\; = 1.757 \text{ V}$$

So $\qquad V_{GSQ} = -I_{DQ} R_S$

$$\Rightarrow \qquad R_S = \frac{V_{GSQ}}{I_{DQ}} = \frac{1.757 \text{ V}}{4 \text{ mA}} = 0.44 \text{ k}\Omega$$

Therefore $\qquad R_D = 3R_S = 3 \times 0.44 = 1.32 \text{ k}\Omega$

Thus, the standard values of

$$R_D \cong 1.33 \text{ k}\Omega \text{ and } R_s \cong 0.43 \text{ k}\Omega$$

Problem 5.19. Design a voltage divider bias network using a depletion type MOSFET with I_{DSS} = 10 mA and V_P = – 4V to have a Q-point at I_{DQ} = 2.5 mA using a supply of 24 V. In addition set V_G = 4 V and use $R_D = 2.5\ R_S$ with R_1 = 22 MΩ. Use standard values.

Solution. The configuration of a voltage divider bias network is drawn below:

Given

I_{DQ} = 2.5 mA

V_{DD} = 24 V

V_G = 4 V

R_D = 2.5 R_S

R_1 = 22 mΩ

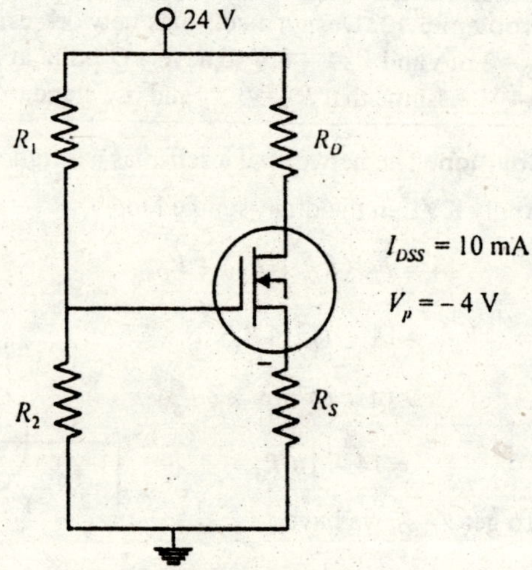

Fig. 5.19

So, we have

$$V_G = \frac{R_2}{R_1 + R_2} V_{DD} = \frac{R_2}{(22 + R_2)} \times 24$$

$$\Rightarrow \qquad 4 = \frac{24\ R_2}{22 + R_2} \qquad \Rightarrow 88 + 4\ R_2 = 24\ R_2$$

$$\Rightarrow \qquad 20\ R_2 = 88$$

$$R_2 = \frac{88}{20} = 4.4\ \text{M}\Omega$$

Now $\qquad V_{GS} = V_G - I_{DQ} R_S$

There $\qquad R_S = \dfrac{V_G - V_{GS}}{I_{DQ}} \qquad\qquad V_{GS} = V_P\left(1 - \sqrt{\dfrac{I_{DQ}}{I_{DSS}}}\right)$

$$= \frac{(4-2)\text{ V}}{2.5\text{ mA}} \qquad\qquad = 4\left(1 - \sqrt{\frac{2.5}{10}}\right)$$

$$= 0.8\ \text{k}\Omega \qquad\qquad = 2\text{ V}$$

Therefore $\qquad R_D = 2.5,\ \ R_S = 2.5 \times 0.8\ \text{k}\Omega = 2\ \text{k}\Omega$

Thus, the standard values, are

$$R_2 = 4.4\ \text{M}\Omega,\ R_S = 0.7\ \text{k}\Omega,\ R_D = 2.2\ \text{k}\Omega$$

Problem 5.20. Design a network enhancement type MOSFET with $V_{GS(th)} = 4$ V, $K = 0.5 \times 10^{-3}$ A/V^2 to have a Q-point of $I_{DQ} = 6$ mA. Use supply of 16 V and standard values.

Solution. The enhancement type MOSFET network is drawn

$$I_{DQ} = 6\text{ mA}$$

$$V_{DD} = 16$$

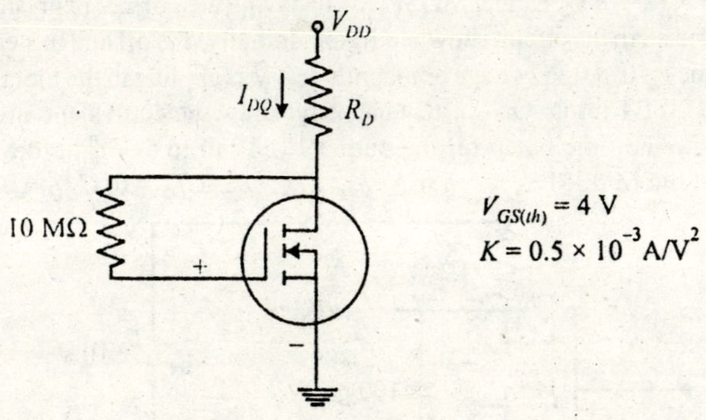

Fig. 5.20

The drain current is

$$I_{DQ} = K(V_{GS} - V_{GS_{th}})^2$$

$$(6\text{ mA}) = 0.5 \times 10^{-3}\ (V_{GS} - 4)^2$$

$$\frac{6\,\text{mA}}{0.5 \times 10^{-3}} = (V_{GS} - 4)^2$$

$$12 = (V_{GS} - 4)^2$$

$$V_{GS} = 4 + \sqrt{12} = 7.464 \text{ V}$$

Therefore $\quad V_{DS} = V_{GS} = 7.464 \text{ V}$

Now $\quad R_D = \dfrac{V_{DD} - V_{DS}}{I_{DQ}} = \dfrac{(16 - 7.464)\text{ V}}{6 \text{ mA}} = 1.423 \text{ k}\Omega$

Thus, a standard value of $R_D = 1.5$ kW.

Problem 5.21. In a test on JFET at a constant gate voltage of – 1 V, an increase in drain voltage from 8 to 10 V caused the drain current to increase from 2.8 to 3 mA. The drain voltage was then held constant and when the gate voltage was increased from – 1 to – 0.5 V the drain current changed from 3 to 4.8 mA. Estimate the values of y_{fs} and y_{os} of the FET.

Solution.

$$r_d = \left.\frac{\Delta V_{DS}}{\Delta I_D}\right|_{V_{GS} = \text{constant}} = \frac{10 - 8}{3 - 2.8} = \frac{2}{0.2 \times 10^{-3}} = 10 \times 10^3$$

$$y_{os} = \frac{1}{10 \times 10^3} = 0.1 \text{ ms}$$

$$y_{fs} = g_m = \left.\frac{\Delta I_D}{V_{GS}}\right|_{V_{DS} = \text{constant}} = \frac{4.8 - 3}{|-1 - (-0.5)|} = \frac{1.8}{0.5} = 3.6 \text{ ms}$$

Problem 5.22. An n-channel MOSFET (T) having a V_T of 2 V [Threshold voltage], is used in the circuit shown in figure. Initially, T is off and in steady state. At time $t = 0$, a step voltage of magnitude 4 V is applied to the input, so that the MOSFET turns 'ON' instantaneously. Draw the equivalent circuit and calculate the time taken for the output V_o to fall to 5 V. The device constant of the MOSFET, $K = 5 \text{ mA/V}^2$, $R_{DS} = \infty$, $C_{DS} = 0$, $C_{DG} = 0$.

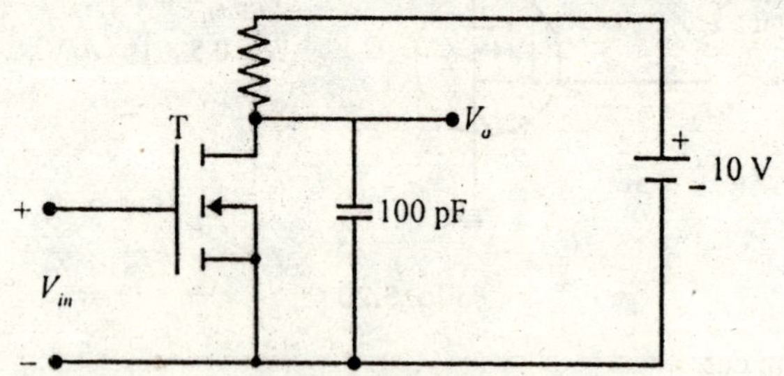

Fig. P. 5.22

Solution. The equivalent circuit is

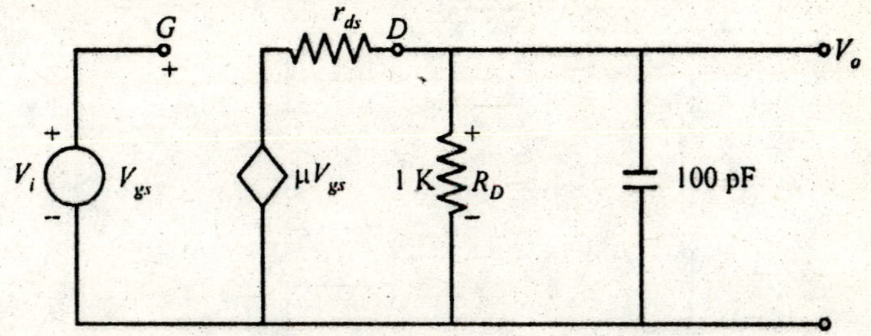

Fig. 5.22 (a)

$$I_D - K(V_{GS} - V_T)^2 - 5(4-2)^2 = 20 \text{ mA}$$

The transistor is off when $V_{GS} < V_T = 2$ V.

So the full voltage of 10 V is on the capacitor by charging. When $V_{GS} = 4 \text{ V} > V_T = 2$ V, the MOSFET is turned on and its resistance becomes very low, $\dfrac{1}{g_m} = \dfrac{1}{2K(V_{GS} - V_T)} = 50\ \Omega$. The low r_{ds} being in shunt with R_D, the capacitor discharges through the transistor.

In the $I_{DS} - V_{DS}$ characteristics of MOSFET, the saturation region is from $V_{GS} - V_T = 4 - 2 = 2$ V and below 2 V, it is the triode or resistor region. So for 5 V, the transistor is still in the saturation region. The capacitor voltage drops in this region by the deposition of charge of opposite polarity on the capacitor by the current I_{DS}.

$$C = \frac{Q}{V} = \frac{It}{V}$$

$$V_C(t) = V_{CO} - \frac{It}{C} = 10 - \frac{20 \times 10^{-3} t}{100 \times 10^{-12}}$$

$$t = \frac{(10-5) \times 100 \times 10^{-12}}{20 \times 10^{-3}} = 25 \text{ ns}$$

Problem 5.23. Calculate the Q-point.

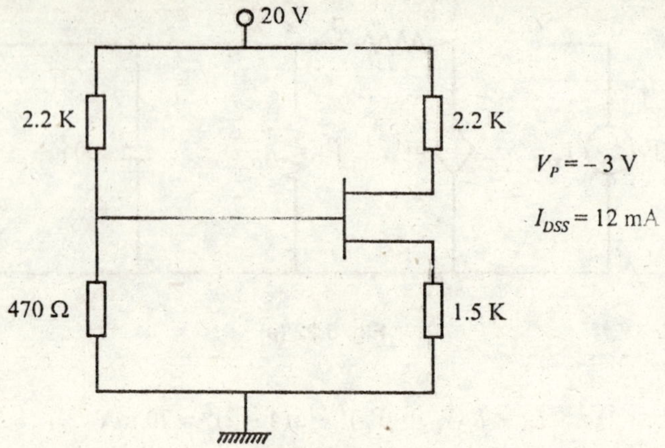

Fig. P. 5.23

Solution.

$$V_{GG} = 20 \times \frac{470}{2670} = 3.52 \text{ V}$$

$$R_G = \frac{0.47 \times 2.2}{2.67} \times 10^3 = 387.27\ \Omega$$

$$I_D = 12 \times 10^{-3} \quad 1 - \frac{[3.52 - I_D(1.5\text{ K})]}{(-3)}^{\,2}$$

$$= 12 \times (1 + 1.173 - 0.5\, I_D)^2$$

$$= 12\,(4.722 - 2.173\, I_D + 0.25\, I_D^2)$$

$$= 56.664 - 27.076\, I_D + 3\, I_D^2 = 0$$

$$I_D = \frac{27.076 \pm \sqrt{733.11 - 680}}{6} = \frac{27.076 \pm 7.28}{6}$$

$$= 5.726 \text{ mA or } 3.299 \text{ mA} \Rightarrow I_D \cong 3.3 \text{ mA}$$

$$V_D = 20 - 5.726 \times 3.7$$

$$= 20 - 21.186 = -1.186 \text{ V (inadmissible)}$$

$$\therefore \quad V_D = 20 - 3.299 \times 3.7$$

$$= 7.79 = 7.8 \text{ V}$$

Problem 5.24. A FET is used as a simple *CS* amplifier. The gain of the amplifier is found to be 40 and 30 with load resistances of 80 kΩ and 40 kΩ respectively. Estimate the drain resistance and trans-conductance of the transistor.

Solution.

$$A_v = g_m \, (R_D \parallel R_L)$$

$$R_D \parallel R_L = \frac{80\, R_D}{80 + R_D}$$

$$40 = g_m \times \frac{80\, R_D}{80 + R_D}$$

$$30 = g_m \times \frac{40\, R_D}{40 + R_D}$$

$$\frac{40}{30} = \frac{80\, R_D}{80 + R_D} \times \frac{40 + R_D}{40\, R_D} = \frac{80 + 2\, R_D}{80 + R_D}$$

$$4(80 + R_D) = 3(80 + 2\, R_D)$$

$$6\, R_D - 4\, R_D = 320 - 240 = 80$$

$$R_D = 40 \text{ K}$$

Now

$$40 = g_m \left(\frac{80 \times 40}{120}\right)$$

$$g_m = \frac{120 \times 40}{80 \times 40} = \frac{120}{80}$$

$$g_m = 1.5 \text{ ms}$$

Reader's Notes

CHAPTER

6

FET-Small Signal Analysis

Introduction

FET amplifier provide an excellent voltage gain with the added feature of high input, impedance. They are also considered low power consumption configuration with good frequency range and minimal size and weight. Both JFET and depletion MOSFET devices can be used to design amplifiers having similar voltage gains. The depletion, MOSFET circuit, however, has a much higher input impedance than a similar JFET configuration.

FET Small Signal Model: The a.c. analysis of FET configuration requires that a small signal a.c. model for the FET be developed. The gate-to-source voltage controls the drain-to-source (channel) current of an FET.

$$\Delta I_D = G_m \ \Delta V_{GS}$$

Where ΔI_D = change in drain-to-source current

ΔV_{GS} = change in gate-to-source voltage

g_m = transconductance

$$\therefore \quad \boxed{g_m = \frac{\Delta I_D}{\Delta V_{GS}}}$$

Also $\boxed{g_m = \frac{2I_{DSS}}{|V_P|}\left[1 - \frac{V_{GS}}{V_P}\right]}$ this equation is obtained by differentiating the Schokley's equation with respect to V_{GS}.

Now, when $V_{GS} = 0$ V $\boxed{g_m = \frac{2I_{DSS}}{|V_P|}}$

$$\therefore \quad \boxed{g_m = g_{m_o}\left[1 - \frac{V_{GS}}{V_P}\right]}$$

From the specification sheets

$$g_m = Y_{fs}$$

FET Input Impedance Z_i: The input impedance of FET is sufficiently assume that the input terminals approximate an open circuit

$$\therefore \quad Z_i(\text{FET}) = \infty\ \Omega$$

practical values for JFET is of $10^9\ \Omega$ and for MOSFET is of $10^{15}\ \Omega$.

FET output impedance (Z_o): The output impedance is defined as

$$Z_o(\text{FET}) = r_d = \frac{1}{Y_{OS}}$$

where, Y_{OS} is a component of an admittance equivalent circuit.

Also $$r_d = \left.\frac{\Delta V_{DS}}{\Delta I_D}\right|_{V_{GS}} = \text{constt.}$$

FET a.c. equivalent circuit: The a.c. equivalent circuit is drawn using input impedance, output impedance g_m. etc.

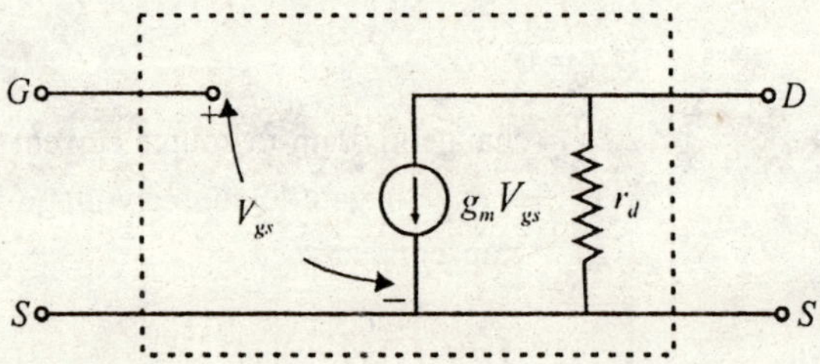

Fig. 6.1

Points to remember

(*i*) On specification sheets, g_m is provided as Y_{fs}.

(*ii*) When V_{GS} is one-half of the pinch off value, g_m is one-half the maximum value.

(*iii*) When I_D is one-fourth the saturation level of I_{DSS}, g_m is one-half the value at saturation.

(*iv*) On specification sheets, the output impedance r_d is provided as $1/Y_{oS}$.

(*v*) The voltage gain for the fixed bias and self-bias JFET is the same (with bypass source capacitance).

(*vi*) The a.c. analysis of JFET and depletion type MOSFET is the same.

(*vii*) The a.c. analysis of an echancement type MOSFET is the same as that of depletion type MOSFET. The only difference is the equation for g_m.

(*viii*) There is no phase shift between input an output for the source follower and common gate configuration. Most others have a 180° phase shift.

(*ix*) The input impedance of FET is quite high. However, it is quite low for the common gate configuration.

Some Important Equations

JFET and MOSFET (r_d^3 10 R_D, r_d^3 10 R_S)

(*a*) Fixed bias

$$Z_i = R_G \quad \text{(Input impedance)}$$

$$Z_o \cong R_D \quad \text{(Output impedance)}$$

$$A_v = -g_m R_D \quad \text{(Voltage gain)}$$

(*b*) Self-bias (Bypassed R_s)

$$Z_i = R_G$$

$$Z_o \cong R_D$$

$$A_v = -g_m R_D$$

(*c*) Self-bias (unpassed R_s)

$$Z_i = R_G \qquad Z_o = R_D$$

$$A_v \cong \frac{-g_m R_D}{1 + g_m R_S}$$

(*d*) Voltage divider bias

$$Z_i = R_1 \parallel R_2 \qquad Z_o = R_s$$

$$A_v = -g_m R_D$$

(*e*) Source followers

$$Z_i = R_G \qquad Z_o = R_s \parallel (1/g_m)$$

$$A_v = \frac{g_m R_s}{1 + g_m R_s}$$

(*f*) Common gate

$$Z_i = R_S \| (1/g_m) \qquad Z_i \cong R_D$$

$$AV = gm\ Rs$$

Enhancement type MOSFET ($R_F >> r_d \| R_d$, $r_d \geq 10\ R_D$)

$$g_m = 2K\ (V_{GSQ} - V_{GS(th)})$$

(*a*) Drain feedback configuration

$$Z_i = \frac{R_F}{1 + g_m + R_D} \qquad Z_o \cong R_D$$

$$A_v \cong -g_m R_D$$

(*b*) Voltage divider bias

$$Z_i = R_1 \| R_2 \qquad Z_o = R_D$$

$$A_V = g_m R_D$$

Problem 6.1. Calculate g_{mo} for a JFET having device parameters $I_{DSS} = 15$ mA and $V_P = -5$ V.

Solution. g_{mo} is the maximum possible value of g_m, which is given by

$$g_{mo} = \frac{I_{DSS}}{|V_P|} = \frac{2 \times 15 \times 10^{-3}}{5}$$

$\Rightarrow \qquad g_{mo} = 6 \times 10^{-3}$ s

$\Rightarrow \qquad g_{mo} = 6$ ms.

Problem 6.2. For JFET having device parameters $g_{mo} = 5$ ms, $V_P = -3.5$ V, what is the device current at $V_{GS} = 0$ V ?

Solution. The current of the device, i.e. drain current at $V_{GS} = 0$ is known as I_{DSS}.

$$g_{mo} = \frac{2\ I_{DSS}}{|V_P|} = \frac{2 \times I_{DSS}}{|V_P|}$$

$$\Rightarrow \qquad I_{DSS} = \frac{g_{mo} \times |V_P|}{2}$$

$$\Rightarrow \qquad I_{DSS} = \frac{5 \times 10^{-3} \times 3.5}{2} = 8.75 \times 10^{-3}\text{ A}$$

$$\Rightarrow \qquad I_{DSS} = 8.75\text{ mA}$$

Problem 6.3. For a JFET having $g_m = 6$ ms at $V_{GSQ} = -1$ V, what is the value of I_{DSS}, if $V_P = -2.5$ V.

Solution. The equation for g_m is given as

$$g_m = g_{mo}\left[1 - \frac{V_{GS}}{V_p}\right], \text{ where } g_{mo} = \frac{2 I_{DSS}}{|V_P|}$$

When $\qquad V_{GS} = -1\text{ V}, \; g_m = 6\text{ ms}$

$$6 \times 10^{-3} = g_{mo}\left[1 + \frac{1}{-2.5}\right]$$

$$\Rightarrow \qquad g_{mo} = \frac{6 \times 10^{-3}}{\left(1 - \frac{1}{2.5}\right)} = \frac{6 \times 10^{-3}}{0.6}$$

$$\Rightarrow \qquad g_{mo} = \frac{6 \times 10^{-3}}{6 \times 10^{-1}} = 10^{-2}\text{ s} = 10\text{ ms}$$

Now $\qquad g_{mo} = \dfrac{2I_{DSS}}{|V_P|}$

$$\Rightarrow \qquad I_{DSS} = \frac{g_{mo} \times |V_P|}{2}$$

$$\Rightarrow \qquad I_{DSS} = \frac{10 \times 10^{-3} \times 2.5}{2} = 12.5\text{ mA}$$

Problem 6.4. Determine the value of g_m for a JFET ($I_{DSS} = 8$ mA, $V_P = -5$ V), when biased at $V_{GSQ} = V_P/4$.

Solution. The equation for g_m is given as

$$g_m = g_{mo}\left(1 - \frac{V_{GS}}{V_P}\right), \quad g_{mo} = \frac{2\, I_{DSS}}{|V_P|}$$

$$\Rightarrow \qquad g_m = 3.2 \times 10^{-3}\left(1 - \frac{V_{P/4}}{V_P}\right) = \frac{2 \times 8 \times 10^{-3}}{5} = 3.2\text{ ms}$$

$$\Rightarrow \qquad g_m = 3.2 \times 10^{-3}\left(1 - \frac{1}{4}\right)$$

$$\Rightarrow \quad g_m = 3.2 \times 10^{-3} \times \frac{3}{4} \quad \Rightarrow \quad g_m = 2.4 \text{ ms}$$

Problem 6.5. For a JFET having specified values of Y_{fs} = 4.5 ms and Y_{os} = 25 μs, determine the output impedance, Z_o, (FET) and device ideal voltage gain A_v (FET).

Solution. We have $g_m = Y_{fs} = 4.5$ ms.

and $$r_d = \frac{1}{Y_{os}} = \frac{1}{25\ \mu s} = \frac{1}{25 \times 10^{-6}}\ \Omega$$

$$\Rightarrow \quad r_d = 40\ \text{k}\Omega$$

The a.c. equivalent model of FET is shown as below:

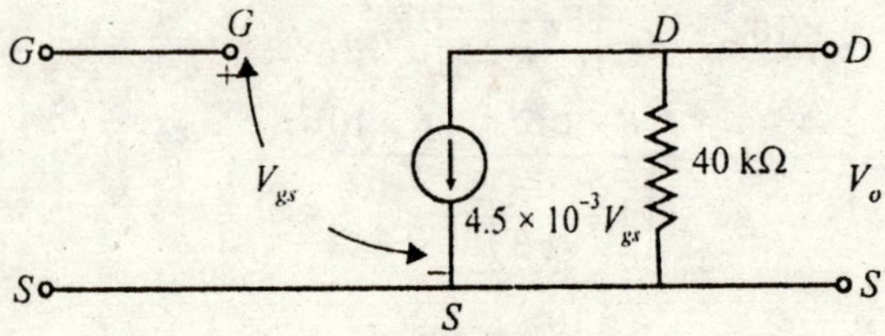

Fig. 6.5

The output impedance, Z_o is given as, Z_o = 40 kΩ

The output voltage, $V_o = -4.5 \times 10^{-3} V_{gs} \times 40$ kΩ

$$V_o = -4.5 \times 10^{-3} V_{gs} \times 40 \times 10^3$$

$$\Rightarrow \quad V_o = -4.5 \times 40 V_{gs}$$

$$\Rightarrow \quad \frac{V_o}{V_{gs}} = -180$$

$$\Rightarrow \quad \text{Voltage gain} = \frac{V_o}{V_{gs}} = -180$$

Problem 6.6. Using the transfer characteristic of Fig. P. 6.6, determine:

(*a*) The value of g_{mo}.

(*b*) g_m at $V_{GS} = -1.5$ V graphically.

(*c*) The value of g_m at $V_{GSQ} = 1.5$ V, by equation.

(*d*) g_m at $V_{GS} = -2.5$ V, graphically.

(*e*) The value of g_m at $V_{GSQ} = -2.5$ V, by equation compare with solution of *d*.

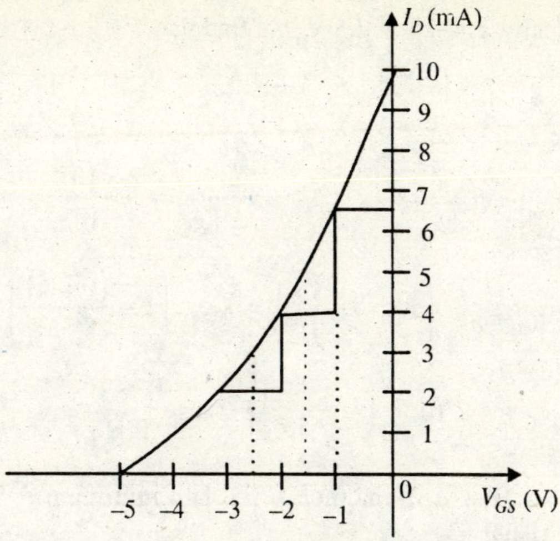

Fig. P. 6.6

Solution.

(*a*) $$g_{mo} = \frac{I_{DSS}}{|V_P|}$$

$I_D = I_{DSS}$ at $V_{GS} = 0$

$\therefore \quad I_{DSS} = 10$ mA

$V_P = V_{GS}$ at $I_D = 0$

i.e. the drain current is zero.

at $V_{GS} = V_P$

$\therefore \quad V_P = -5$ V

Put the values of $I_{DSS} = 10$ mA

and $V_P = -5$ V

$$g_{mo} = \frac{2 \times 10}{5}$$

$$g_{mo} = 4 \text{ ms}$$

(*b*) $$g_m = \frac{2\,I_{DSS}}{|V_P|}\left(1 - \frac{V_{GS}}{V_P}\right) = g_{mo}\left(1 - \frac{V_{GS}}{V_P}\right)$$

$$= 4 \times 10^{-3}\left(1 - \frac{(-1.5)}{-5}\right) = 4 \times 10^{-3}\left(1 - \frac{15}{50}\right) = 2.8 \text{ ms}$$

(*c*) Graphically, we have at $V_{GSQ} = -1.5$ V

$$\Delta I_D = 6.5 - 4 = 2.5 \text{ mA}$$

$$\Delta V_{gs} = 1.9 - 1 = 0.9 \text{ V}$$

$$\therefore \quad g_m = \frac{\Delta I_D}{\Delta V_{GS}} = \frac{2.5 \times 10^{-3}}{0.9} = 2.77 \times 10^{-3} = 2.8 \text{ ms}$$

(*d*) Graphically, at $V_{GS} = -2.5$ V, we find that

$$\Delta I_D = 4 - 2 = 2 \text{ mA}$$

$$\Delta V_{GS} = 3 - 2 = 1 \text{ V}$$

$$\therefore \qquad (1/g_m) = \frac{\Delta V_{GS}}{\Delta I_D} \Rightarrow \quad g_m = \frac{\Delta I_D}{\Delta V_{gs}} = \frac{2 \times 10^{-3}}{1} = 2 \text{ ms}$$

(*e*)

$$g_m = g_{mo}\left(1 - \frac{V_{GS}}{V_P}\right) = 4 \times 10^{-3}\left(1 - \frac{(-2.5)}{(-5)}\right)$$

$$= 4 \times 10^{-3}\left(1 - \frac{2.5}{50}\right) = 2 \text{ ms}$$

Problem 6.7. For a *n*-channel JFET [y_{fs} minimum = 750 μs, y_{os} (maximun) = 10 μs]

(*a*) What is the value of g_m?

(*b*) What is the value of r_d?

Solution.

(*a*) The value of $g_m = y_{fs} = 750$ μs = 0.75 ms.

(*b*) The value of $r_d = \dfrac{1}{y_{os}} = \dfrac{1}{10 \times 10^{-6}} = 100 \text{ k}\Omega$. The a.c. equivalent of FET

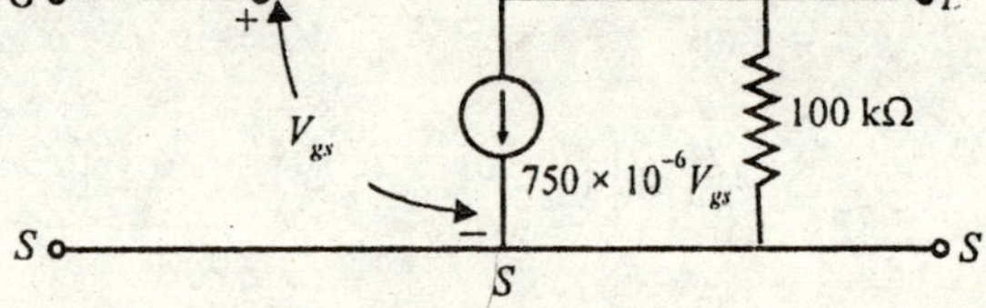

Fig. 6.7

Problem 6.8. Sketch the a.c. equivalent model of a JFET, if $Y_{fs} = 5.6$ ms and $Y_{os} = 15$ μs.

Solution. $g_m = y_{fs} = 5.6$ ms and $r_d = \dfrac{1}{y_{DS}}$

$$= \frac{1}{15 \times 10^{-6}} = \frac{1000 \times 10^3}{15} = 66.67 \text{ k}\Omega$$

The a.c. equivalent of JFET is shown below:

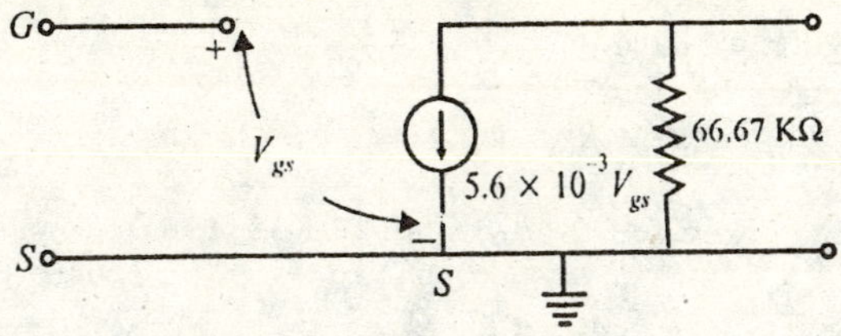

Fig. 6.8

Problem 6.9. Determine Z_i , Z_o and A_v for the network shown in Fig. P. 6.9, if I_{DSS} = 10 mA , V_P = – 4 V and r_d = 40 kΩ.

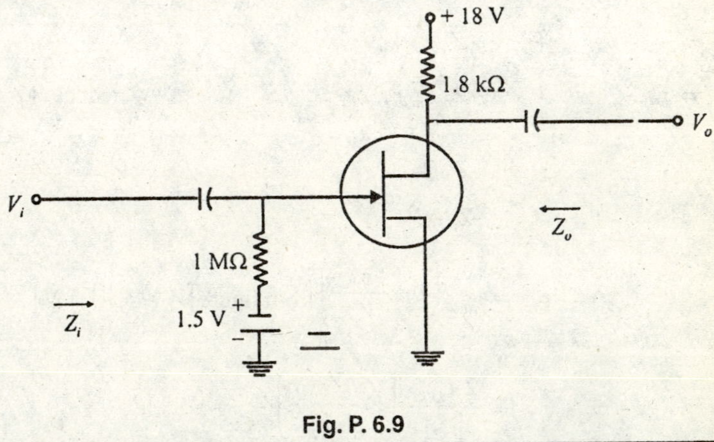

Fig. P. 6.9

Solution. Note that the two capacitors have the short-circuit equivalent, because the reactance $X_e = \dfrac{1}{2\pi fc}$ is sufficiently small compared to other impedance levels of the network, and the d.c. batteries V_{GG} and V_{DD} are also set zero volts by a short-circuit. Equivalent thus, the a.c. equivalent network for the given figure is drawn as below:

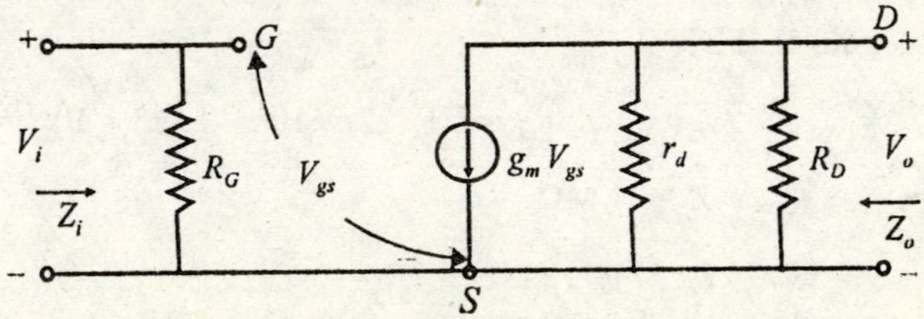

Fig. 6.9 (a)

$R_G = 1\ \text{m}\Omega$ $\quad r_d = 40\ \text{k}\Omega$ $\quad V_P = -4\text{V}$

$R_D = 1.8\ \text{k}\Omega$ $\quad I_{DSS} = 10\ \text{mA}$

From the figure, we have

$$Z_i = R_G = 1\text{m}\Omega$$

$$Z_o = r_d \parallel R_D = (40 \parallel 1.8)\ \text{k}\Omega$$

$$= \frac{40 \times 1.8}{41.8}\ \text{k}\Omega = 1.722\ \text{k}\Omega$$

The output voltage

$$V_o = -(r_d \parallel R_D)\, g_m V_{gs}$$

$$\Rightarrow \quad V_o = -Z_o\, g_m V_{gs}$$

$$\Rightarrow \quad V_o = -Z_o\, g_m V_i \qquad \text{(Since, } V_i = V_{gs}\text{)}$$

$$\Rightarrow \quad \frac{V_o}{V_i} = -Z_o g_m$$

$$\Rightarrow \quad A_v = -Z_o\, g_m$$

$$g_m = g_{mo}\left(1 - \frac{V_{GSQ}}{V_P}\right) \text{ from figure } [V_{GS} = -1.5\ \text{V}]$$

$$= \frac{2\, I_{DSS}}{|V_P|}\left(1 - \frac{V_{GSQ}}{V_P}\right)$$

$$= \frac{2 \times 10 \times 10^{-3}}{4}\left(1 - \frac{(-1.5)}{(-4)}\right)$$

$$= 5 \times 10^{-3}\left(1 - \frac{1.5}{4}\right) = 3.125 \times 10^{-3} = 3.125\ \text{ms.}$$

Hence, gain of JFET,

$$A_v = -Z_o\, g_m = -1.722 \times 10^3 \times 3.125 \times 10^{-3} = -5.38$$

Hence $\quad Z_i = 1\ \text{M}\Omega$

$\quad Z_o = 1.722\ \text{k}\Omega$

Voltage gain $\quad A_v = -5.38$

Problem 6.10. Determine Z_i , Z_o and A_v for the network shown in Fig. P. 6.10, if y_{fs} = 3000 μs and y_{os} = 50 μs.

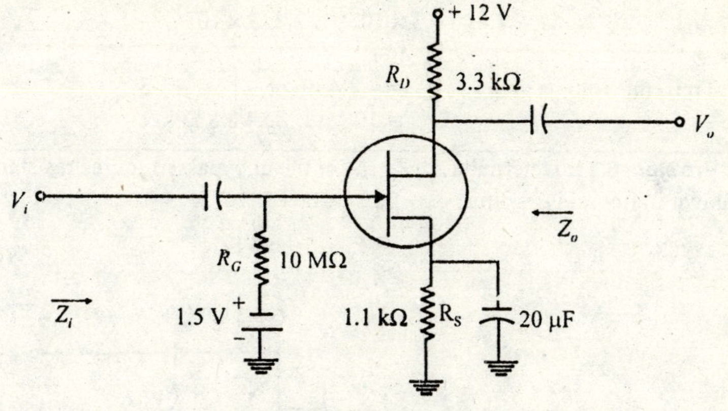

Fig. P. 6.10

Solution. From the figure, we have R_s is bypassed as the capacitor 20 μF is short circuit equivalent. The a.c. equivalent network for the above circuit is drawn below:

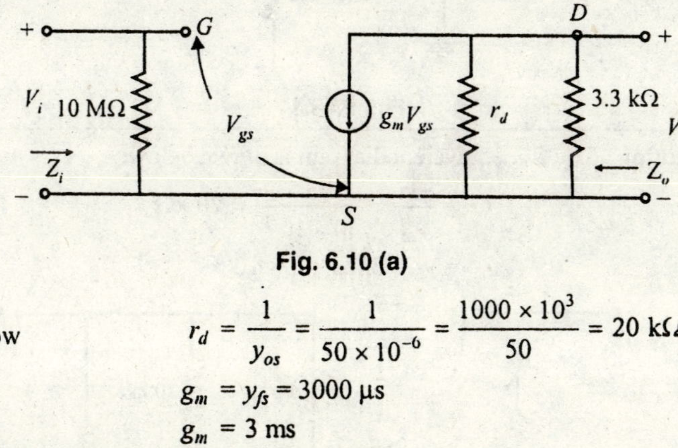

Fig. 6.10 (a)

Now $$r_d = \frac{1}{y_{os}} = \frac{1}{50 \times 10^{-6}} = \frac{1000 \times 10^3}{50} = 20 \text{ k}\Omega$$

$$g_m = y_{fs} = 3000 \text{ μs}$$

$\Rightarrow$ $$g_m = 3 \text{ ms}$$

From the circuit, we have

Input impedance Z_i = 10 MΩ

Output impedance $Z_o = r_d \parallel 3.3 \text{ K} = 20 \text{ K} \parallel 3.3 \text{ K}$

$$= \frac{20 \times 3.3}{23.3} \text{ k}\Omega = 2.833 \text{ k}\Omega$$

Output voltage $V_o = -g_m V_{gs} Z_o$

$\Rightarrow$ $V_o = -g_m Z_o V_i$ $\{\because V_i = V_{gs}\}$

$$\Rightarrow \qquad \frac{V_o}{V_i} = -g_m Z_o$$

$$\Rightarrow \qquad A_v = -3 \times 10^{-3} \times 2.833 \times 10^{-3}$$

$$\Rightarrow \qquad A_v = -8.499$$

Thus, the voltage gain $A_v = -8.499$ or $A_v \simeq -8.5$

$$Z_i = 10\ \text{M}\Omega,\ Z_o = 2.833\ \text{k}\Omega$$

Problem 6.11. Determine Z_i, Z_o, A_v for the unbypassed source resistance shown in figure y_{fs} = 3 ms, f_{os}= µs:

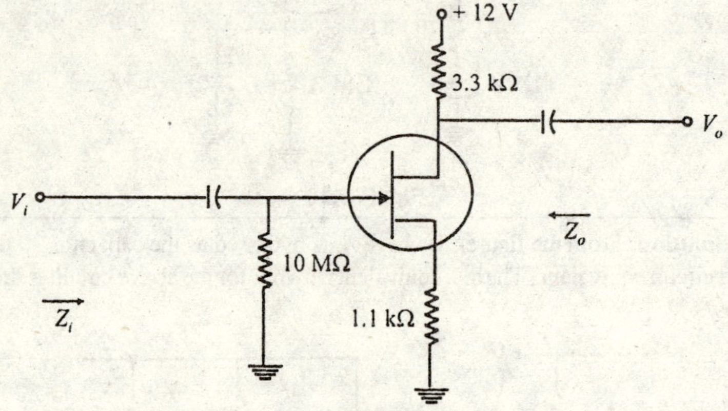

Fig. P. 6.11

Solution. The a.c. equivalent diagram is drawn below:

$$r_d = \frac{1}{y_{os}} = \frac{1}{50 \times 10^{-6}} = 20\ \text{k}\Omega$$

$$g_m = y_{fs} = 3\ \text{ms} = 3 \times 10^{-3}\ \text{s}$$

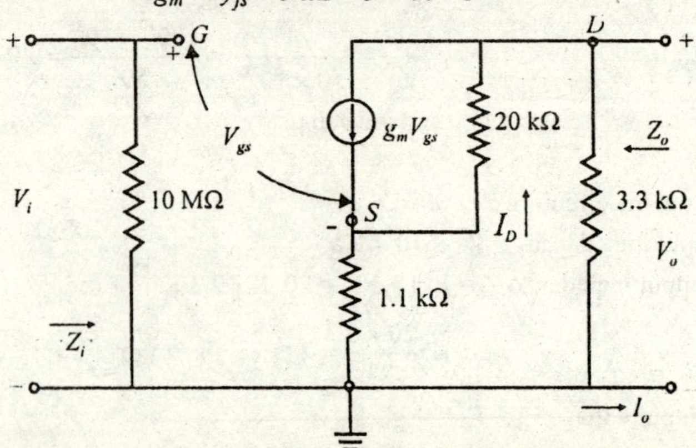

Fig. 6.11 (a)

The input impedance of the self-bias, JFET configuration is given

$$Z_i = 10\ \text{M}\Omega$$

The output impedance of the configuration

$$Z_o = \frac{\left[1 + g_m\ R_s + \dfrac{R_S}{r_d}\right]}{\left[1 + g_m\ R_s + \dfrac{R_S}{r_d} + \dfrac{R_D}{r_d}\right]} \times R_D$$

Here $g_m = 3 \times 10^{-3}$ s , $R_s = 1.1$ kΩ

$R_D = 3.3$ kΩ , $r_d = 20$ kΩ

$$Z_o = \frac{\left[1 + \left(3 \times 10^{-3} \times 1.1 \times 10^3\right) + \dfrac{1.1}{20}\right] \times 3.3\ \text{K}}{\left[1 + \left(3 \times 10^{-3} \times 1.1 \times 10^3\right) + \dfrac{1.1}{20} + \dfrac{3.3}{20}\right]}$$

$$\Rightarrow \quad Z_o = \frac{[1 + 3.3 + 0.055]}{[1 + 3.3 + 0.55 + 0.165]} \times 3.3\ \text{K}$$

$$\Rightarrow \quad Z_w = \frac{4.355}{4.52} \times 3.3\ \text{K} \quad \Rightarrow \quad Z_o = 3.179\ \text{k}\Omega$$

We get, the formula of output impedance is given by appearing Kirchoff's law

$$I_o = g_m\ V_{gs} + I_{rd} - I_D$$

$$V_{rd} = V_o + V_{gs}$$

$$\Rightarrow \quad I_o = g_m\ V_{gs} + \frac{V_o + V_{gs}}{r_d} - I_D$$

Now $V_{gs} = -(I_D + I_o)\ R_S$

$$I_o = -\left(g_m + \frac{1}{r_d}\right)(I_D + I_o) R_S - \frac{I_D\ R_D}{r_d} - I_D$$

$$I_o = \frac{-I_D\left[1 + g_m\ R_S + \dfrac{R_S}{r_d} + \dfrac{R_D}{r_d}\right]}{\left(1 + g_m\ R_S + \dfrac{R_S}{r_d}\right)}$$

Now $$Z_o = \frac{V_o}{I_o} = \frac{-I_D\ R_D}{I_o}$$

Put the values of the desired component, get the output impedance, Z_o.
The voltage gain is given by

$$\Rightarrow \qquad A_v = \frac{V_O}{V_I} = \frac{-g_m R_D}{1 + g_m R_s + \dfrac{R_D + R_S}{r_d}}$$

$$\Rightarrow \qquad A_v = -\frac{3 \times 10^{-3} \times 3.3 \times 10^3}{1 + (3 \times 10^{-3} \times 1.1 \times 10^3) + \dfrac{(3.3 + 1.1)\text{K}}{20\text{K}}}$$

$$= \frac{-9.9}{1 + 3.3 + \left(\dfrac{4.4}{20}\right)} = \frac{-9.9}{4.3 + 0.22} = 2.19$$

Problem 6.12. Determine Z_i, Z_o and V_o for the network shown in Fig. P. 6.12, if $V_i = 20$ mV.

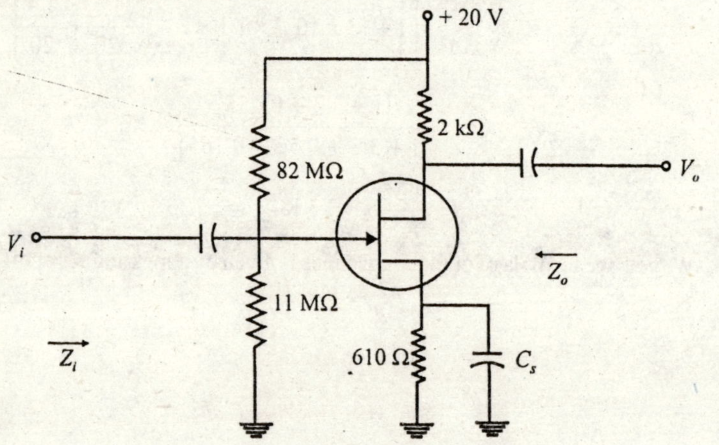

Fig. P. 6.12

$I_{DSS} = 12$ mA, $V_P = -3$ V, $r_d = 100$ kΩ, $V_{GSQ} = 0.925$ V

Solution. The a.c. equivalence of the above network is drawn below:

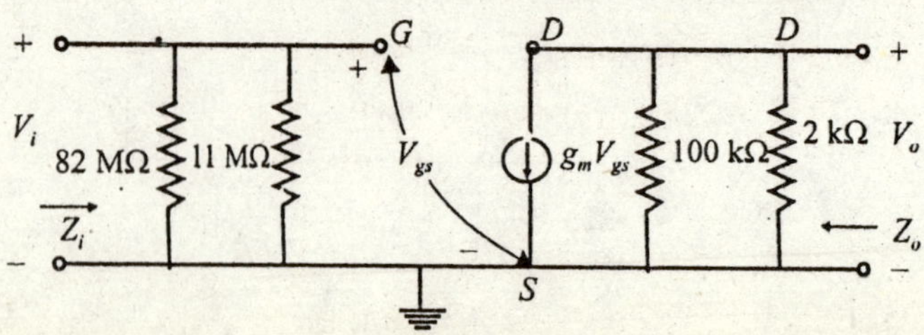

Fig. 6.12 (a)

The input impedance, Z_i is given as

$$Z_i = (82\ \text{M}\Omega) \parallel (\parallel \text{M}\Omega) = \frac{82 \times 11}{82 + 11}\ \text{M}\Omega$$

$$= 9.699\ \text{M}\Omega = 9.7\ \text{M}\Omega$$

The output impedance $\quad Z_o = (100\ \text{k}\Omega) \parallel (2\ \text{k}\Omega)$

$$\Rightarrow \quad Z_o = \frac{100 \times 2}{100 + 2} \quad \Rightarrow \quad Z_o = 1.96\ \text{k}\Omega$$

Voltage gain $\quad A_v = \dfrac{V_o}{V_i} = \dfrac{-g_m V_{gs} \times Z_o}{V_i} = \dfrac{-g_m Z_o V_i}{V_i}$

$$\Rightarrow \quad Av = -g_m Z_o$$

Now $\quad g_{mo} = \dfrac{2 I_{DSS}}{|V_P|} = \dfrac{2 \times 12}{3} = 8\ \text{ms}$

$$g_m = g_{mo}\left(1 - \frac{V_{GSQ}}{V_P}\right) = 8 \times 10^{-3}\left(\frac{1-(-0.9525)}{(-3)}\right)$$

$$= 5.46\ \text{ms}$$

The voltage gain $\quad A_v = -g_m Z_o = -5.46 \times 10^{-3} \times 1.96 \times 10^3 = -10.701$

Therefore, output voltage

$$V_o = A_v \times V_i = -10.701 \times 20 = -214.03\ \text{mV}$$

Problem 6.13. For a JFET voltage divider configuration determine Z_o, Z_i and V_o, if $V_i = 20$ mV and $r_d = 20$ kΩ, $I_{DSS} = 12$ mA, $V_P = -3$ V, $V_{GSQ} = -0.9525$ V.

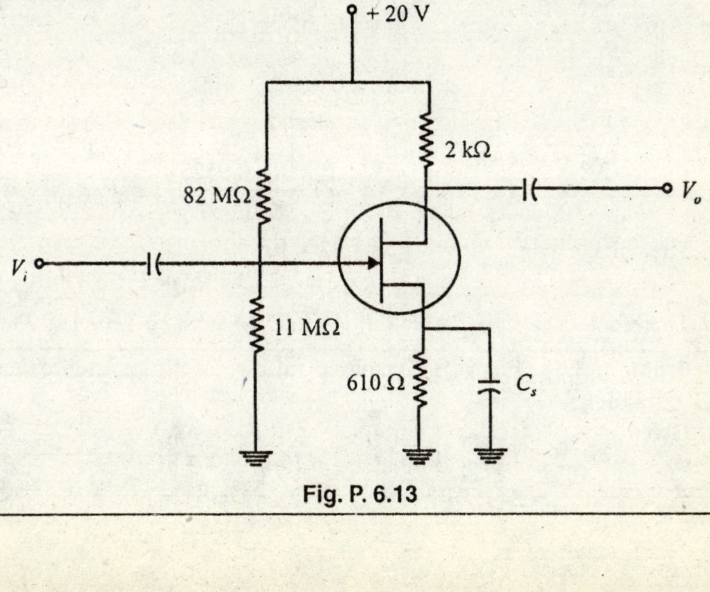

Fig. P. 6.13

Solution. The a.c. equivalent of the JFET voltage divider configuration is drawn below:

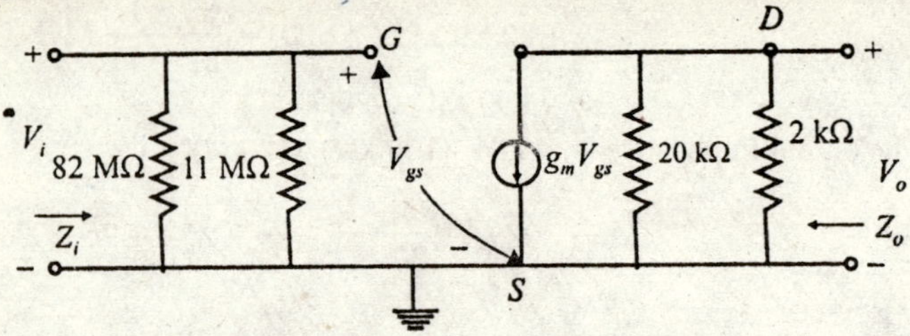

Fig. 6.13 (a)

The input impedance

$$Z_i = (82\ \text{M}\Omega) \parallel (11\ \text{M}\Omega)$$

$$= \frac{82 \times 11}{82 + 11} = 9.7\ \text{M}\Omega$$

The output impedance

$$Z_o = (20\ \text{k}\Omega) \parallel (2\ \text{k}\Omega) = \frac{20 \times 2}{22}\ \text{k}\Omega$$

$$= 1.818\ \text{k}\Omega \simeq 1.82\ \text{k}\Omega$$

Voltage gain $\quad A_v = \dfrac{V_o}{V_i} = \dfrac{-g_m V_{gs} (r_d \parallel R_D)}{V_{gs}}$

$\Rightarrow \quad A_v = -g_m Z_o$

$\Rightarrow \quad A_v = -g_m \times 1.82 \times 10^3$

Now, to calculate g_m, we have

$$g_{mo} = \frac{2 I_{DSS}}{|V_P|} = \frac{2 \times 12}{3} = 8\ \text{ms}$$

So $\quad g_m = g_{mo}\left(1 - \dfrac{V_{GSQ}}{V_P}\right)$

$$= 8 \times 10^{-3}\left(1 - \frac{(-0.9525)}{(-3)}\right) = 5.46\ \text{ms}$$

The voltage gain $\quad A_v = -g_m \times 1.82 \times 10^3$

$$= -5.46 \times 10^{-3} \times 1.82 \times 10^3 = -9.937$$

Therefore $\quad V_o = A_v \times V_i = -9.937 \times 20 \times 10^{-3} = -198.74\ \text{mV}.$

Problem 6.14. For a JFET source follower configuration calculate Z_i, Z_o and A_v:

Given $\quad I_{DSS} = 9\ \text{mA} \quad V_P = -4.5\ \text{V}$

$\quad r_d = 40\ \text{k}\Omega \quad V_{GSQ} = -2.818\ \text{V}$

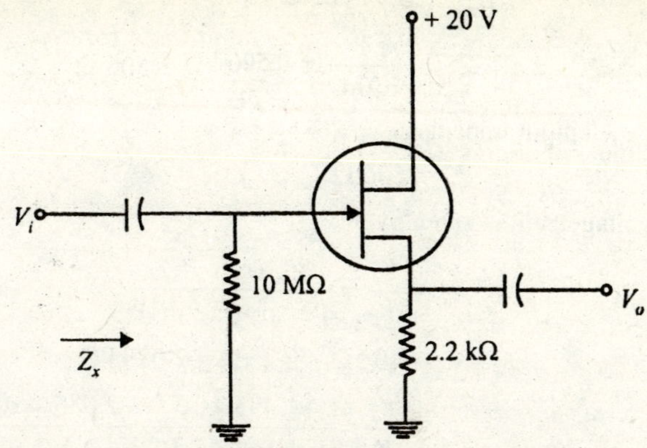

Fig. P. 6.14

Solution. For the JFET source followers configuration the small signal model is drawn below by having capacitors short-circuit equivalent. The JFET source follower is a common drain configuration from the figure, we have the input impedance.

$$Z_i = 10\ \text{M}\Omega$$

The output impedance is given by

$$Z_o = r_d \parallel 2.2\ \text{k}\Omega \parallel (1/g_m)$$

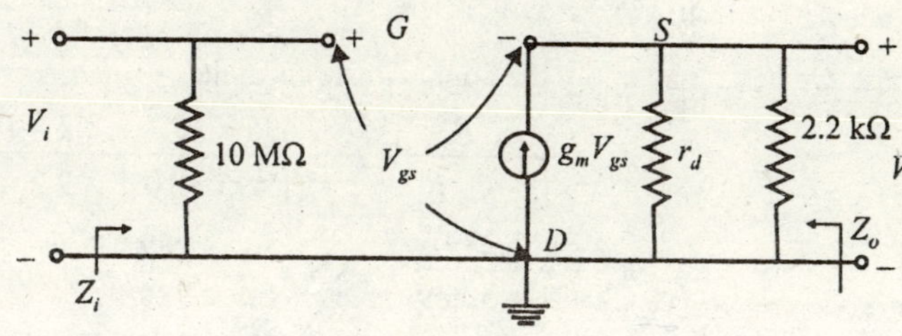

Fig. 6.14 (a)

Now $\quad r_d = 40\ \text{k}\Omega$

$$g_{mo} = \frac{2\ I_{DSS}}{|V_P|} = \frac{2 \times 9 \times 10^{-3}}{4.5} = 4\ \text{ms}$$

Now $$g_m = g_{mo}\left(1 - \frac{V_{GSQ}}{V_P}\right) = 4 \times 10^{-3}\left(1 - \frac{(2.818)}{(-4.5)}\right)$$

$$= 4 \times 10^{-3}\ (1 - 0.6262) = 1.495 \times 10^{-3} s = 1.495\ \text{ms}$$

$$Z_o = 40\ \text{k}\Omega \parallel 2.2\ \text{k}\Omega \parallel 668.8\ \Omega$$

$$\Rightarrow \quad Z_o^{-1} = (40 \times 10^3)^{-1} + (2200)^{-1} + (668.8^{-1})$$

$$\Rightarrow \qquad Z_o^{-1} = 1.9747 \text{ ms}$$

$$\Rightarrow \qquad Z_o = \frac{1}{1.9747} = 0.506 \text{ k}\Omega \simeq 506\ \Omega$$

Thus, the output impedance

$$Z_o = 506\ \Omega$$

The voltage gain is given by

$$A_v = \frac{V_o}{V_i} = \frac{g_m\ (r_d \parallel R_s)}{1 + g_m\ (r_d \parallel R_s)}$$

Now $\qquad r_d \parallel R_s = 40 \text{ k}\Omega \parallel 2.2 \text{ k}\Omega = 2.085 \text{ k}\Omega$

$$A_v = \frac{V_o}{V_i} = \frac{1.495 \times 10^{-3} \times 2.085 \times 10^3}{1 + (1.495 \times 10^{-3} \times 2.085 \times 10^3)}$$

$$= \frac{3.0075}{1 + 3.1175} = \frac{3.1175}{4.1175} = 0.7571$$

Another Method

It $10R_s \le r_d$, then $A_v = \dfrac{V_o}{V_i} = \dfrac{g_m R_s}{1 + g_m R_s}$

Here $\qquad r_d = 40 \text{ k}\Omega$ and $10\ R_s = 10 \times 2.2 \text{ k}\Omega = 22 \text{ k}\Omega$

Hence $\qquad 40 \text{ k}\Omega \ge 22 \text{ k}\Omega$

Thus, we have may apply the approximate formula

$$A_v = \frac{g_m R_s}{1 + g_m R_s} = \frac{1.495 \times 10^{-3} \times 2.2 \times 10^3}{1 + (1.495 \times 10^{-3} \times 2.2 \times 10^3)}$$

$$\Rightarrow \qquad A_v = \frac{3.289}{1 + 3.289} \quad \Rightarrow \quad A_v = \frac{3.289}{4.289} = 0.7668$$

Hence, we find that the voltage gain by approximate formula than the actual value calculated.

Problem 6.15. Determine the pinch-off a JFET with g_{mo} = 10 ms and I_{DSS} = 12 mA.

Solution. The relation between g_{mo} and pinch-off voltage (V_P) given by

$$g_{mo} = \frac{2\ I_{DSS}}{|V_P|}$$

$$|V_P| = \frac{2\ I_{DSS}}{g_{mo}} = \frac{2 \times 12 \times 10^{-3}}{10 \times 10^{-3}} = 2.4 \text{ volt}$$

For *n*-channel JFET $V_P = -2.4$ V

For *p*-channel JFET $V_P = -2.4$ V

Problem 6.16. Calculate the value of g_m for a JFET at a bias point of $V_{GS} = -1$ V. Given that $I_{DSS} = 10$ mA and $V_P = -3$ V.

Solution. The g_{mo} of the JFET is given by

$$g_{mo} = \frac{2\,I_{DSS}}{|V_P|} = \frac{2 \times 12 \times 10^{-3}}{3} = 8 \text{ ms}$$

Now
$$g_m = g_{mo}\left(1 - \frac{V_{GSQ}}{V_P}\right)$$

$$\Rightarrow \quad g_m = 8 \times 10^{-3}\left(1 - \frac{(-1)}{(-3)}\right)$$

$$\Rightarrow \quad g_m = 8 \times 10^{-3}\left(1 - \frac{1}{3}\right)$$

$$\Rightarrow \quad g_m = 8 \times 10^{-3} \times \frac{2}{3}$$

$$\Rightarrow \quad g_m = 5.33 \text{ ms}$$

Problem 6.17. A JFET ($I_{DSS} = 10$ mA, $V_P = -5$ V) is biased at $I_D = I_{DSS}/4$. What is the value of 'g_m' at that bias point.

Solution. The drain current I_D of a JFET is given by

$$I_D = \left(1 - \frac{V_{GS}}{V_P}\right)^2$$

$$\Rightarrow \quad \frac{I_{DSS}}{4} = I_{DSS}\;1 - \frac{V_{GS}}{V_P}^{\;2}$$

$$\Rightarrow \quad \sqrt{\frac{1}{4}} = \left(1 - \frac{V_{GS}}{V_P}\right)^2$$

$$\Rightarrow \quad \frac{V_{GS}}{V_P} = 1 - \frac{1}{2} = \frac{1}{2}$$

$$\Rightarrow \quad V_{GS} = \frac{1}{2}\,V_P = \frac{1}{2} \times -5$$

$$\Rightarrow \quad V_{GS} = -2.5 \text{ V}$$

$$g_{mo} = \frac{2\,I_{DSS}}{|V_P|}$$

$$\Rightarrow \quad g_{mo} = \frac{2 \times 10 \times 10^{-3}}{5}$$

$$\Rightarrow \quad g_{mo} = 4 \text{ ms}$$

Now $$g_m = g_{mo}\left(1 - \frac{V_{GS}}{V_P}\right)$$

$$= 4 \times 10^{-3}\left(1 - \frac{2.5}{5}\right) = 4 \times 10^{-3} \times \frac{1}{2} = 2 \text{ ms.}$$

Problem 6.18. A specification sheet provides the following data (at listed drain-source current) y_{fs} = 4.5 ms, y_{os} = 25 μS. At the listed drain-source current, determine:

(*a*) g_m (*b*) r_d?

Solution. $$g_m = y_{fs} = \frac{\Delta I_D}{\Delta V_{GS}} = 4.5 \text{ ms}.$$

$$r_d = \frac{1}{y_{os}} = \left.\frac{\Delta V_{DS}}{\Delta I_D}\right|_{V_{GS} = \text{constant}}$$

$$\Rightarrow \quad r_d = \frac{1}{25 \times 10^{-6}} = \frac{100 \times 10^4}{25}$$

$$\Rightarrow \quad r_d = 40 \text{ k}\Omega$$

(*a*) g_m = 4.5 ms (*b*) r_d = 40 kΩ

Problem 6.19. If a JFET having a specified value of r_d = 100 kΩ has an ideal voltage gain of A_v (FET) = – 200. What is the value of g_m.

Solution. The ideal voltage gain of a JFET is given by

$$A_v = -r_d g_m \Rightarrow A_v = g_m r_d$$

$$\Rightarrow \quad g_m = \frac{-A_v}{r_d}$$

$$\Rightarrow \quad g_m = \frac{-(-200)}{100 \text{ k}\Omega} \Rightarrow g_m = 2 \text{ ms}$$

Problem 6.20. Sketch the a.c. equivalent model for a JFET, if I_{DSS} = 10 mA, V_p = – 4 V, V_{GSQ} = – 2 V and y_{os} = 25 μs.

Solution. $$g_{mo} = \frac{2 I_{DSS}}{|V_P|} = \frac{2 \times 10 \times 10^{-3}}{4} = 5 \text{ ms}$$

$$g_m = g_{mo}\left(1 - \frac{V_{GSQ}}{V_P}\right) = 5 \times 10^{-3}\left(1 - \frac{(-2)}{(-4)}\right)$$

$$\Rightarrow \quad g_m = 5 \times 10^{-3}\left(1 - \frac{1}{2}\right) = 2.5 \text{ ms}$$

$$r_d = \frac{1}{y_{os}} = \frac{1}{25 \times 10^{-6}} = \frac{1000 \times 10^3}{25} = 40 \text{ k}\Omega.$$

Thus, the equivalent a.c. network is given by

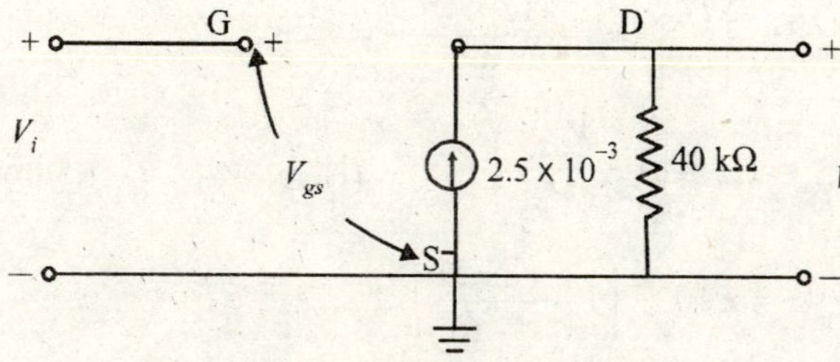

Fig. 6.20

Problem 6.21. Determine Z_i, Z_o and A_v for the network JFET. Fixed biased configuration, if $I_{DSS} = 12$ mA, $V_P = -6$ V and $y_{os} = 40$ μs:

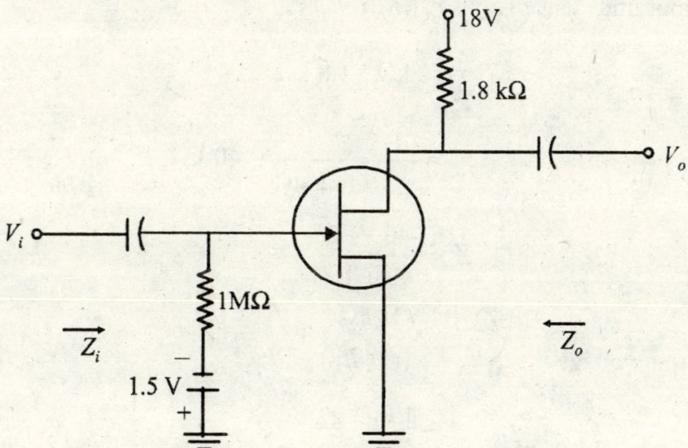

Fig. P. 6.21

Solution. The small signal model of the above configurations is drawn below:

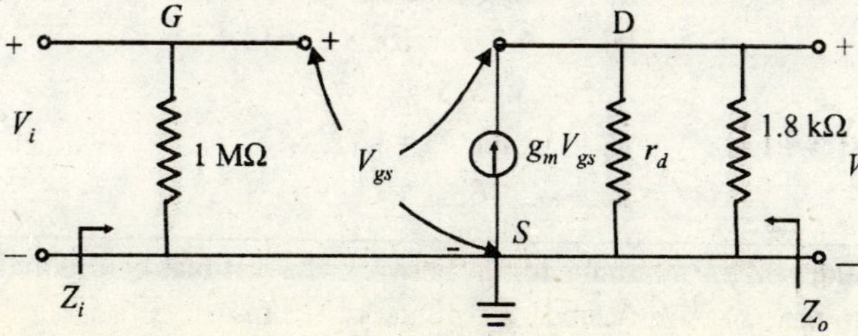

Fig. 6.21 (a)

From the figure, we have

Input impedance $Z_i = 1\ \text{M}\Omega$

$$g_{mo} = \frac{2\ I_{DSS}}{|V_P|} = \frac{2 \times 12 \times 10^{-3}}{6} = 4\ \text{ms}$$

Now $g_m = g_{mo} = \left(1 - \frac{V_{GSQ}}{V_P}\right)$ (Here, $V_{GSQ} = -1.5$ from figure)

$$\Rightarrow \quad g_m = 4 \times 10^{-3} \left(1 - \frac{(-1.5)}{(-6)}\right)$$

$$\Rightarrow \quad g_m = 4 \times 10^{-3} \left(1 - \frac{1}{4}\right)$$

$$\Rightarrow \quad g_m = 3\ \text{ms}$$

The output impedance $Z_o = r_d \parallel R_D$

$$\Rightarrow \quad Z_o = 40\ \text{k}\Omega \parallel 1.8\ \text{k}\Omega$$

Here $$r_d = \frac{1}{y_{os}} = \frac{1}{25\ \mu\text{s}} = 40\ \text{k}\Omega$$

$$\Rightarrow \quad Z_o = \frac{40 \times 1.8}{40 + 1.8}$$

$$\Rightarrow \quad Z_o = 1.722\ \text{k}\Omega$$

The output voltage $= V_o = -(r_d \parallel R_D)\, g_m\, V_{gs}$

$$\Rightarrow \quad \frac{V_o}{V_{gs}} = -(r_d \parallel R_D)\, g_m$$

$$\Rightarrow \quad A_v = \frac{V_o}{V_{gs}} = -Z_o g_m$$

Hence, the voltage gain

$$A_v = -Z_o\, g_m$$

$$\Rightarrow \quad A_v = -1.722 \times 10^3 \times 3 \times 10^{-3}$$

$$\Rightarrow \quad A_v = -5.1675$$

Thus, we have $Z_i = 1\ \text{M}\Omega,\ Z_o = 1.722\ \text{k}\Omega$

$$A_v = -5.168$$

Problem 6.22. Determine for the network of a self-bias configuration, if $I_{DSS} = 6$ mA, $V_P = -6$ V, and $y_{os} = 40\ \mu$s, $y_{fs} = 3$ ms:

(*a*) Z_i (*b*) Z_o (*c*) A_v

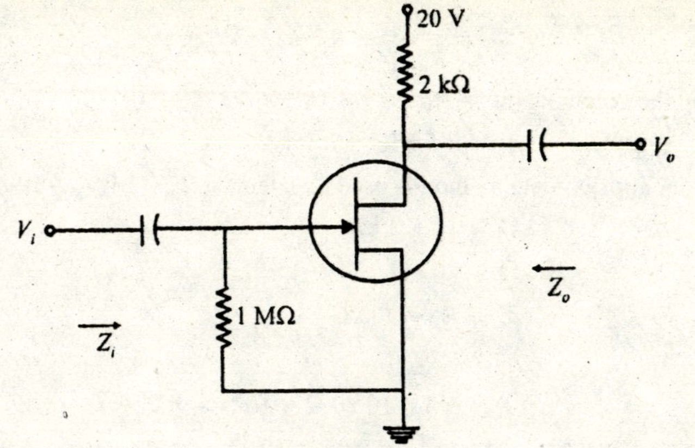

Fig. P. 6.22

Solution.

$$g_{mo} = \frac{2\,I_{DSS}}{|V_P|} = \frac{2 \times 6 \times 10^{-3}}{6} = 2 \text{ ms}$$

$$r_d = \frac{1}{y_{os}} = \frac{1}{40 \times 10^{-6}} = \frac{1000 \times 10^3}{40} = 25 \text{ k}\Omega$$

The input impedance $Z_i = 1 \text{ M}\Omega$.

The small model signal diagram of the selfbias configuration is shown below:

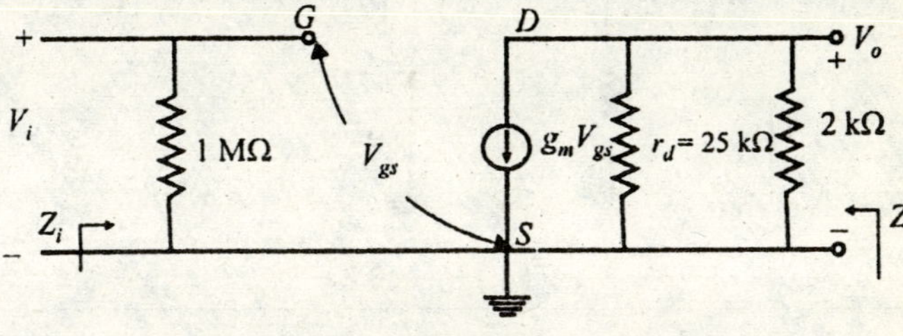

Fig. 6.22 (a)

The output impedance

$$Z_o = r_d \parallel R_D$$

$$\Rightarrow \quad Z_o = 25 \text{ k}\Omega \parallel 2 \text{ k}\Omega$$

$$\Rightarrow \quad Z_o = \frac{25 \times 2}{25 + 2} = 1.852 \text{ k}\Omega$$

$$g_m = g_{mo}\left(1 - \frac{V_{GSQ}}{V_P}\right) = y_{fs} = 3 \text{ ms}$$

The output voltage $V_o = -g_m V_{gs} (r_d \| R_D)$

$$\Rightarrow \quad A_v = \frac{V_o}{V_{gs}} = -g_m Z_o$$

Thus, the voltage gain $= A_v = -3 \times 10^{-3} \times 1.852 \times 10^3$

$$A_v = -5.556$$

Another approximate method is used to calculate Z_o, A_v, if $r_d \geq 10 R_D$.

$$25 \text{ k}\Omega > 10 \times 2 \text{ k}\Omega$$

$$\therefore \quad 25 \text{ k}\Omega > 20 \text{ k}\Omega$$

$$Z_o = R_D = 2 \text{ k}\Omega$$

$$A_v = -g_m R_D$$

$$= -3 \times 10^{-3} \times 2 \times 10^3 = -6$$

Problem 6.23. For the network of JFET common gate configuration determine: (*a*) input impedance Z_i, (*b*) output impedance z_o, (*c*) output voltage V_o

If $V_i = 10^{-4}$ V $\quad r_d = 25$ kΩ

$I_{DSS} = 8$ mA, $\quad V_P = -2.8$ V, $\quad V_{gsQ} = -1.75$ V

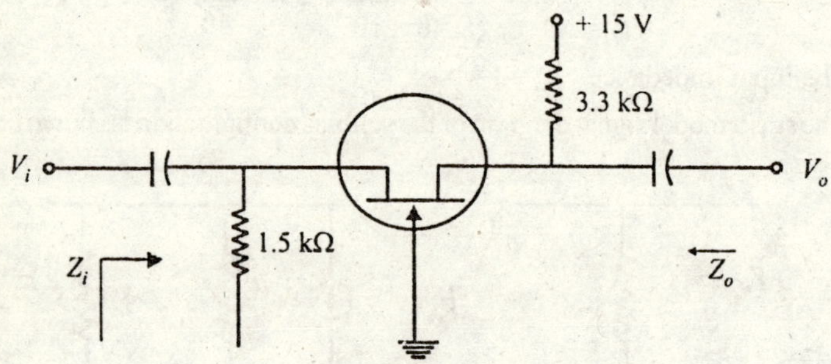

Fig. P. 6.23

Solution. $\quad r_d = 2.5$ kΩ

$$g_{mo} = \frac{2 I_{DSS}}{|V_P|} = \frac{2 \times 8 \times 10^{-3}}{2.8}$$

$$\Rightarrow \quad g_{mo} = 5.71 \text{ ms}$$

$$g_m = g_{mo}\left(1 - \frac{V_{GSQ}}{V_P}\right)$$

$$= 5.71 \times 10^{-3}\left(1 - \frac{(-1.75)}{(-2.8)}\right)$$

$$= 2.14 \times 10^{-3} = 2.14 \text{ ms}$$

The small signal model configuration is drawn below:

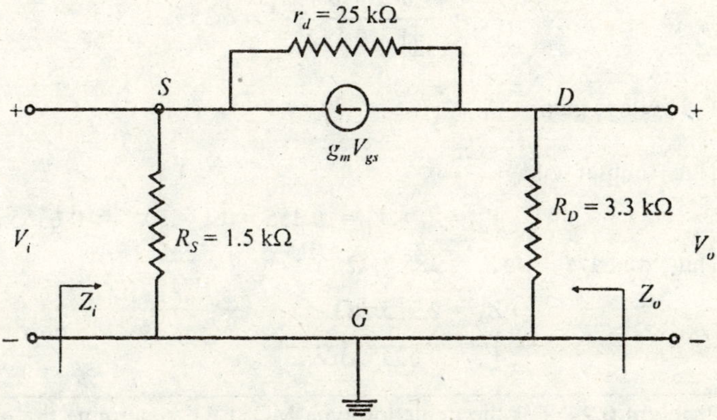

Fig. 6.23 (a)

The imput impedance $Z_i = R_s \parallel \left(\dfrac{r_d + R_D}{1 + g_m r_d}\right)$

Now $\dfrac{r_d + R_D}{1 + g_m r_d} = \dfrac{(25 + 3.3) \text{ k}\Omega}{1 + 2.14 \times 10^{-3} \times 25 \times 10^3}$

$$= \frac{28.3}{54.3} \text{ k}\Omega = 0.5192 \text{ k}\Omega$$

Now $Z_i = 1.5 \text{ k}\Omega \parallel 0.5192 \text{ k}\Omega$

$$= \frac{1.5 \times 0.5192}{1.5 + 0.5192} \text{ k}\Omega = 0.38569 \text{ k}\Omega = 385.7 \ \Omega$$

The output impedance $Z_o = r_d \parallel R_D = 25 \parallel 3.3 = \dfrac{25 \times 3.3}{28.3} \text{ k}\Omega = 2.915 \text{ k}\Omega$

The voltage gain $A_v = \dfrac{\left[g_m R_D + \dfrac{R_D}{r_d}\right]}{\left[1 + \dfrac{R_D}{r_d}\right]}$

$$\Rightarrow \qquad A_v = \frac{\left[2.14 \times 10^{-3} \times 3.3 \times 10^3 + \frac{3.3}{25}\right]}{\left[1 + \frac{3.3}{25}\right]}$$

$$\Rightarrow \qquad A_v = \frac{[7.062 + 0.132]}{[1 + 0.132]} = 6.355$$

The voltage gain $A_v = \frac{V_o}{V_i}$

Thus, output voltage

$$V_o = A_v \times V_i = 6.355 \times 0.1 \times 10^{-3} = 0.6355 \text{ mV}$$

Thus, we have $Z_i = 385.7 \ \Omega$

$Z_o = 2.915 \ \text{k}\Omega$

$V_o = 0.6355 \ \text{mV}$

Problem 6.24. For the depletion type MOSFET, determine the output voltage, if $y_{os} = 20$ μs:

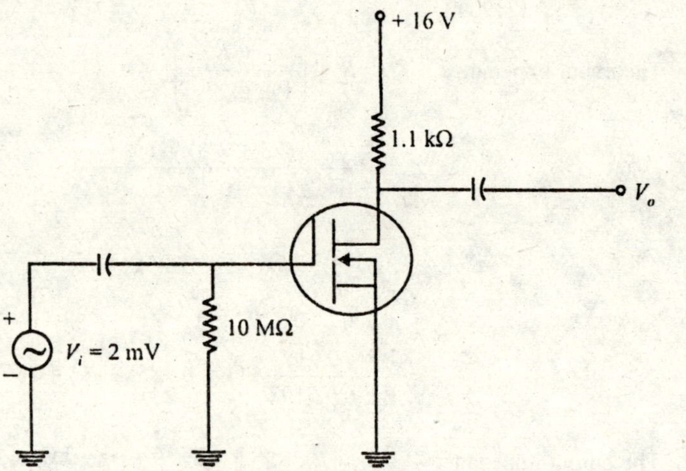

Fig. P. 6.24

$I_{DSS} = 8$ mA, $V_p = -3$ V, $V_{gs} = +1$ mV

Solution. $$r_d = \frac{1}{y_{os}} = \frac{1}{20 \times 10^{-6}} = \frac{1000 \times 10^3}{20} = 50 \text{ k}\Omega$$

$$g_{mo} = \frac{2\,I_{DSS}}{|V_P|} = \frac{2\times 8\times 10^{-3}}{3} = 5.33 \text{ ms}$$

$$g_m = g_{mo}\left(1-\frac{V_{gs}}{V_P}\right) = 5.33\times 10^{-3}\left(1-\frac{0.001}{-3}\right)$$

$$= 5.33\times 10^{-3}\,(1.000333) = 5.3318 \text{ ms}$$

The input impedance $Z_i = 10\ \text{M}\Omega$.

The small signal diagram of configuration shown in Fig. 6.24 (a):

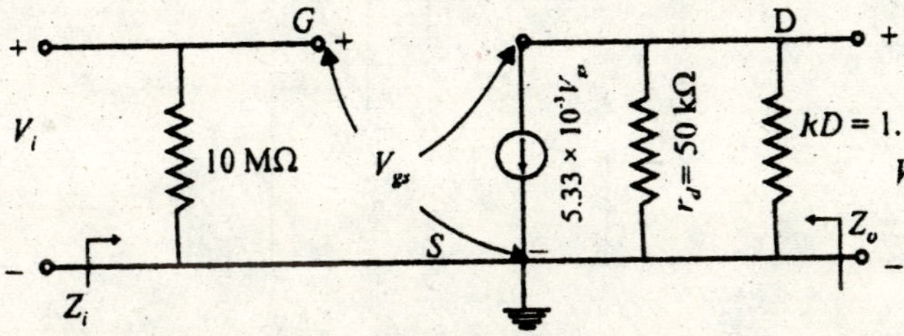

Fig. 6.24 (a)

The input impedance

$$Z_o = r_d \,||\, R_D = 50\ \text{k}\Omega \,||\, 1.1\ \text{k}\Omega$$

$$= \frac{50\times 1.1}{51.1}\ \text{k}\Omega = 1.076\ \text{k}\Omega$$

The voltage gain $A_v = -g_m\,(r_d \,||\, R_D)$

$\Rightarrow \quad \dfrac{V_o}{V_i} = -g_m Z_o$

$\Rightarrow \quad V_o = -g_m Z_o V_i$

$\Rightarrow \quad V_o = -5.33\times 10^{-3}\times 1.076\times 10^{3}\times 2\times 10^{-3}$

$\Rightarrow \quad V_o = -11.473\ \text{mV}$

Another method of approximation can be used since

$$r_d = 10\,R_d \geq 50\ \text{k}\Omega > 1.1\times 10\ \text{k}\Omega$$

$$50\ \text{k}\Omega > 11\ \text{k}\Omega$$

$$A_v = -g_m R_D$$

$$= -5.33\times 10^{-3}\times 1.1\times 10^{3} = -5.863$$

Output voltage $V_o = A_v \times V_i = -5.863\times 2\times 10^{-3} = -11.72\text{υ}\ \text{mV}$

Problem 6.25. Determine the output of the depletion type MOSFET, if input voltage is 4 mV:

$y_{os} = 35\ \mu s$

$y_{fs} = 6000\ \mu s$

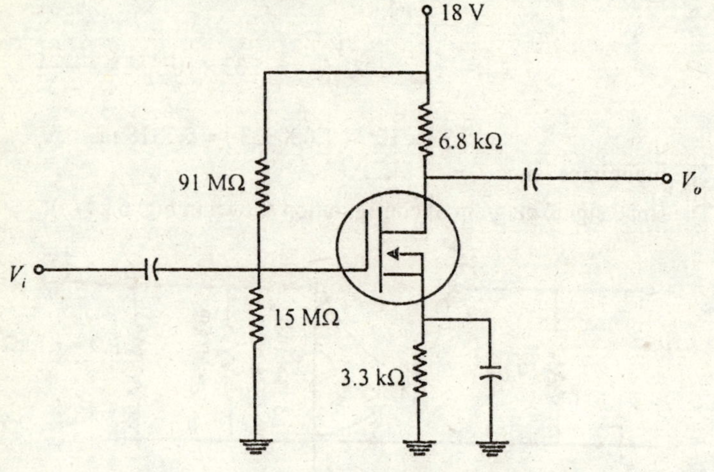

Fig. P. 6.25

Solution.

$$r_d = \frac{1}{y_{os}} = \frac{1}{35 \times 10^{-6}} = \frac{1000 \times 10^3}{35} = 28.57\ \text{k}\Omega$$

$$g_m = y_{fs} = 6000 \times 10^{-6} = 6\ \text{ms}$$

The small signal model is drawn below:

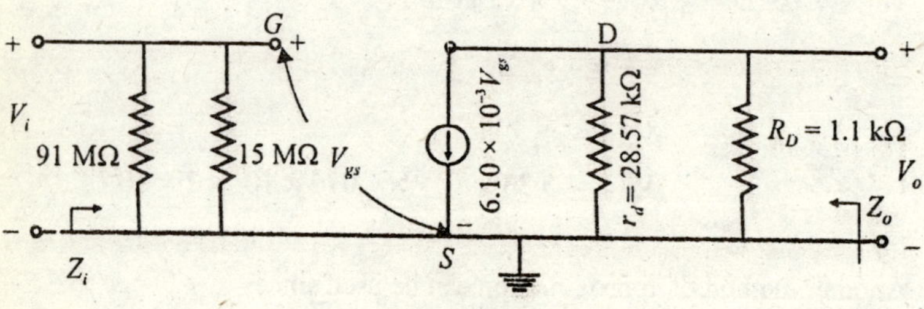

Fig. 6.25 (a)

The input impedance

$$Z_i = 91\ \text{M}\Omega \parallel 15\ \text{M}\Omega$$

$$\Rightarrow \qquad Z_i = \frac{91 \times 15}{91 + 15} = 12.88\ \text{M}\Omega$$

The output impedance

$$Z_o = R_D \parallel r_d = 6.8\ \text{k}\Omega \parallel 28.57\ \text{k}\Omega = 5.493\ \text{k}\Omega$$

The output voltage $V_o = A_v V_i$

$$\Rightarrow \quad V_o = -g_m (r_d \| R_D) V_i$$

$$\Rightarrow \quad V_o = -g_m Z_o V_i$$

$$\Rightarrow \quad V_o = -6 \times 10^{-3} \times 5.493 \times 10^3 \times 4 \times 10^{-3}$$

$$\Rightarrow \quad V_o = -131.82 \text{ mV}$$

Problem 6.26. Determine g_m for a MOSFET of enhancement type, if $V_{gs(th)} = 3$ V, and it is biased at $V_{GSQ} = 8$V. Assume $k = 0.3 \times 10^{-3}$.

Solution. For an enhancement MOSFET, the relationship between the output drain current and controlling voltage is defined by

$$I_D = K(V_{GS} - V_{GS(th)})^2$$

Since

$$g_m = \frac{\Delta I_D}{\Delta V_{GS}} = \frac{d I_D}{d V_{GS}}$$

$$\Rightarrow \quad g_m = 2\text{K}(V_{GS} - V_{GS(th)})$$

$$g_m = 2 \times 0.3 \times 10^{-3} (8 - 3)$$

$$\Rightarrow \quad g_m = 2 \times 0.3 \times 10^{-3} \times 5$$

$$\Rightarrow \quad g_m = 3 \text{ ms}$$

Problem 6.27. For an enhancement MOSFET drain feedback configuration determine, Z_i, Z_o and A_v. If $K = 0.3 \times 10^{-3}$:

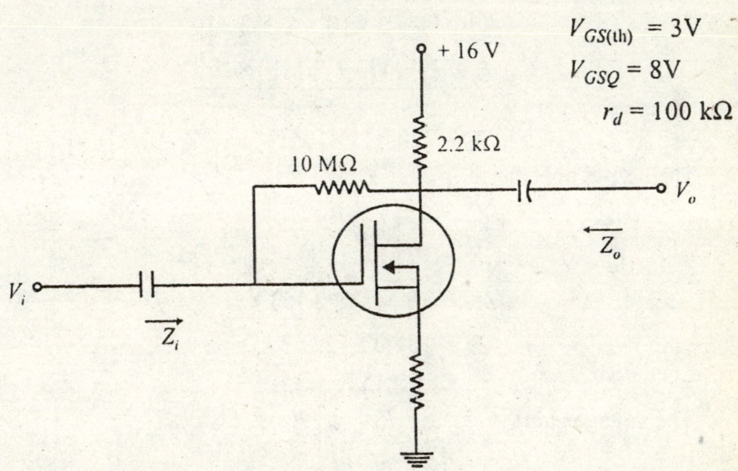

Fig. P. 6.27

Solution. $g_m = 2\text{ K}\,(V_{GSQ} - V_{GS(th)})$

The capacitors are short-circuit equivalent in the a.c. equivalent model.

$$= 2 \times 0.3 \times 10^{-3}\,(8 - 3)$$

$$= 2 \times 0.3 \times 10^{-3} \times 5 = 3\text{ ms}$$

$$r_d = 100\text{ k}\Omega$$

The small signal model is drawn below:

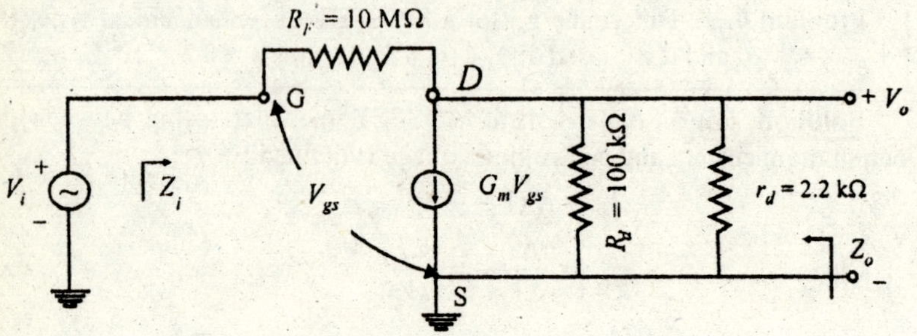

Fig. 6.27 (a)

The input impedance

$$Z_i = \frac{R_F + r_d \parallel R_D}{1 + g_m\,(r_d \parallel R_D)}$$

Now $\quad r_d \parallel R_D = 100\text{ k}\Omega \parallel 2.2\text{ k}\Omega = 2.153\text{ k}\Omega$

$$\Rightarrow \quad Z_i = \frac{10 \times 10^6 + 2.153 \times 10^3}{1 + 3 \times 10^{-3}\; 2.153 \times 10^3}$$

$$\Rightarrow \quad Z_i = \frac{(10000 + 2.513) \times 10^3}{7.458}$$

$$\Rightarrow \quad Z_i = 1341.13 \times 10^3\,\Omega \;\Rightarrow\; Z_i = 1.341\text{ M}\Omega$$

The output impedance

$$Z_o = R_F \parallel r_d \parallel R_D$$

$$\Rightarrow \quad Z_o = 10\text{ M}\Omega \parallel 100\text{ k}\Omega \parallel 2.2\text{ k}\Omega$$

$$\Rightarrow \quad Z_o = 10\text{ M}\Omega \parallel 2.153\text{ k}\Omega$$

$$\Rightarrow \quad Z_o = 10^4\text{ k}\Omega \parallel 2.153\text{ k}\Omega$$

$$\Rightarrow \quad Z_o = 2.1525\text{ k}\Omega$$

The voltage gain $A_v = -g_m\,(R_F \parallel r_d \parallel R_D)$

$$\Rightarrow \quad A_v = -g_m Z_o$$

$$\Rightarrow \quad A_v = -3 \times 10^{-3} \times 2.1525 \times 10^3$$

$$\Rightarrow \quad A_v = -6.4576.$$

Problem 6.28. Determine V_o of E-MOSFET, if $V_i = 4$ mV, $V_{GS(th)} = 4$ V and $I_{D(on)} = 4$ mA with $V_{GS(on)} = 7$ V and $y_{os} = 20\ \mu$s.

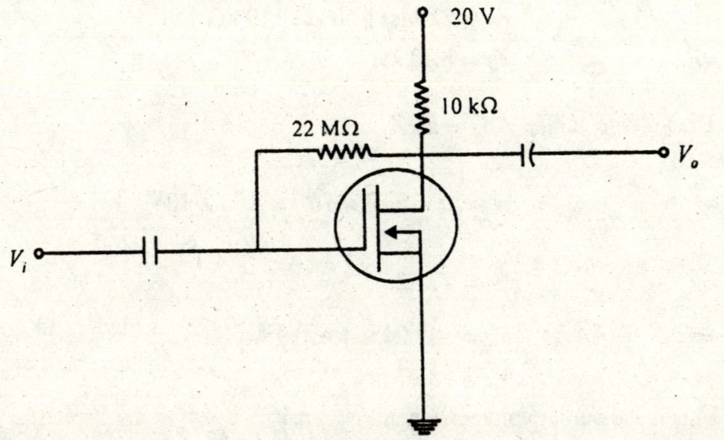

Fig. P. 6.28

Solution. $I_D = K(V_{GSQ} - V_{GS(th)})^2 \qquad g_m = 2\,K(V_{GSQ} - V_{GS(th)})$

$$\Rightarrow \qquad 4 \times 10^{-3} = K(7-4)^2 \qquad = 2 \times 0.44 \times 10^{-3} \times 3 = 2.667 \text{ ms.}$$

$$\Rightarrow \qquad K = \frac{4 \times 10^{-3}}{9} = 0.44 \times 10^{-3}$$

$$r_d = \frac{1}{y_{os}} = \frac{1}{20 \times 10^{-6}} = \frac{1000 \times 10^3}{20} = 50 \text{ k}\Omega$$

The small signal model is shown in Fig. 6.28 (a).

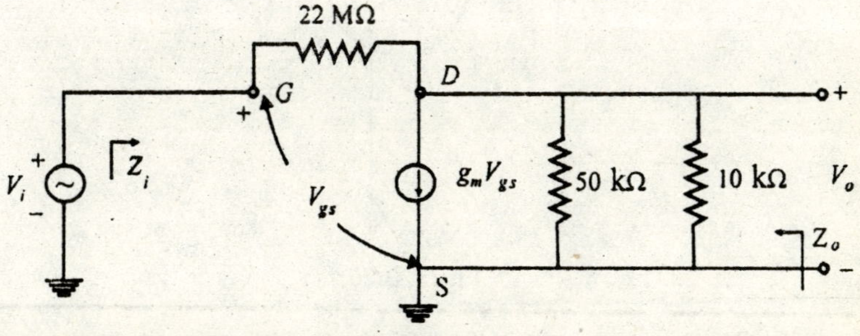

Fig. 6.28 (a)

The output impedance

$$Z_o = R_F \| R_D \| r_d$$

$$\Rightarrow \quad Z_o = 22\text{ M}\Omega \| 50\text{ k}\Omega \| 10\text{ k}\Omega$$

$$\Rightarrow \quad Z_o = 8.33\text{ k}\Omega$$

The voltage gain, $A_v = -g_m Z_o$

$$\Rightarrow \quad Av = \frac{V_o}{V_i} - 2.67 \times 10^{-3} \times 8.33 \times 10^3$$

$$\Rightarrow \quad \frac{V_o}{V_i} = -22.24$$

$$\Rightarrow \quad V_o = -22.24 \times 4 \times 10^{-3}$$

$$\Rightarrow \quad V_o = 88.97\text{ mV}$$

This is the output voltage, when $V_i = 4$ mV

Problem 6.29. Determine the fixed bias metwork to have an a.c. gain of that is determine the value of R_D at $V_{GSQ} = \frac{1}{3} V_P$:

$I_{DSS} = 12$ mA, $V_P = -3$ V, $r_d = 40$ kΩ

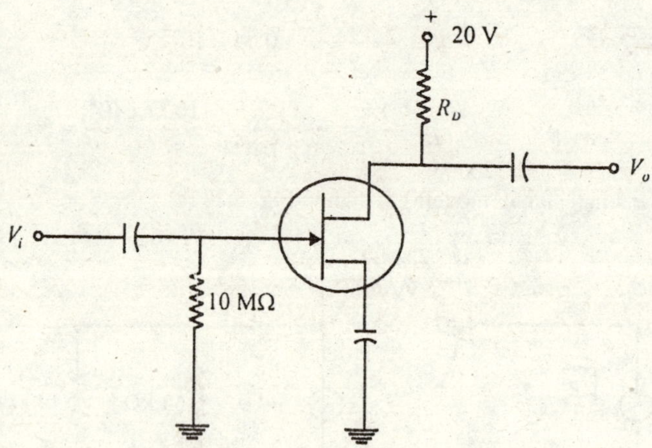

Fig. P. 6.29

Solution. $$g_{mo} = \frac{2I_{DSS}}{|V_P|} = \frac{2 \times 12 \times 10^{-3}}{3} = 8\text{ ms}$$

$$r_d = 40 \text{ k}\Omega$$

$$g_m = g_{mo}\left(1 - \frac{V_{GSQ}}{V_P}\right) = 8 \times 10^{-3}\left(1 - \frac{1}{3}\right) = 5.33 \text{ ms}$$

Gain

$$A_v = -g_m (R_D \parallel r_d)$$

$$-10 = -5.33 \times 10^{-3} (R_D \parallel r_d)$$

$$\Rightarrow \quad R_D \parallel r_d = \frac{-10}{-5.33 \times 10^{-3}} = 1.876 \text{ k}\Omega$$

$$\Rightarrow \quad (R_D)^{-1} + (40)^{-1} = (1.876)^{-1}$$

$$\Rightarrow \quad (R_D)^{-1} = (1.876)^{-1} - (40)^{-1}$$

$$\Rightarrow \quad (R_D)^{-1} = (0.508)$$

$$\Rightarrow \quad R_D = (0.508)^{-1} = \frac{1}{0.508} = 1.968 \text{ k}\Omega$$

$$\therefore \quad R_D \simeq 2 \text{ k}\Omega$$

Problem 6.30. Choose the values of R_D and R_s for the network shown in Fig. P. 6.30, that will result in a gain of 8 using relatively high of g_m for this device defined at $V_{GS} = \frac{1}{4} V_P$.

$$I_{DSS} = 10 \text{ mA} \qquad V_P = -4 \text{ V}$$

$$y_{os} = 20\ \mu\text{s} \qquad g_{mo} = 5 \text{ ms}$$

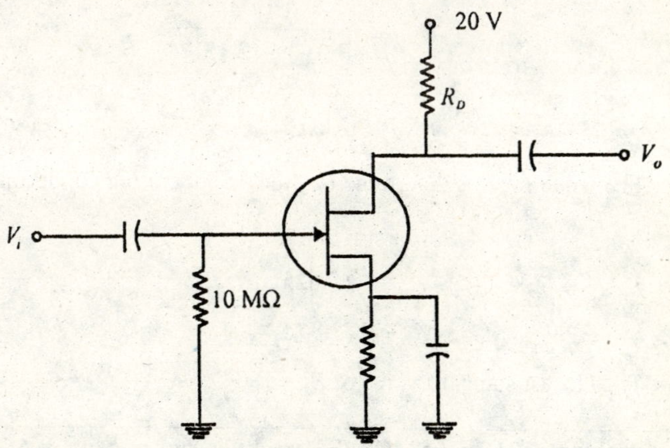

Fig. P. 6.30

Solution.

$$g_{mo} = \frac{2I_{DSS}}{|V_P|} = \frac{2 \times 10}{4} = 5 \text{ ms}$$

$$r_d = \frac{1}{y_{os}} = \frac{1}{20 \times 10^{-6}} = \frac{1000 \times 10^3}{20} = 50 \text{ k}\Omega$$

$$I_D = I_{DSS}\left(1 - \frac{V_{GSQ}}{V_P}\right)^2$$

$$= 10 \times 10^{-3}\left(1 - \frac{1}{4}\right)^2 = \frac{9}{16} \times 10 \times 10^{-3} = 5.625 \text{ mA}$$

$$g_m = g_{mo}\left(1 - \frac{V_{GSQ}}{V_P}\right) = 5 \text{ ms}\left(1 - \frac{1}{4}\right)$$

The magnitude of $A_v = g_m (R_D \| r_d)$

$\Rightarrow$ $8 = 3.75 \times 10^{-3} (R_D \| r_d)$

$\Rightarrow$ $R_D \| r_d = 2.13 \text{ k}\Omega$

$\Rightarrow$ $R_D \| 50 \text{ k}\Omega = 2.13 \text{ k}\Omega$

$R_D \| 50 \text{ kW} = 2.13 \text{ kW}$

$\Rightarrow$ $$R_D = \frac{1}{-(50)^{-1} + (2.13)^{-1}}$$ $$V_{GSQ} = \frac{1}{4}V_P$$

$\Rightarrow$ $R_D = 2.2 \text{ k}\Omega$ $\quad \frac{1}{4}(\times 4) = -1 \text{ V}$

The value of R_s is determined by the d.c. operating condition as follows:

$$V_{GSQ} = -I_D R_s$$

$\Rightarrow$ $-1 = -(5.625 \text{ mA}) R_s$

$\Rightarrow$ $R_s = 177.8 \ \Omega$

The closest standard value of $R_s = 180 \ \Omega$.

Problem 6.31. In the MOSFET amplifier shown in Fig. P. 6.31, the transistor has $\mu = 50$, $r_d = 10 \text{ k}\Omega$. $C_{gs} = 5$ pF, $C_{gd} = 1$ pF and $C_{ds} = 2$ pF. Draw a small signal equivalent circuit for the amplifier for midband frequencies and calculate its midband voltage gain:

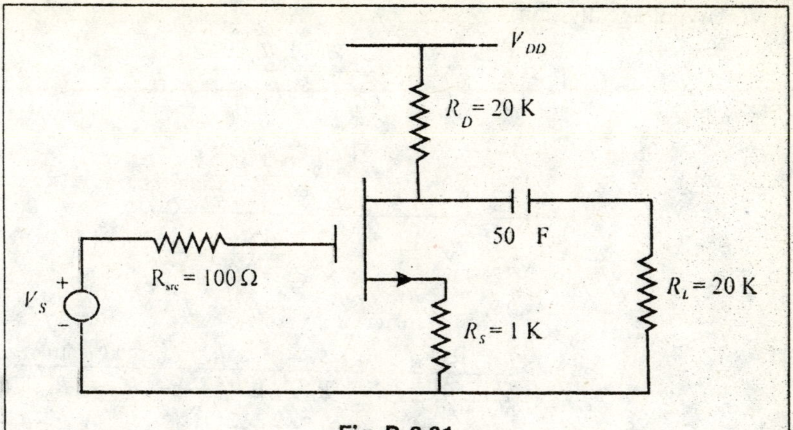

Fig. P. 6.31

Solution. All the FET capacitances come into effect only at high frequencies and coupling capacitor 50 μF at low frequencies. At mid frequency, the equivalent circuit is:

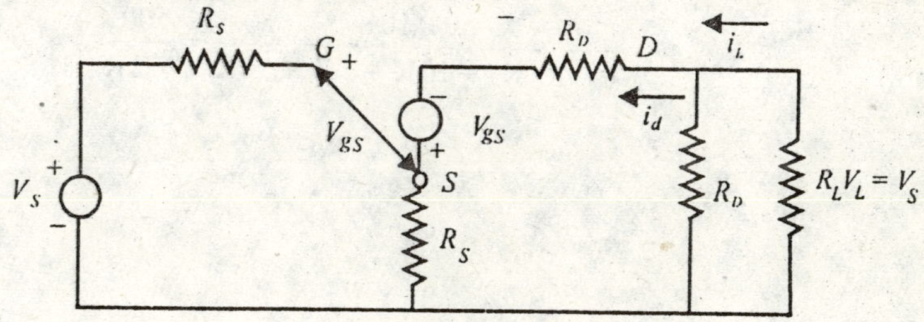

Fig. 6.31 (a)

$$V_{GS} = V_s - i_d R_S$$

$$i_d = \frac{\mu V_{gs}}{R_S + r_d (R_D \parallel R_L)}$$

$$\therefore \qquad V_{gs} = V_S - \left(\frac{\mu V_{gs}}{R_S + r_d + \dfrac{R_D R_L}{R_D + R_L}} \right) R_S$$

$$V_{gs} = R_S - \left(\frac{\mu V_{gs}}{(R_S + r_d) + (R_D \parallel R_L)} \right) V_S$$

By voltage division

$$V_o = V_L = \mu V_{gs} \times \frac{(R_D \parallel R_L)}{r_d + R_s + (R_D \parallel R_L)}$$

$$A_v = \frac{V_o}{V_s} = \frac{\dfrac{\mu V_{gs}(R_D \parallel R_L)}{r_d + R_s + (R_D \parallel R_L)}}{V_{gs}\left(1 + \dfrac{\mu R_s}{r_d + R_s + (R_D \parallel R_L)}\right)}$$

$$= \frac{\mu(R_D \parallel R_L)}{r_d + R_s + (R_D \parallel R_L)} \times \frac{r_d + R_s + (R_D \parallel R_L)}{\mu R_s + r_d + R_S + (R_D \parallel R_L)}$$

$$= \frac{\mu(R_D \parallel R_L)}{\mu R_s + r_d + R_s + (R_D \parallel R_L)}$$

$$(R_D \parallel R_L) = 10\,\text{K}$$

$$A_v = \frac{50 \times 10\text{K}}{(50 \times 1\text{K}) + 10\text{K} + 1\,\text{K} + 10\text{K}} = \frac{500\text{K}}{71\text{K}} = 7.04$$

CHAPTER

7

Compound Configuration

Introduction

The configuration in which more than one devices (FET, BJT, MOSFET) are connected together is called *compound configuration*. There are different types of compound configurations, they are as follows.

(*i*) **Cascade Connection:** A cascade connection is a series connection with the output of one stage applied as the input to the second stage. The cascade connection provides a multiplication of the gain of each stage for a larger overall gain.

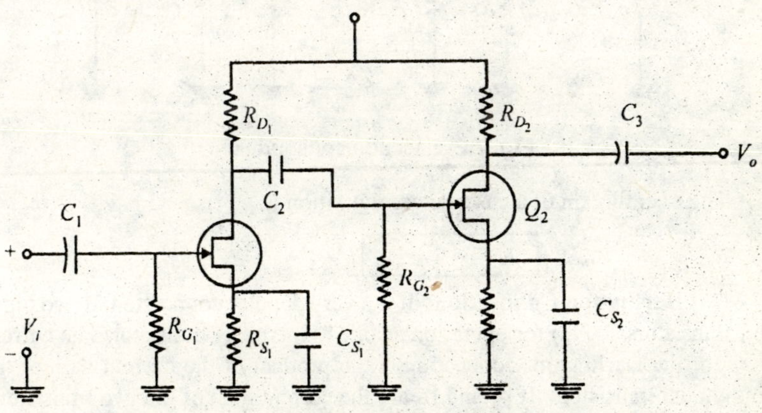

Fig. 7.1 Cascaded FET amplifier

The gain of overall cascade amplifier is the product of stage gains Av_1 and Av_2, i.e.

$$A_v = A_{v1} \cdot Av_2 = (-g_{m1}\, R_{D1})\,(-g_{m2}\, R_{D2})$$

The input impedance of the cascade amplifier is that of stage 1, i.e.

$$Z_i = R_{G1}$$

The output impedance of the cascade amplifier is that of stage 2

$$Z_0 = R_{D2}$$

The main function of cascading stages is the larger overall gain achieved.

(*ii*) **Cascade Configuration:** A cascade configuration has one transistor on top (in series with) another. Fig. 7.2 is a cascade with a common-emitter stage is feeding a common-base stage. This arrangement is designed to provide high input impedance with low voltage gain to ensure that it provides high frequency operation.

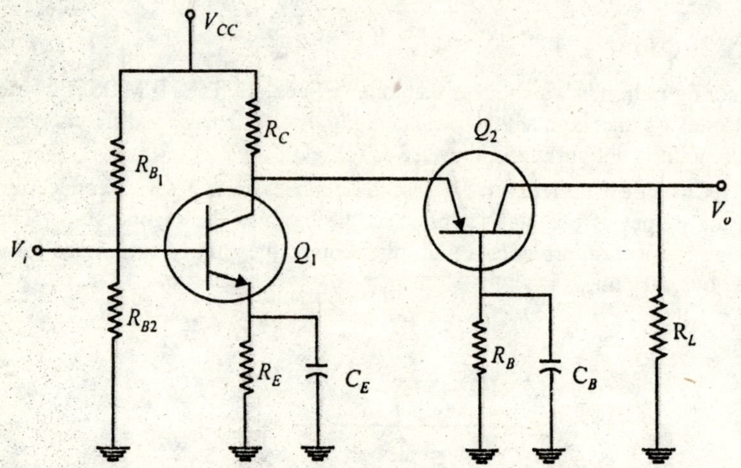

Fig. 7.2 Cascade configuration

The overall gain of a cascade configuration is

$$\boxed{A_v = A_{v1} \cdot A_{v2}}$$

(*iii*) **Darlington Connection:** It is a very popular connection of two bipolar transistors (BJTs) for operation as one "superbeta" transistor. The current gain of the Darlington connection is the product of the current gain of the individual transistors. If β_1 and β_2 are the current gain of the two transistors, then the Darlington connection provides the current gain of $\beta_D = \beta_1 \cdot \beta_2$.

If the $\qquad \beta_1 = \beta_2 = \beta.$

Then $\qquad \boxed{\beta_D = \beta^2}$

Thus, the Darlington connection provides a transistor having very large current gain, typically a few thousand.

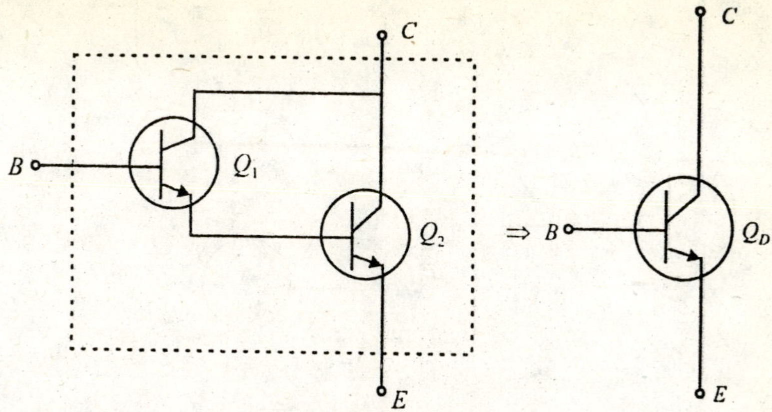

Fig. 7.3 Darlington connection

The input impedance $\boxed{Z_i = R_B \parallel (r_i + \beta_D R_E)}$

The current gain $\boxed{A_i = \beta_D \dfrac{R_B}{(R_B + \beta_D R_E)}}$

The output impedance $\boxed{Z_o = R_E \parallel r_i \parallel \dfrac{r_i}{\beta_D} \approx \dfrac{r_i}{\beta_D}}$

The voltage gain $\boxed{A_v = \dfrac{V_o}{V_i} = \dfrac{R_E + \beta_D R_E}{r_i + (R_E + \beta_D R_E)} \approx 1}$

(*iv*) **Feedback Pair:** The feedback pair connection is a two transistor circuit that operates like the Darlington circuit. It uses a *pnp*-transistor to derive an *npn*-transistor. It also has very high current gain being product of the current gains of the individual transistors. The input impedance:

$$\boxed{Z_i = R_B \parallel (r_i + \beta_1 \beta_2 R_C)}$$

The a.c. current gain $\boxed{A_i = \beta_1 \beta_2 \dfrac{R_B}{R_B + Z_i}}$

The output impedance $\boxed{Z_o = \dfrac{r_{i1}}{\beta_1 \beta_2}}$

The voltage gain of the circuit $\boxed{A_v = \dfrac{\beta_1 \beta_2 R_c}{(\beta_1 \beta_2 R_c + r_{i1})}}$

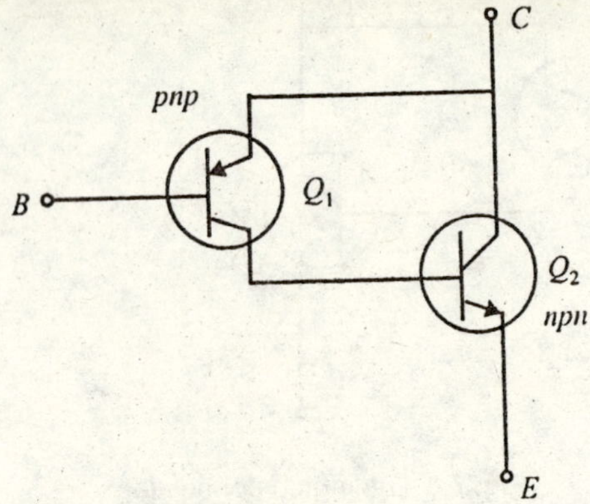

Fig. 7.4 Feedback pair connection

(*v*) **Differential Amplifier Circuit:** This circuit is extremely popular in IC units. The main feature of the differential amplifier is the very large gain, when opposite signals are applied to the inputs as compared to the very small gain resulting from the common inputs. The ratio of this difference gain to the common gain is called *common mode rejection* (CMMR).

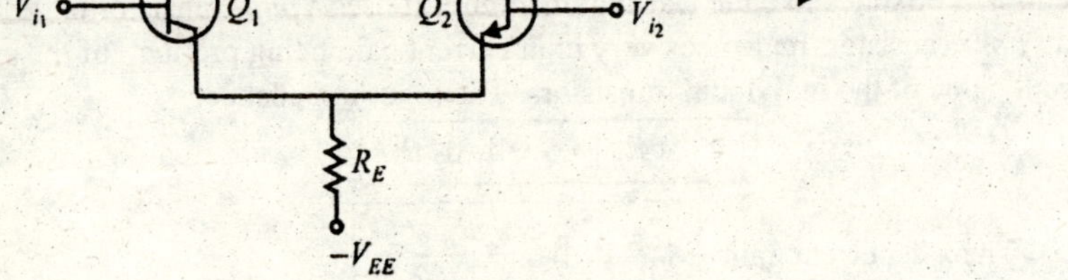

Fig. 7.5 Basic differential amplifier circuit

$$A_v = \frac{V_o}{V_i} = \frac{R_c}{2\, r_e}$$

$$A_c = \frac{\beta R_c}{r_i + 2\,(\beta + 1)\, R_E}$$

$$A_d = \frac{V_o}{V_i} = \frac{\beta\, R_c}{2\, r_i}$$

Problem 7.1. For the circuit shown in Fig. P. 7.1, calculate the d.c. bias voltage. V_{E2} and emitter current I_{E2}:

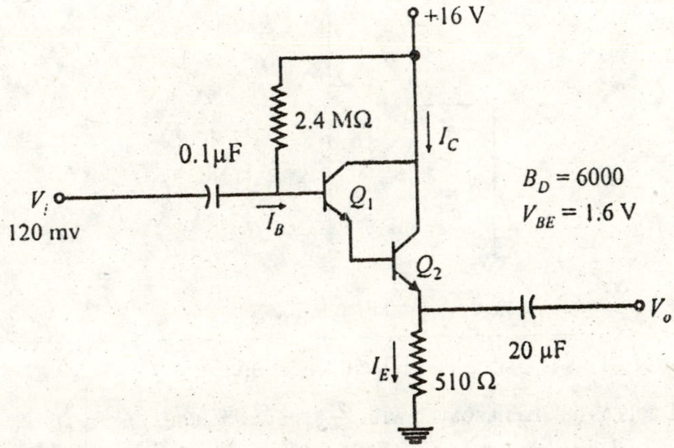

Fig. P. 7.1

Solution. The base current

$$I_B = \frac{V_{CC} - V_{BE}}{R_B + B_D\ R_E}$$

This is a circuit of Darlington pair

$$\therefore \qquad I_B = \frac{16\text{ V} - 1.6\text{ V}}{2.4\text{ M}\Omega + 6000\ (510\ \Omega)} = \frac{14.4\text{ V}}{5.46\text{ M}\Omega}$$

or $\qquad I_B = 2.637\ \mu\text{A} \approx 2.64\ \mu\text{A}$

The emitter current $I_E = (\beta + 1)\ I_B \approx 6000 \times 2.64\ \mu\text{A}$

or $\qquad I_E = 15840\ \mu\text{A} = 15.84\text{ mA}$

The emitter d.c. voltage is

$$V_{E2} = V_E = I_E \times R_E$$

or $\qquad V_{E2} = (15.84\text{ mA}) \times (510\ \Omega) = 8.0784\text{ V} \approx 8.08\text{ V}$

The base voltage is $V_B = V_E + V_{BE}$

or $\qquad V_B = 8.08\text{ V} + 1.6\text{ V} = 9.68\text{ V}$

The emitter current $I_{E2} = I_E = 15.84$ mA.

Problem 7.2. Calculate the current through the 2 kΩ load in the circuit shown in Fig. P. 7.2.

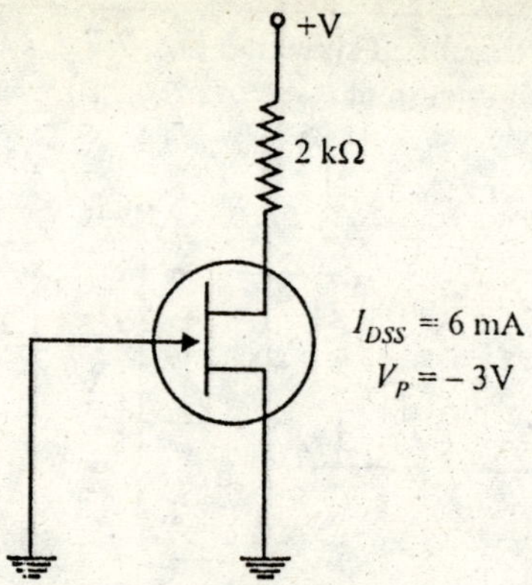

Fig. P. 7.2

Solution. This is a simple JFET current source.

JFET acts as a current source with V_{GS} set to 0 V, and the drain current is fixed at

$$I_D = I_{DSS}$$

Since $\quad V_{GS} = 0, I_D = I_{DSS} = 6 \text{ mA}$

JFET as a constant current source can be equivalent shown as

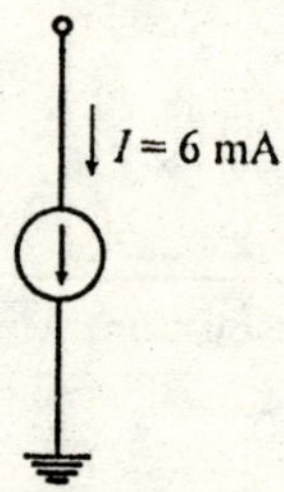

Fig. 7.2 (a)

Problem 7.3. For the circuit given below, calculate the current I:

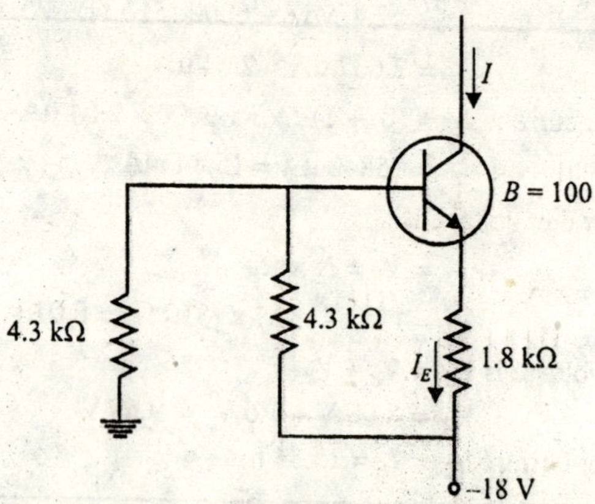

Fig. P. 7.3

Solution. This is a BJT constant source

Given $V_{EE} = 18\text{ V},\ R_1 = R_2 = 4.3\text{ k}\Omega$

Assuming that the base input impedance is much larger than R_1 and R_2

$$V_B = \frac{R_1}{R_1 + R_2}(-V_{EE})$$

or $$V_B = \frac{4.3\text{ k}\Omega}{4.3\text{ k}\Omega + 4.3\text{ k}\Omega}(-18\text{ V}) = -\frac{4.3\text{ k}\Omega \times 18\text{ V}}{8.6\text{ k}\Omega}$$

or $$V_B = -9\text{ V}$$

and $$V_E = V_B - V_{BE} = -9\text{ V} - 0.7\text{ V} = -9.7\text{ V}$$

Now $$I_E = \frac{V_E - (-V_{EE})}{R_E} \approx I_C = I$$

or $$I = \frac{V_E + V_{EE}}{R_E} = \frac{-9.7\text{ V} + 18\text{ V}}{1.8\text{ k}\Omega}$$

or required circuit $$I = \frac{8.3\text{ V}}{1.8\text{ k}\Omega} = 4.611\text{ mA}$$

Problem 7.4. Calculate the current I is the circuit of given Fig. P. 7.4·

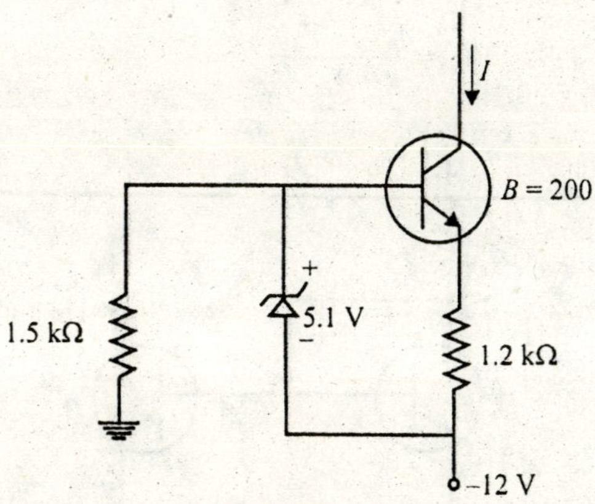

Fig. P. 7.4

Solution. Applying KVL to the base emitter loop, I_E can be obtained as

$$I_E = \frac{V_z - V_{BE}}{R_E}$$

Since, $I_C = I \approx I_E$, we get the value of I as

$$I_E = \frac{V_z - V_{BE}}{R_E} = \frac{5.1\text{ V} - 0.7\text{ V}}{1.2\text{ k}\Omega} = \frac{4.4\text{ V}}{1.5\text{ k}\Omega}$$

or, required $I = 3.67$ mA.

Problem 7.5. Calculate the mirrored current I is the circuit of Fig. P. 7.5.

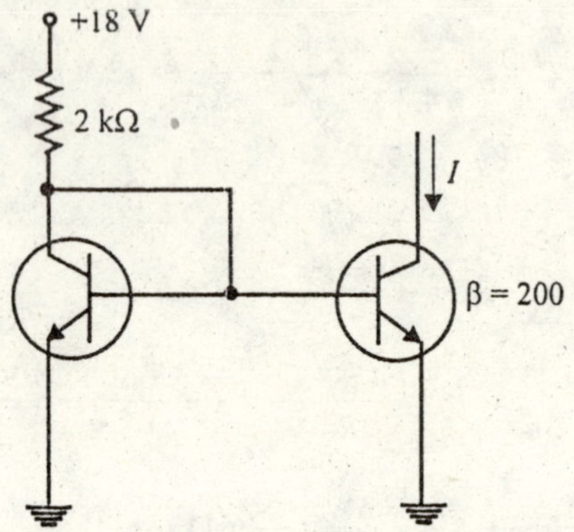

Fig. P. 7.5

Solution. The constant current provided at the collector of Q_2 mirror is that of Q_1.

Since $$I_x = \frac{V_{CC} - V_{BE}}{R_x}$$

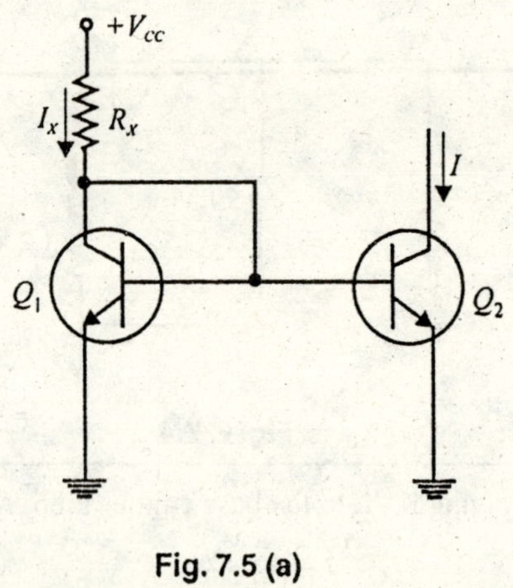

Fig. 7.5 (a)

The current I_x set by V_{CC} and R_x is mirrored is the current into the collector of Q_2.

Therefore $$I = I_x = \frac{V_{CC} - V_{BE}}{R_x}$$

or $$I = \frac{18\text{ V} - 0.7\text{ V}}{2\text{ k}\Omega} = \frac{17.3\text{ V}}{2\text{ k}\Omega} = 8.65\text{ mA}$$

Problem 7.6. Calculate the current I, through each of the transistors Q_2 and Q_3 in the following circuit.

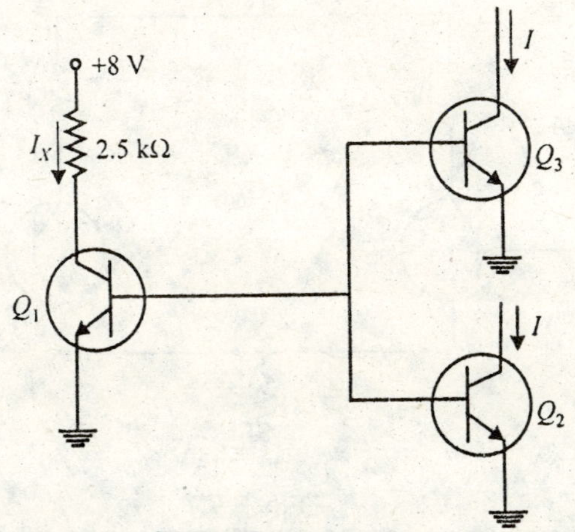

Fig. P. 7.6

Solution. In current mirror circuits, we assume that the emitter current I_E for all the transistors are the same (Q_1, Q_2, and Q_3 being fabricated near eacn other on the same chip).

The three base currents are approximately

$$I_B = \frac{I_E}{\beta + 1} \approx \frac{I_E}{\beta}$$

The collector current of each transistor is then

$$I_C \approx I_E$$

finally, the current through resistor R_x

I_x is $$I_x = I_E + 3\,I_B$$

or $$I_x = I_E + 3\frac{I_E}{\beta} = \frac{\beta + 3}{\beta} I_E \approx I_E$$

Therefore $$I = I_x = \frac{V_{CC} - V_{BE}}{R_x}$$

or $$I = \frac{8\text{ V} - 0.7\text{ V}}{2.5} = \frac{7.3\text{ V}}{2.5\text{ k}\Omega}$$

or $$I = 2.92\text{ mA}$$

Problem 7.7. Calculate d.c. bias values of I_c and V_c for the watched transistor of figure given below :

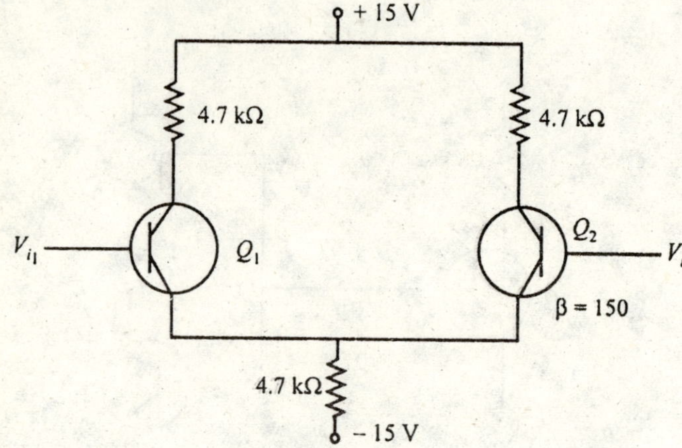

Fig. P. 7.7

Solution.

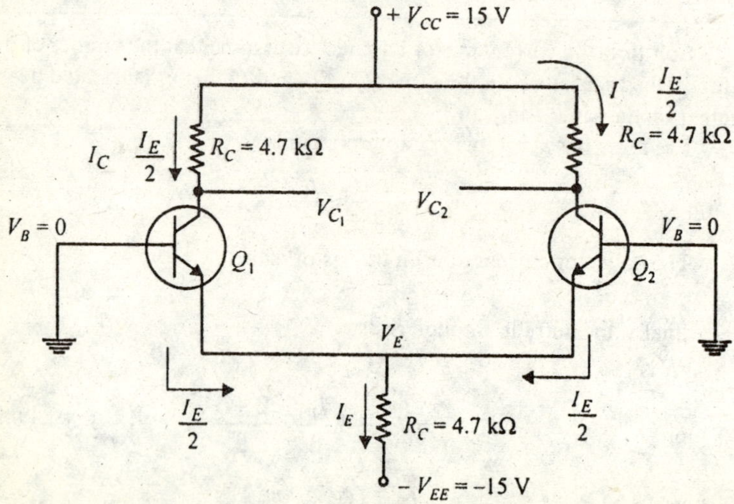

Fig. 7.7 (a)

Now $V_E = V_B - V_{BE} = 0 - 0.7 \text{ V}$

or $V_E = -0.7 \text{ V}$

The emitter d.c. bias current is then

$$I_E = \frac{V_E - (-V_{EE})}{R_E} \approx \frac{V_{EE} - 0.7 \text{ V}}{R_E}$$

or $$I_E = \frac{15 - 0.7 \text{ V}}{4.7 \text{ k}\Omega} = \frac{14.3 \text{ V}}{4.7 \text{ k}\Omega} = 3.0425 \text{ mA}$$

The collector current is then

$$I_C = \frac{I_E}{2} = \frac{3.0425 \text{ mA}}{2} = 1.521 \text{ mA}$$

i.e. $I_{C1} = I_{C2} = I_C = 1.521 \text{ mA}$

Resulting in a collector voltage of

$V_{C1} = V_{C2} = V_C$, which equals from diagram to

$V_C = V_{CC} - I_C R_C = 15 \text{ V} - (1.521 \text{ mA})(4.7 \text{ k}\Omega)$

or $V_C = 15 \text{ V} - 7.15 \text{ V} = 7.85 \text{ V}$

The common emitter voltage is thus – 0.7 V, while the collector bias voltage is 7.85 V for both outputs.

Problem 7.8. Calculate the d.c. bias values of I_C and V_C for the transistor shown in Fig. P. 7.8.

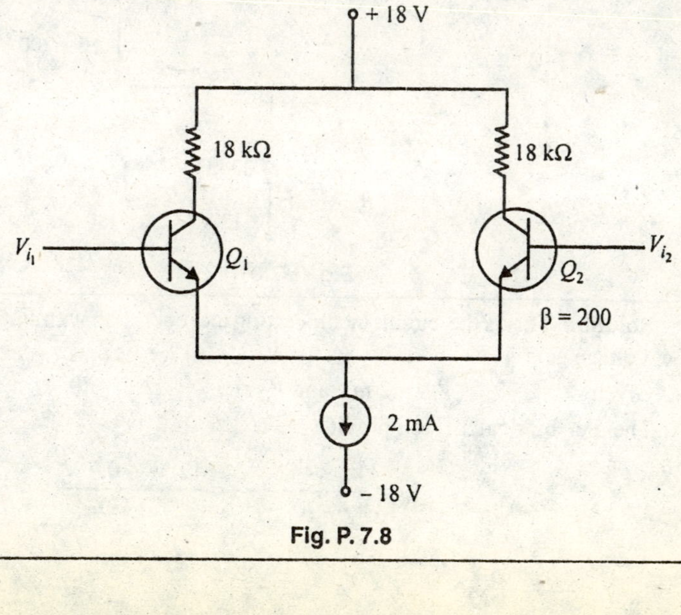

Fig. P. 7.8

Solution. Figure P. 7.8 is a differential amplifier circuit.

As is shown in the complete diagram of a differential amplifier circuit is the last problem, Given, $R_C = 18\ k\Omega$, $V_{CC} = +18V$, $V_{EE} = -18\ V$, $I_E = 2\ mA$

As we know

$$I_C \approx \frac{I_E}{2}$$

$$\therefore \qquad I_{C1} = I_{C2} = I_C = \frac{2\ mA}{2} = 1\ mA$$

Now, collector voltage

$$V_{C1} = V_{C2} = V_C$$

$$\Rightarrow \qquad V_C = V_{CC} - I_C R_C = 18V - (1mA)(18\ k\Omega)$$

$$\text{or} \qquad V_C = 18\ V - 18\ V = 0\ V.$$

Problem 7.9. Calculate V_o in the circuit is shown in Fig. P. 7.9:

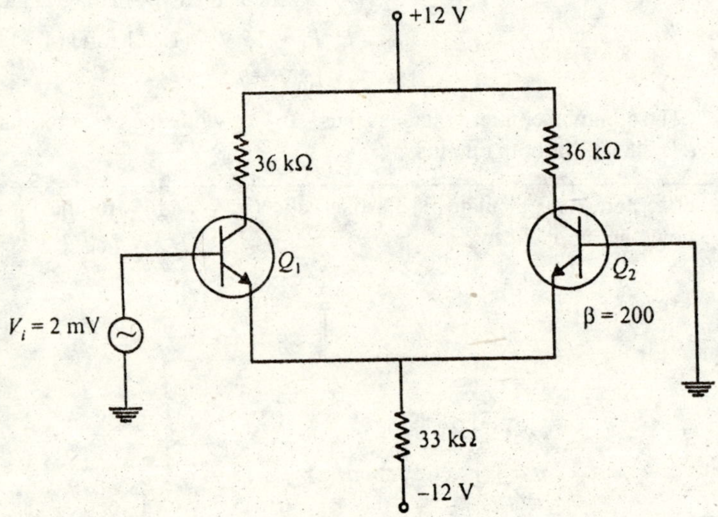

Fig. P. 7.9

Solution. This is the circuit incorporating a.c. operation with single-ended a.c. voltage gain

Given $\qquad V_{i1} = 2\ mV$ and $V_{i2} = 0$

The d.c. bias calcualtion provide

$$I_E = \frac{V_{EE} - 0.7\ V}{R_E} = \frac{12\ V - 0.7\ V}{33\ k\Omega}$$

or $$I_E = \frac{11.3\text{ V}}{33\text{ k}\Omega} = 342.42\ \mu\text{A}$$

The collector d.c. current is then

$$I_C = \frac{I_E}{2} = 171.212\ \mu\text{A}$$

Now, collector voltage

$$V_C = V_{CC} - I_C R_C = 12\text{ V} - (171.212\ \mu\text{A})(36\text{ k}\Omega)$$

or $$V_C = 12\text{ V} - 6.164\text{ V} = 5.836\text{ V}$$

The value of r_e is $r_e = \dfrac{26\text{ mV}}{(I_E/2)}$, because, $\dfrac{I_E}{2}$ is the emitter current for each transistor

or $$r_e = \frac{26\text{ mV}}{0.1712\text{ mA}} = 151.86\ \Omega$$

The a.c. voltage gain magnitude can be calculated using the relation

$$A_v = \frac{V_o}{V_{i1}} = \frac{R_C}{2\,r_e} \quad \text{or} \quad A_v = \frac{36\text{ k}\Omega}{2 \times 151.36\ \Omega} = 118.53$$

Hence, output as voltage $V_o = A_v V_{i1} = (118.53)(2\text{ mV}) = 237\text{ mV}$

or required $V_o = 237\text{ mV} = 0.237\text{ V}$

Problem 7.10. For the differential amplifier circuit shown in Fig. P. 7.10, calculate the differential gain, the common mode gain and the CMRR.

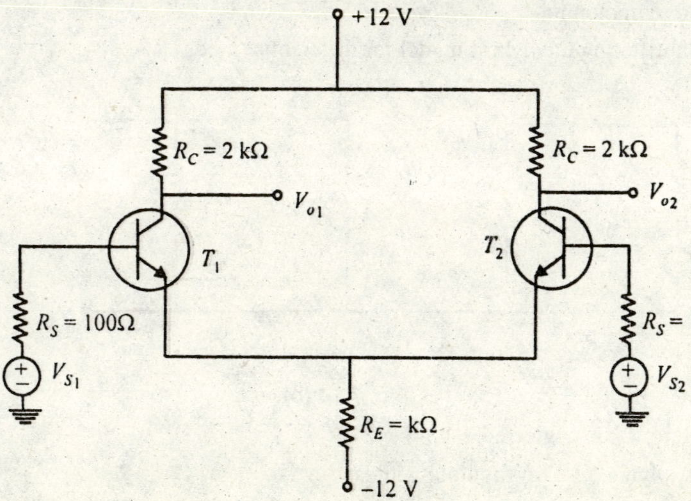

Fig. P. 7.10

Given that T_1 and T_2 are identical with the h-parameters h_{ie} = 2 K, $h_{re} = 8 \times 10^{-4}$, h_{fe} = 100, h_{oe} is negligible. [GATE' 1997]

Solution: Given V_{CC} = + 12 V, V_{EE} = – 12 V

Circuit shown is of emitter-coupled differential amplifier.

The circuit is bisected for two differential modes are shown:

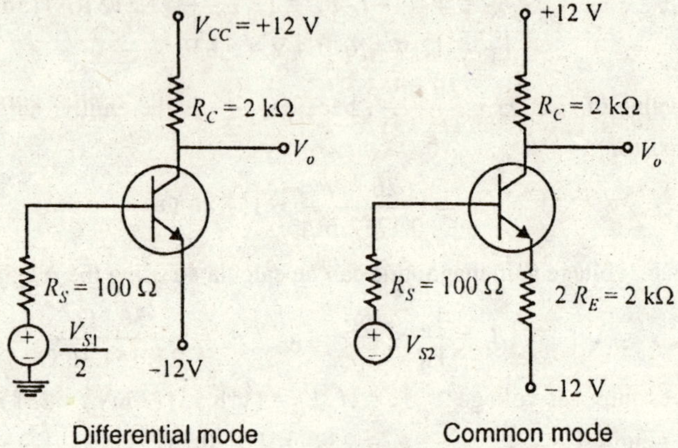

Fig. 7.10 (a)

For differential mode, V_S becomes $\frac{V_S}{2}$, and

For common mode, R_E becomes 2 R_E.

Substituting the hybrid model for differential mode

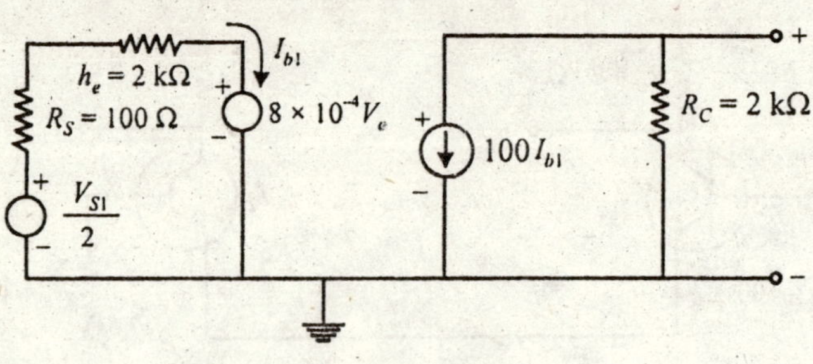

Fig. 7.10 (b)

Here, h_{oe} is neglected.

Applying KVL to first loop yields

$$\frac{V_{S1}}{2} = (2.1 \text{ k}\Omega)\, I_{b1} + 8 \times 10^{-4}\, V_o \qquad \text{... (1)}$$

and

$$V_o = -\,(2 \text{ k}\Omega)\, 100\, I_{b1} = -\,200\, I_{b1} \qquad \text{... (2)}$$

Putting value of V_o from (2) in Eq. (1), we get

$$\frac{V_{S1}}{2} = 2.1\, I_{b1} - 8 \times 10^{-4} \times 200\, I_{b1} = 1.94\, I_{b1}$$

$$\Rightarrow \quad V_{S1} = 2 \times 1.94\, I_{b1} = 3.88\, I_{b1} \quad \text{... (3)}$$

From equations (2) divided by (3), we get

$$\frac{V_o}{V_{S1}} = -\frac{200}{3.88}$$

or differential gain $A_d = -\dfrac{200}{3.88} = -51.55$

Substituting the hybrid model for common mode

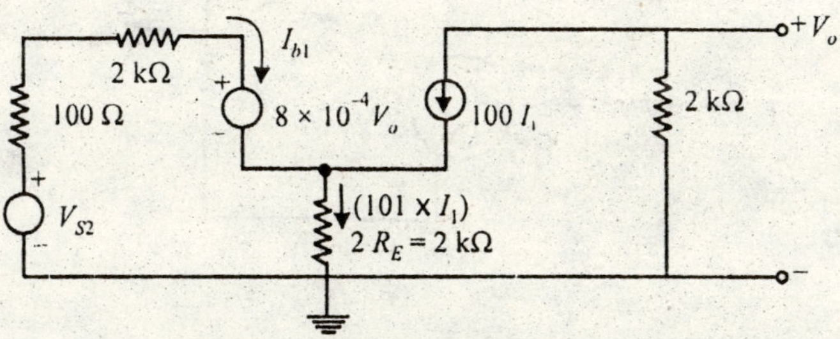

Fig. 7.10 (c)

From KVL

$$V_{S2} = 2.1 I_1 + 8 \times 10^{-4}\, V_o + (2\text{ k}\Omega)(101\, I_1)$$

$$\Rightarrow \quad V_{S2} = 204.1\, I_1 + 8 \times 10^{-4}\, V_o \quad \text{... (4)}$$

Also $\quad V_o = -200\, I_1 \quad$ ... (5)

Putting the value of V_o from Eqs (5) into (4) yields

$$V_{S2} = 204.1\, I_1 + 8 \times 10^{-4} (-200\, I_1)$$

$$\Rightarrow \quad V_{S2} = 204.1\, I_1 - 0.16\, I_1 = 203.94\, I_1 \quad \text{... (6)}$$

From Eqs (5) divided by (6), we get

$$\frac{V_o}{V_{S2}} = \frac{-200\, I_1}{203.94\, I_1} = -\frac{200}{203.94}$$

or Common mode gain

$$A_C = \frac{V_o}{V_{S2}} = -\frac{200}{203.94} = -0.981$$

Now, common mode rejection ratio is given by

$$\text{CMRR} = \left|\frac{A_d}{A_c}\right| = \frac{51.5}{0.981} = 52.5$$

Problem 7.11. For the cascaded amplifier shown, find input impedance, output impedance, current gain and voltage gain for individual stage as well as for overall amplifier.

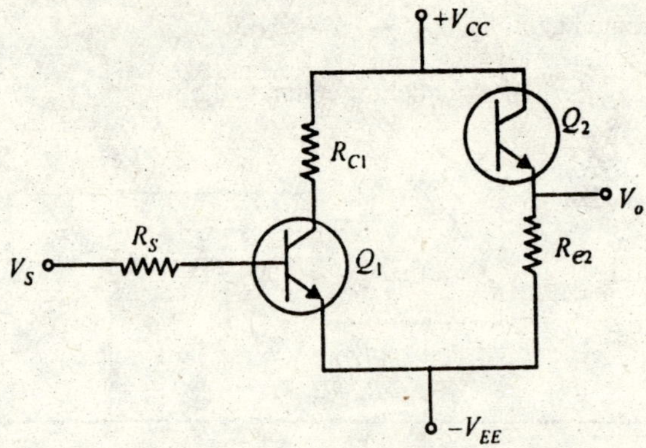

Fig. P. 7.11

Take $R_S = 1\ \text{k}\Omega$, $R_{C1} = R_{e2} = 2\ \text{k}\Omega$, and use standard values of the h-parameters.

Solution. The circuit shown is a $CE - CC$ configuration

The h-parameters are:

	CE	CC
h_i	1.1 kΩ	1.1 kΩ
h_r	2.5×10^{-4}	1
h_f	50	− 51
h_o	25 μs	25 μs

For second stage amplification (CC stage):

current gain $\quad A_{i2} = \dfrac{-h_{fc}}{1 + h_{oc} R_{e2}} = \dfrac{51}{1 + 25 \times 10^{-6} \times 2 \times 10^{3}}$

or $\quad A_{i2} = 48.57$

Input impedance $R_{i2} = h_{ic} + h_{rc} A_{i2} R_{e2} = 1.1 + 1 \times 48.57 \times 2$

$$= 98.24 \text{ k}\Omega$$

voltage gain $A_{v2} = Ai_2 \times \dfrac{R_{e2}}{R_{i2}}$

or $$A_{v2} = \frac{48.57 \times 2 \text{ k}\Omega}{98.24 \text{ k}\Omega} = 0.99 \approx 1$$

For first stage (*CE* stage):

Effective load $R_{L1} = \dfrac{R_{C1}\ R_{i2}}{R_{C1} + R_{i2}} = \dfrac{(2 \text{ k}\Omega) \times (98.24 \text{ k}\Omega)}{100.24 \text{ k}\Omega}$

$$= 1.96 \text{ k}\Omega$$

current gain $A_{i1} = \dfrac{-h_{fe}}{1 + h_{oe}\ R_{L1}} = \dfrac{-50}{1 + 25 \times 10^{-6} \times 1.96 \times 10^{3}}$

or $A_{i1} = -47.66$

input impedance

$$R_{i1} = h_{ie} + h_{re} A_{i1} R_{L1}$$

or $R_{i1} = 1.1 \text{ k}\Omega + 2.5 \times 10^{-4} (-47.46)(1.96) = 1.08 \text{ k}\Omega$

voltage gain $A_{v1} = A_{i1} \times \dfrac{R_{L1}}{Ri_1}$

$$= (-47.46) \times \frac{1.96 \text{ k}\Omega}{1.08 \text{ k}\Omega} = -47.46 \times 1.815$$

or $A_{v1} = -86.13$

output Impedance $R_{o1} = \left(h_{oe} - \dfrac{h_{fe}\ h_{re}}{h_{ie} + R_S} \right)^{-1}$

or $$R_{o1} = \left(25 \times 10^{-6} - \frac{50 \times 2.5 \times 10^{-4}}{1100 + 100} \right)^{-1}$$

$$= 68.59 \text{ k}\Omega$$

output impedance taking R_{C1} into account is

$$R'_{o1} = R_{C1} \parallel R_{o1} = (69.59 \text{ k}\Omega) \,||\, (2 \text{ k}\Omega)$$

or $$R'_{o1} = \frac{68.59 \times 2}{68.59 + 2} = 1.94 \text{ k}\Omega$$

For overall cascade amplifier

Output impedance

$$R_o = \left(h_{oc} - \frac{h_{fc}\, h_{rc}}{h_{ic} + R'_{S2}} \right)^{-1}, \quad R'_{S2} = R'_{o1}$$

or $$R_o = \left(25 \times 10^{-4} - \frac{1 \times (-51)}{1100 + 1940} \right)^{-1} = 49.94 \ \Omega$$

Output impedance taking R_{e2} into account is

$$R'_o = R_o \,||\, R_{e2} = (49.94 \ \Omega) \,||\, (2000 \ \Omega)$$

$$= \frac{49.94 \times 2000}{2049.94} = 48.72 \ \Omega$$

current gain $$A_i = A_{i1} \ A_{i2} \frac{R_{C1}}{R_{i2} + R_{e1}}$$

$$= (-47.66)(48.57) \frac{2}{2 + 98.24} = -46.19$$

voltage gain $$A_v = A_{v1} \times A_{v2} = (-86.13)(0.99)$$

$$= -85.27$$

Problem 7.12. Two transistors are direct coupled with the first emitter connected to the second base, and the two collectors are connected as indicated in Fig. P. 7.12. For this circuits shown, $\alpha_1 = 0.99$, $\alpha_2 = 0.98$, $V_{CC} = 22$ V, $R_C = 120 \ \Omega$ and $I_E = -120$ mA.

Neglect reverse saturation current and assume both transistors in the active region.

Find (*a*) the currents I_{C1}, I_{B1}, I_{E1}, I_{C2}, I_{B2}, and I_C.

(*b*) the current ratios $\frac{I_C}{I_B}, \frac{I_C}{I_E}$.

(*c*) the voltage V_{CE}.

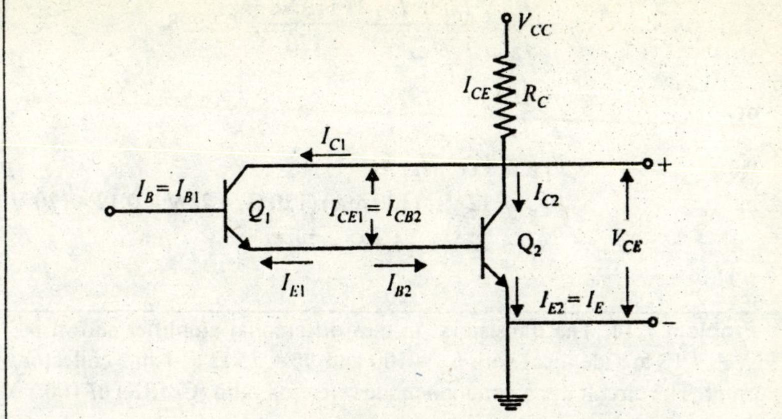

Fig. P. 7.12

Solution. Given $\alpha_1 = 0.99$ and $\alpha_2 = 0.98$

Therefore $$\beta_1 = \frac{\alpha_1}{1-\alpha_1} = \frac{0.99}{1-0.99} = \frac{0.99}{0.01} = 99$$

and $$\beta_2 = \frac{\alpha_2}{1-\alpha_2} = \frac{0.98}{0.02} = 49$$

(*a*) Now given , $I_{E2} = I_E = 120$ mA

or $$(\beta + 1)\, I_{B2} = 120 \times 10^{-3}$$

$$\Rightarrow \quad I_{B2} = \frac{120 \times 10^{-3}}{(49+1)} = \frac{120 \times 10^{-3}}{50} = 2.4 \text{ mA}$$

$$I_{E1} = -I_{B2} = -2.4 \text{ mA}$$

$$I_{C2} = \beta_2 I_{\beta 2} = 49 \times 2.4 \text{ mA} = 117.6 \text{ mA}$$

$$(\beta_1 + 1)\, I_{B1} = I_{E1}$$

or $$I_{B1} = \frac{I_{E1}}{\beta_1 + 1} = \frac{2.4 \text{ mA}}{99+1} = \frac{2.4 \text{ mA}}{100}$$

or $$I_{B1} = 2.4 \text{ mA} \times 10^{-2} = 0.024 \text{ mA}$$

$$I_{C1} = \beta_1 \times I_{B1} = 99 \times 0.024 \text{ mA} = 2.376 \text{ mA}$$

Now $$I_C = I_{C1} + I_{C2}$$

or $$I_C = 2.376 \text{ mA} + 117.6 \text{ mA} = 119.98 \text{ mA}$$

(*b*) $$\frac{I_C}{I_B} = \frac{I_{C1} + I_{C2}}{I_{B2}} = \frac{119.98}{0.024} = 5000$$

$$\frac{I_C}{I_E} = \frac{I_{C1} + I_{C2}}{I_{E2}} = \frac{119.98}{120} = 0.9998$$

or
$$\frac{I_C}{I_E} \approx 1$$

Now
$$V_{CE} = V_{CC} - I_C R_C$$

or
$$V_{CE} = 22\text{ V} - (120\text{ mA})(120\,\Omega) = 22\text{ V} - 0.12 \times 120\text{ V}$$
$$= 22\text{ V} - 14.4\text{ V} = 7.6\text{ V}$$

(*c*)

Problem 7.13. The transistors in the differential amplifier shown in Fig. P. 7.13 are identical with $h_{fe} = 100$ and $r_e = 25\,\Omega$ at 1 mA collector current. The circuit has a common mode rejection ratio (CMRR) of 100:

(*a*) What is the differential gain $\dfrac{V_o}{(V_1 - V_2)}$ of the circuit?

(*b*) What is the common mode gain of the circuit ?

(*c*) If d.c. voltages of 1010 mV and 990 mV are applied to inputs 1 and 2 respectively with reference to ground, what will be the output voltage V_o?

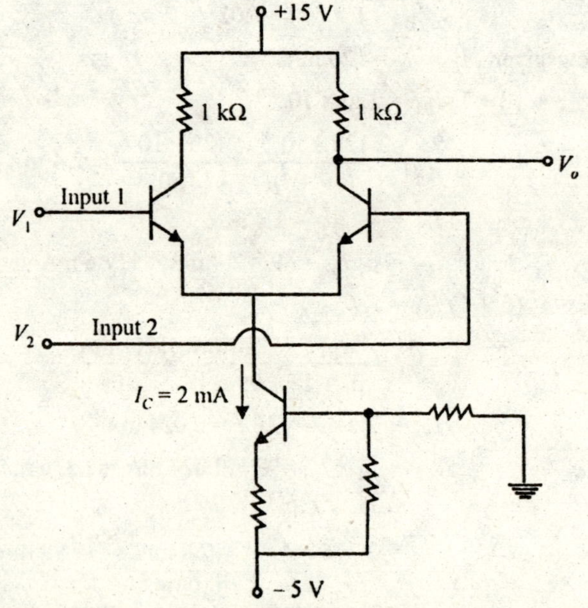

Fig. P. 7.13

Solution. Given $I_C = 2$ mA and $h_{fe} = 100$

from relation $I_C = h_{fe} I_B$

we have $$I_B = \frac{I_C}{h_{fe}} = \frac{2 \times 10^{-3}}{100} = 20\ \mu\text{A} = 0.02\ \text{mA}$$

and $$I_E = \frac{(\beta + 1)}{\beta} I_C = \frac{(h_{fe} + 1)}{h_{fe}} I_C = \frac{101}{100} \times 2\ \text{mA}$$

or $$I_E = 2.02\ \text{mA}$$

Assuming base of current source transistor at earth

(*a*) $$R_E = \frac{5 - 0.6}{2.02 \times 10^{-3}} = \frac{4.4}{2.02 \times 10^{-3}} = 2.178\ \text{k}\Omega$$

or $$R_E \approx 2.2\ \text{k}\Omega$$

Differential gain $$A_d = -\frac{R_C}{2\left(r_e + \dfrac{R_B}{h_{fe} + 1}\right)}$$

Neglecting biasing and source resistance, we get

$$A_d = -\frac{R_C}{2\, r_e} = -\frac{1000}{2 \times 25}$$

or $$A_d = -\frac{1000}{50} = -20$$

(*b*) Given CMRR = 100

But $$\text{C0MRR} = \frac{A_d}{A_c},\ A_c = \text{ common mode gain}$$

or $$A_c = \frac{A_d}{\text{CMRR}} = -\frac{20}{100}$$

or common mode gain

$$A_c = -0.2$$

Alternatively $$A_c \approx \frac{R_c}{2R_E + r_e}$$

or $$A_c = \frac{1000}{4400 + 25} = \frac{1000}{4425} = 0.226$$

(c) $$V_o = V_{o2} = -A_d V_d + A_c V_c$$

as A_d and A_c are minus.

$$V_{o2} = A_d V_d - A_c V_c$$

For input V_1, output V_o is without phase reversal and for input V_2, phase reversed

Now $$V_d = V_1 - V_2 = 1010 \text{ mV} - 990 \text{ mV}$$

or $$V_d = 20 \text{ mV}$$

$$V_c = \frac{V_1 + V_2}{2} = \frac{1010 \text{ mV} + 990 \text{ mV}}{2} = \frac{2000 \text{ mV}}{2}$$

or $$V_c = 1000 \text{ mV} = 1 \text{ V}$$

Hence, we have final output V_o as

$$V_o = 2.0 \times (20 \times 10^{-3}) \text{ V} - 0.2 \times 1 \text{ V}$$

$$= 0.4 - 0.2 \text{ V} = 0.2 \text{ V}$$

or $$V_o = 0.2 \text{ V} = 200 \text{ mV}$$

Problem 7.14. In the circuit shown in Fig. P. 7.14, there are two identical transistors T_1 and T_2 with current gain β. Another transistor T_3 has current gain β'.

Establish a relation between I_2 and I_o in terms of β and β'.

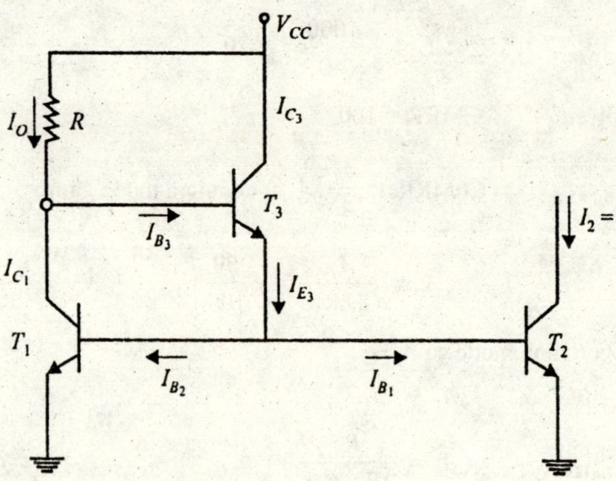

Fig. P. 7.14

Solution. Given $\beta_1 = \beta_2 = \beta$

and $\beta_3 = \beta'$

Since, the two transistors are identical, hence

$$I_{C1} = I_{C2} = I_2$$

and $I_{B1} = I_{B2} = I_B$

Now, from figure

$$I_0 = I_{C1} + I_{B3}$$

or $I_0 = I_2 + \dfrac{I_{E3}}{\beta' + 1}$, But $I_{E3} = I_{B1} + I_{B2} = 2\, I_B$

or $$I_0 = I_2 + \frac{2I_B}{\beta' + 1}$$

$$I_{B2} = \frac{I_{C2}}{\beta} = \frac{I_2}{\beta}$$

or $$I_B = \frac{I_2}{\beta}$$

Putting I_B in the expression of I_0, we get

$$I_0 = I_2 + \frac{2}{\beta' + 1} \times \left(\frac{I_2}{\beta}\right) = I_2 + \frac{2I_2}{\beta\,(\beta' + 1)}$$

or $$I_0 = I_2 + \left(1 + \frac{2}{\beta\,(\beta' + 1)}\right)$$

or $$I_0 = I_2 \left(\frac{\beta\,(\beta' + 1) + 2}{\beta\,(\beta' + 1)}\right)$$

or $$\frac{I_2}{I_0} = \frac{\beta\,(\beta' + 1)}{\beta\,(\beta' + 1) + 2}$$

Problem 7.15. A differential amplifier uses JFETs as a source coupled pair. Given g_m = 1m mho, r_d = 50 kΩ, the element values are R_D = 30 kΩ, R_s = 40 kΩ.

Determine ADM, ACM and CMRR.

Solution. The circuit diagram for differential amplifier using JFETs is shown in Fig. 7.15.

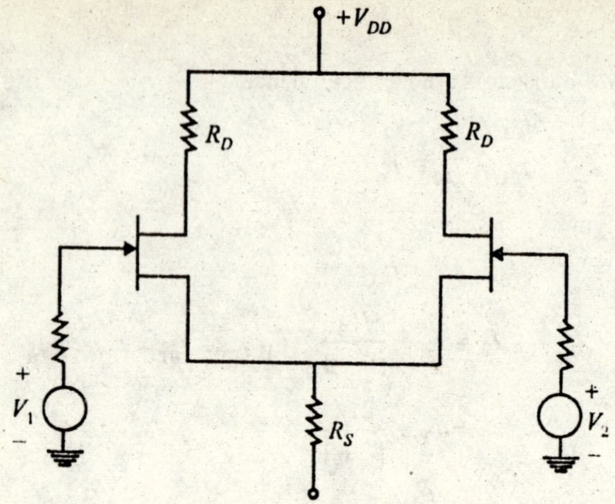

Fig. 7.15

amplification factor $\mu = g_m r_d$

or $\mu = 1\text{m}℧ \times 50\text{ k}\Omega = 50$

or $\mu = 50$

Differential mode gain

$$\text{ADM} = \frac{-\mu R_D}{r_d + R_D}$$

or $$\text{ADM} = \frac{-50 \times 30\text{ k}\Omega}{30\text{k}\Omega + 50\text{ k}\Omega} = \frac{-50 \times 30}{80}$$

or $$\text{ADM} = \frac{-1500}{80} = -18.75$$

$$\text{ACM} = \frac{-\mu R_D}{r_d + R_D + 2R_s(1+\mu)}$$

or $$\text{ACM} = \frac{-50 \times 30}{50 + 30 + 2 \times 40 \times 51} = \frac{-1500}{80 + 80 \times 51}$$

or $$\text{ACM} = \frac{-1500}{4160} = -0.3606$$

$$\text{CMRR} = \left|\frac{\text{ADM}}{\text{ACM}}\right|$$

or $$\text{CMRR} = \frac{18.75}{0.3606} = 52$$

Problem 7.16. For the Darlington amplifier shown in Fig. P. 7.16, obtain its *z*-matric for identical T_1 and T_2 having $h_{fe} = 100$, $R_E = h_{ie} = 1$ K. **[IES' 2001]**

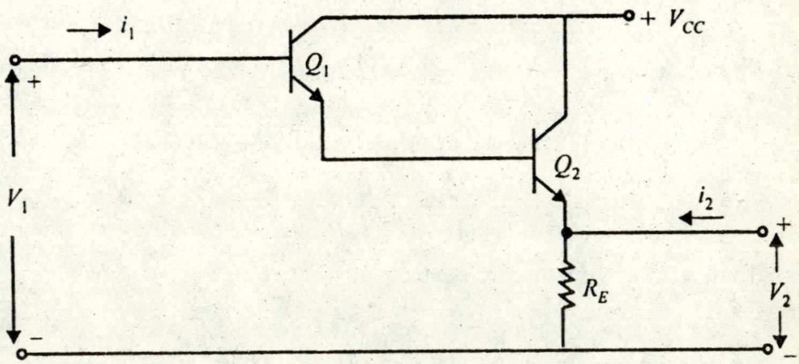

Fig. P. 7.16

Solution. These type network rather most of the analog circuits should be solved with the help of their equivalent network and from KVL.

Drawing the hybrid model of two transistors together.

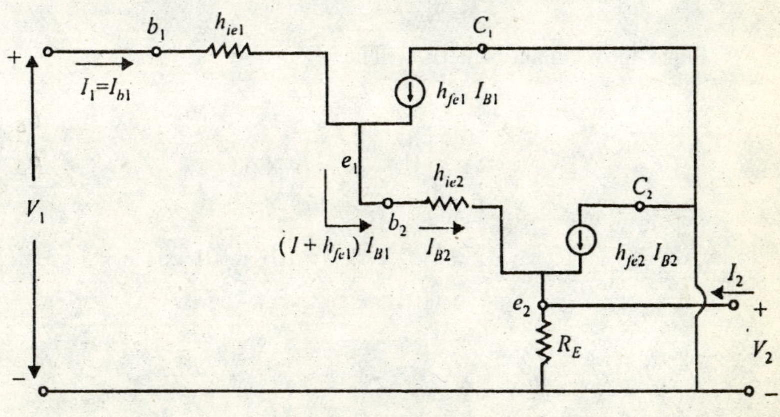

Fig. 7.16 (a)

Given $h_{ie1} = h_{ie2} = h_{ie} = 1\ k\Omega,\ R_E = 1\ k\Omega$

$h_{fe1} = h_{fe2} = h_{fe} = 100$

From the circuit $I_{B2} = (1 + h_{fe})\ I_1$

Current through $R_E = I_{B2} + h_{fe}\ I_{B2} + I_2$

Now, applying KVL to the loop containing V_1, R_E, etc. we get

$$\begin{aligned} V_1 &= h_{ie1}\ I_{B1} + h_{ie2}\ I_{B2} + (I_{B2} + h_{fe2}\ I_{B2} + I_2)\ R_E \\ &= h_{ie}\ I_1 + h_{ie}\ (1 + h_{fe})\ I_1 + (1 + h_{fe})\ I_{B2}\ R_E + I_2\ R_E \\ &= h_{ie}\ I_1 + h_{ie}\ (1 + h_{fe})\ I_1 + (1 + h_{fe})^2\ I_1 R_E + I_2\ R_E \end{aligned}$$

or $$V_1 = [h_{ie} + h_{ie}\ (1 + h_{fe}) + (1 + h_{fe})^2\ R_E]\ I_1 + R_E\ I_2 \quad \text{... (1)}$$

Similarly from KVL to the Loop containing V_2, we get

$$\begin{aligned} V_2 &= (I_{B2} + h_{fe}\ I_{B2} + I_2)\ R_E \\ &= (1 + h_{fe})^2\ I_1\ R_E + R_E\ I_2 \end{aligned} \quad \text{... (2)}$$

To determine z-matrix are have equations

$$V_1 = z_{11} I_1 + z_{12}\ I_2$$

and $$V_2 = z_{21} I_1 + z_{22}\ I_2$$

Hence, from the equations (1) and (2), we get

$$\begin{aligned} z_{11} &= h_{ie} + h_{ie}\ (1 + h_{fe}) + (1 + h_{fe})^2\ R_E \\ z_{12} &= R_E \\ z_{21} &= (1 + h_{fe})^2\ R_E \\ z_{22} &= R_E \end{aligned}$$

Since, the resultant z-matrix will be as

$$\begin{bmatrix} z_{11} & z_{12} \\ z_{21} & z_{22} \end{bmatrix} = \begin{bmatrix} h_{ie} + h_{ie}\ (1 + h_{fe}) + (1 + h_{fe})^2\ R_E & R_E \\ (1 + h_{fe})^2\ R_E & R_E \end{bmatrix}$$

Substituting the given values, we get

$$\begin{aligned} z_{11} &= 1k\Omega + 1k\Omega \times 101 + 101^2 \times 1\ K = 10.3\ M\Omega \\ z_{21} &= (101)^2 \times 1\ k\Omega = 10.2\ M\Omega \\ z_{12} &= z_{22} = R_E = 1\ k\Omega \end{aligned}$$

CHAPTER

8

Feedback and Oscillator Circuits

Feedback: In a feedback network a fraction of output is fed back to the input to get a desired value of output. There are two types of feedback depending upon the polarity:

(*i*) ***Positive feedback:*** The feedback network in which the polarity of the feedback signal is same as that of the input signal is called *positive feedback.*

(*ii*) ***Negative feedback:*** The feedback network in which the polarity of the feedback signal is opposite to that of the input signal is called *negative feedback.*

A typical feedback connection is shown below:

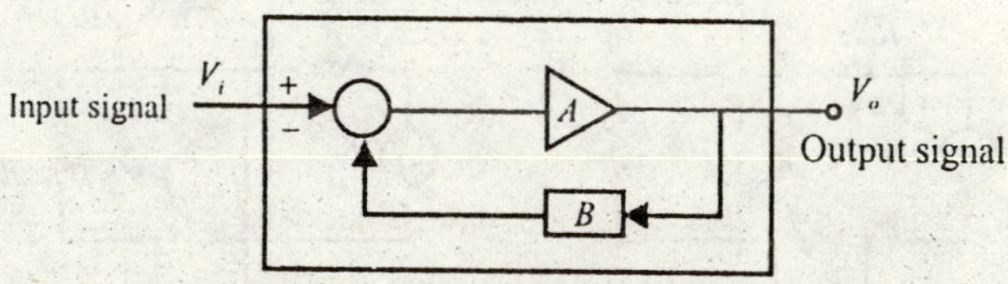

Fig. 8.1 Feedback amplifier

Types of Feedback Connections: There are basically four ways of connecting the feedback signals. Both voltage and current can be fed back to the input either in series and parallel. Specifically there can be:

(*i*) Voltage series feedback (*ii*) Voltage shunt feedback
(*iii*) Current series feedback (*iv*) Current shunt feedbac:

(*i*) ***Voltage series feedback***

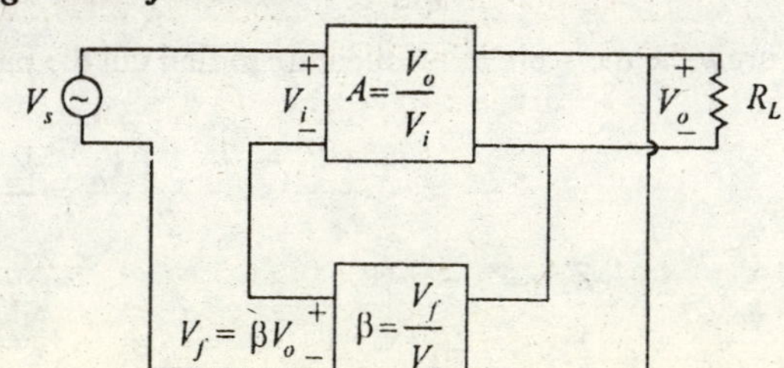

Fig. 8.2

If there is no feedback ($V_f = 0$), the voltage gain of the amplifier stage is given by

$$A = \frac{V_o}{V_i} = \frac{V_o}{V_s}$$

If the feedback signal is connected in series with the input then

$$V_i = V_s - V_f$$

Since $V_o = AV_i = A(V_s - V_f)$

$\Rightarrow$ $V_o = AV_s - AV_f = AV_s - A(BV_o)$

$\Rightarrow$ $(1 + A\beta)V_o = AV_s$

$$\Rightarrow \quad \frac{V_o}{V_s} = \frac{A}{1 + \beta A}$$

Thus, the overall gain with feedback is

$$\boxed{\frac{V_0}{V_s} = \frac{A}{1 + \beta A}}$$

Thus, we find that the gain with feedback is the amplifier gain reduced by the factor $(1 + \beta A)$.

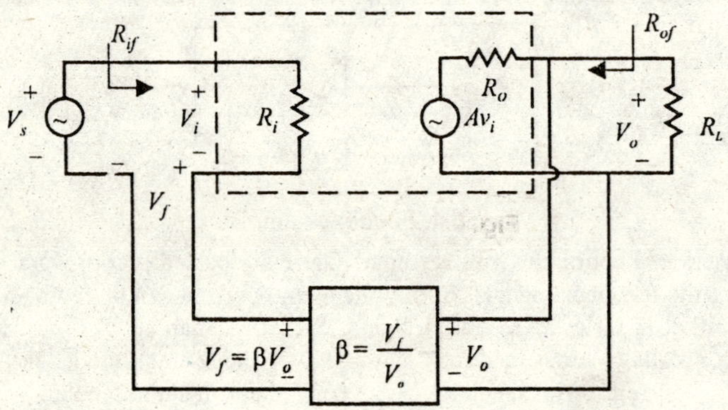

Fig. 8.3

Now, we draw the more elaborate diagram to find out the input and output impedances.

$$I_i = \frac{V_i}{Z_i} = \frac{V_s - V_f}{Z_i} = \frac{V_s - \beta V_o}{Z_i} = \frac{V_s - \beta AV_i}{Z_i}$$

$\Rightarrow$ $I_i Z_i = V_s - \beta AV_i$

$\Rightarrow$ $V_s = I_i Z_i + \beta AV_i$ Now $V_i = I_i Z_i$

$$\Rightarrow \qquad Z_{if} = \frac{V_s}{I_i} = Z_i + (\beta A) Z_i = Z_i (1 + \beta A)$$

$$\Rightarrow \qquad \boxed{Z_{if} = (1 + \beta A) Z_i}$$

Thus, the input impedance with feedback is increased by a factor of $(1 + \beta A)$.

Similarly, for output impedance

$$V = IZ_o + AV_i$$

For $\qquad V_s = 0,\ V_i = -V_f$

So that $\qquad V = IZ_o - AVl = IZ_o - A(\beta V)$

$$IZ_o = V + A\beta V$$

Now $\qquad \boxed{Z_{0f} = \frac{V}{I} = \frac{Z_o}{(1 + \beta A)}}$

Thus, the output impedance with the feedback is reduced by a factor of $(1 + \beta A)$.

(*ii*) ***Voltage shunt feedback:*** The circuit diagram of voltage shunt feedback is drawn below:

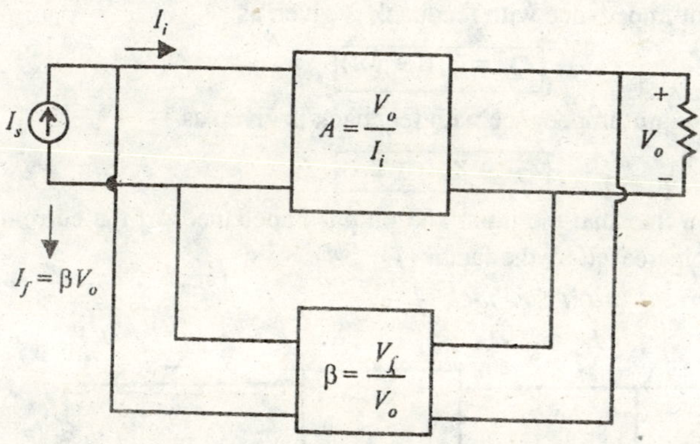

Fig. 8.4

Similarly, we can get the voltage gain with feedback is

$\boxed{A_f = \frac{A}{1 + \beta A}}$, where A is the voltage gain without feedback.

The input impedance with feedback is given by

$$\boxed{Z_{if} = \frac{Z_i}{1 + \beta A}}$$

The output impedance with feedback is given as

$$\boxed{Z_{of} = \frac{Z_o}{1 + \beta A}}$$

Thus, we have seen that the inptu as well as output impedances of voltage shunt feedback are reduced by a factor of $(1 + \beta A)$.

(iii) ***Current series feedback***

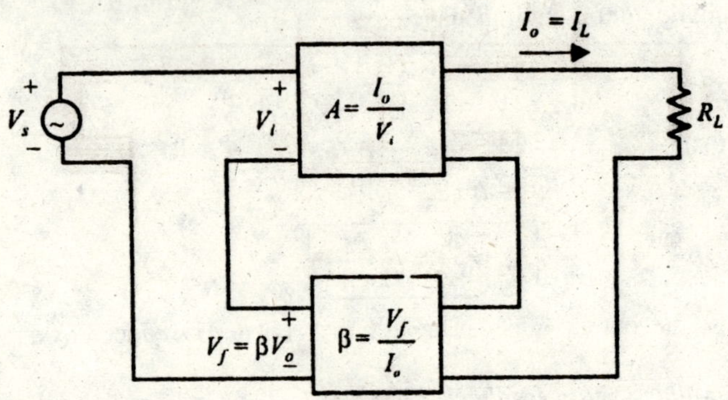

Fig. 8.5

The input impedance with feedback is given as

$$\boxed{Z_{if} = Z_i (1 + \beta A)}$$

and the output impedance with feedback is given as

$$\boxed{Z_{of} = Z_o (1 + \beta A)}$$

Thus, we find that the input and output impedances of the current series feedback is increased by the factor $(1 + \beta A)$.

(*iv*) ***Current shunt feedback***

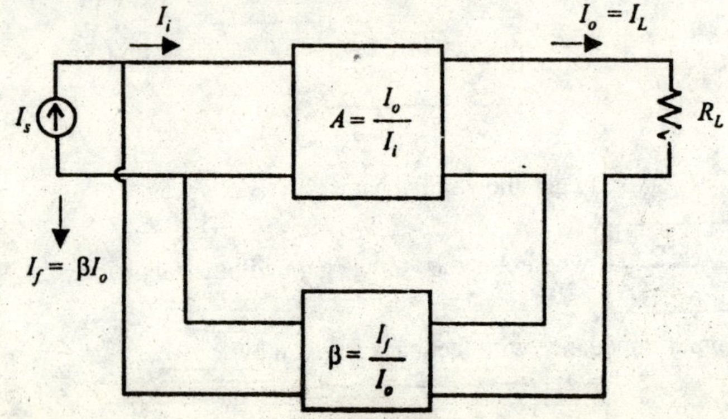

Fig. 8.6

The input impedance with feedback $$Z_{if} = \frac{Z_i}{1+\beta A}$$

and the output impedance with feedback $$Z_{of} = Z_o\left(1+\beta A\right)$$

Thus, we can summarise in the following table:

	Voltage series	*Voltage shunt*	*Current series*	*Current shunt*
Gain without feedback (A)	V_o/V_i	V_o/I_i	I_o/V_i	I_o/I_i
Feedback (β)	I_f/V_o	I_f/V_o	V_f/I_o	I_f/I_o
Gain with feedback (A_f)	V_o/V_s	V_o/I_s	I_o/V_s	I_o/I_s
Input Impedance (Z_{if})	$Z_i(1+\beta A)$ (increased)	$Z_i(1+\beta A)$ (increased)	$Z_i/(1+\beta A)$ (decreased)	$Z_i/(1+\beta A)$ (decreased)
Output impedance (Z_{of})	$\frac{Z_0}{(1+\beta A)}$ (decreased)	$Z_0(1+\beta A)$ (increased)	$Z_i/(1+\beta)$ (decreased)	$Z_i/(1+\beta A)$ (increased)

Effect of Negative Feedback: While the negative feedback results in reduced overall voltage gain but a number of improvements are obtained that is:

(*i*) Higher input impedance.

(*ii*) Better stabilized voltage gain.

(*iii*) Improved frequency response.

(*iv*) Lower output impedance.

(*v*) Reduced noise.

(*vi*) More linear operation.

Oscillator Operation: The use of positive feedback that results in a feedback amplifier having closed loop gain $|A_f|$, greater than 1 and satisfies the phase condition will result in operation as an oscillator circuit. An oscillator circuit then provides a varying output signal. If the output signal varies sinusoidally, the circuit is referred to as a *sinusoidal oscillator.*

If the output voltage rises quickly to one voltage level and later drops quickly to another voltage level, the circuit is generally referred to as a *pulse or square wave oscillator.* The circuit diagram of oscillator is Fig. 8.7.

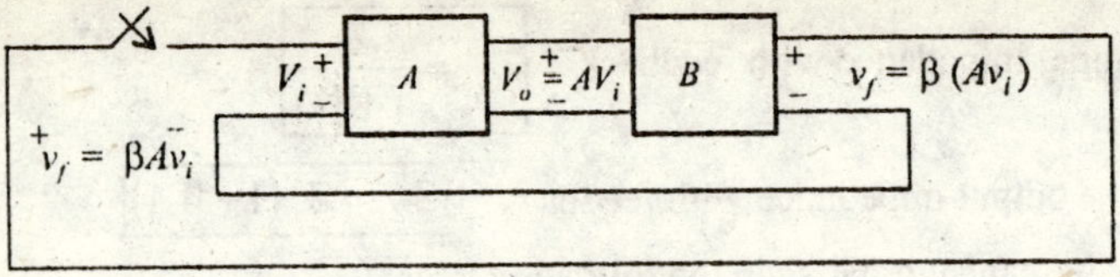

Fig. 8.7

The output waveform will exist after the switch is closed if the condition

$$\boxed{\beta A = 1 \quad \text{is met}}$$

This is known as the *Barkhausen criterion* for oscillation. The quantity βA is known as *loop* gain.

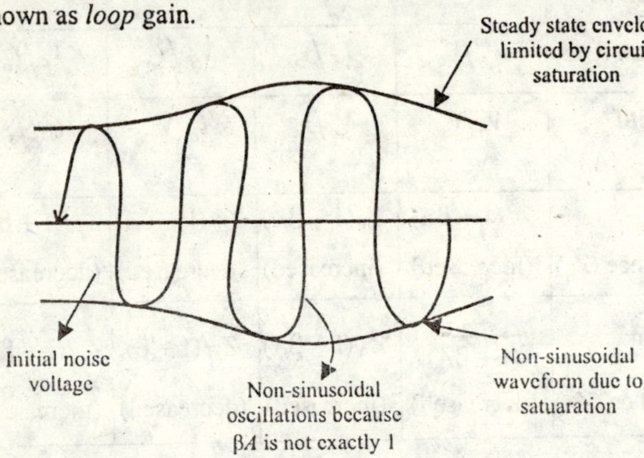

Fig. 8.8

Build-up of Steady State Oscillations: Other condition to be met is that there must bea a phase shift of 180° to maintain the oscillations.

There are different types of oscillations:

() ***Phase-shift oscillator***

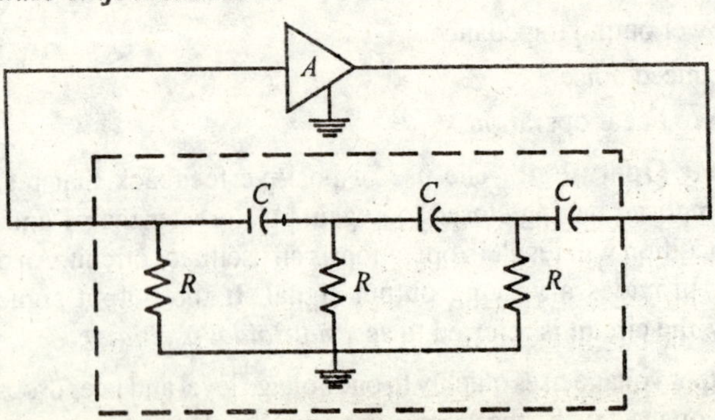

Fig. 8.9 Feedback network

Idealized Phase Shift Oscillator: The requirements for oscillations are that the loop gain βA is greater than unity and that the phase shift around the feedback network is 180° (providing positive feedback). For a phase shift network the frequency, at which the phase shift is exactly 180° is given by

$$\boxed{f = \frac{1}{2\pi RC\sqrt{6}}}$$

and

$$\boxed{\beta = \frac{1}{2q}}$$

For the loop gain $\beta A \geq 1$, the gain of the amplifier stage must be greater than $1/\beta$ or $2q$.

i.e. $$\boxed{A > 2q}$$

(*ii*) ***Colpitts oscillator:*** The op-amp provides the basic amplification needed while the oscillator frequency is set by an *LC* feedback network.

Wien Bridge Oscillator: The basic diagram of a Wien Bridge oscillator is drawn below:

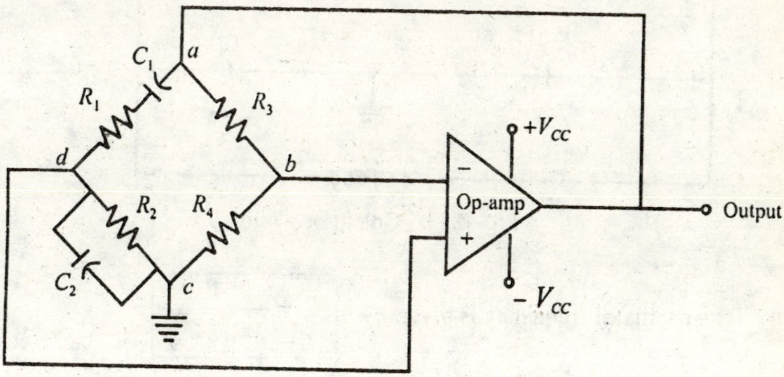

Fig. 8.10

Neglecting the loading effects of the op-amp input and output impedances, the analysis of the bridge circuit results as

$$\boxed{\frac{R_3}{R_4} = \frac{R_1}{R_2} + \frac{C_2}{C_1}}$$

and frequency of oscillation of Wien Bridge is given by

$$\boxed{f_o = \frac{1}{2\pi\sqrt{R_1C_1\,R_2C_2}}}$$

It $R_1 = R_2 = R$ and $C_1 = C_2 = C$, then

$$\boxed{f_o = \frac{1}{2\pi RC}}$$

and $\boxed{\frac{R_3}{R_4} = 2}$; thus the ratio of R_3 to R_4 greater than 2 will provide sufficient loop gain for the circuit to oscillate at frequency f_o. Colpitt network is drawn below:

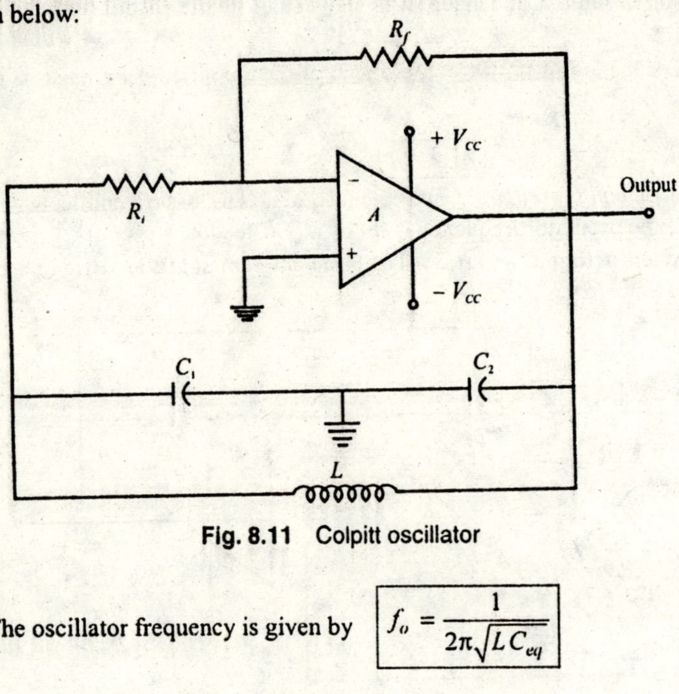

Fig. 8.11 Colpitt oscillator

The oscillator frequency is given by $\boxed{f_o = \frac{1}{2\pi\sqrt{LC_{eq}}}}$

Where $$C_{eq} = \frac{C_1 C_2}{C_1 + C_2}$$

(*iii*) **Hartley oscillator:** If the elements in the basic resonant circuit of the given figure are X_1 and X_2 as inductors and X as capacitor, the circuit is called *Hartley oscillator*.

The frequency of oscillation is given as $\boxed{f = \frac{1}{2\pi\sqrt{L_{eq}C}}}$

Where $$L_{eq} = L_1 + L_2 + 2M.$$

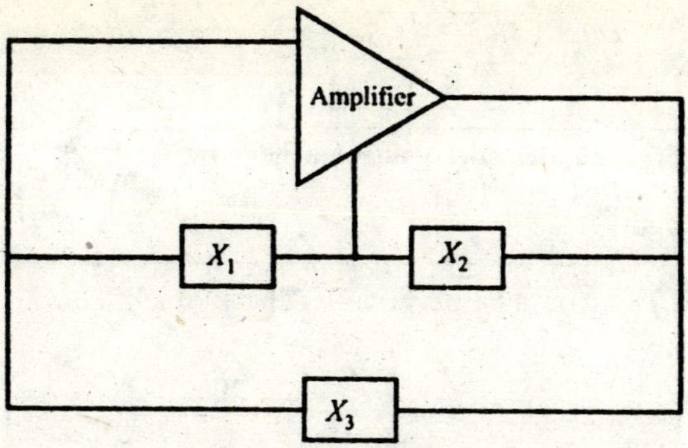

Fig. 8.12

Unijunction Oscillator: This type of oscillator is used in relaxation oscillator. The diagram is drawn below:

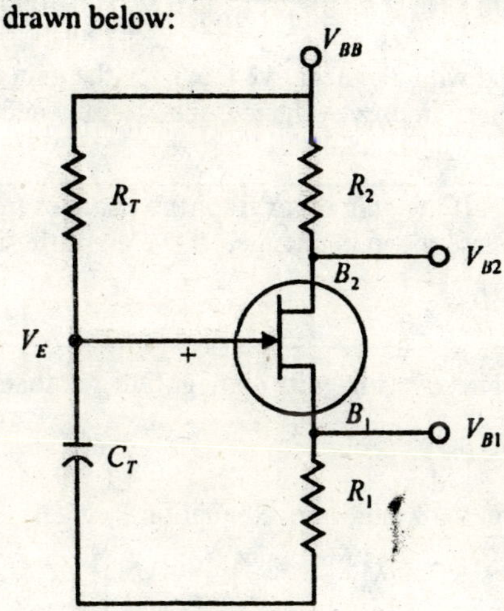

Fig. 8.13

Resistor R_T and capacitor C_T are the timing components that set the circuit oscillating rate. The oscillating frequency is calenlated using unijunction transistor intrinsic stand-off ratio η, as

$$f_o = \frac{1}{R_T C_T \ln\left[\frac{1}{(1-\eta)}\right]}$$

For a unijunction transistor η varies from 0.4 to 0.6. Take η = 0.5.

$$f_o = \frac{1}{R_T C_T \ln\left[\frac{1}{(1-0.5)}\right]} = \frac{1.44}{R_T C_T \ln 2}$$

$$\Rightarrow \qquad \boxed{f_o \cong \frac{1.5}{R_T C_T}}$$

Problem 8.1. Calculate the gain of a negative feedback amplifiere has $A = -2000$ and $\beta = \frac{-1}{10}$.

Solution. The gain of a negative feedback amplifier with feedback is given as:

$$A_f = \frac{A}{1 + \beta A} = \frac{-2000}{1 + 2000 \times \frac{1}{10}}$$

$$\Rightarrow \qquad A_f = \frac{-2000}{1 + 200} = \frac{-2000}{201} = -9.95$$

Thus, we find with the negative feedback the gain reduces dractically but many other factors improves, hence, inspite of reduction in gain it is used frequently.

Problem 8.2. If the gain of an amplifier changes from a value of – 1000 by 10%, calculate the gain change, if the amplifier is used in a feedback circuit having $\beta = \frac{-1}{20}$.

Solution. The overall gain of negative feedback system is given by

$$A_f = \frac{A}{1 + \beta A}$$

Now, differentiating the above equation, we get

$$\left|\frac{dA_f}{A_f}\right| = \frac{1}{|1 + \beta A|}\left|\frac{dA}{A}\right|$$

For $\quad \beta A \gg 1, \quad \left|\frac{dA_f}{A_f}\right| = \frac{1}{|\beta A|}\left|\frac{dA}{A}\right|$

Thus, substituting the values

$$\left|\frac{dA}{A}\right| = 10, \ \beta = \frac{-1}{20}, \text{ and } A = -1000, \text{ we get}$$

$$\left|\frac{dA_f}{A_f}\right| = \left|\frac{1}{\frac{1}{20} \times 1000}\right| \times 10 = \frac{10}{50} = 0.2\%$$

Thus, the gain change in a feedback amplifier is found to be 0.2%, which is quite small as compared without feedback.

Problem 8.3. Calculate the gain with and without feedback for an FET amplifier shown for circuit values $R_1 = 800\ k\Omega$, $R_2 = 200\ \Omega$, $R_o = 40\ k\Omega$, $R_D = 8\ k\Omega$ and $g_m = 5000\ \mu s$.

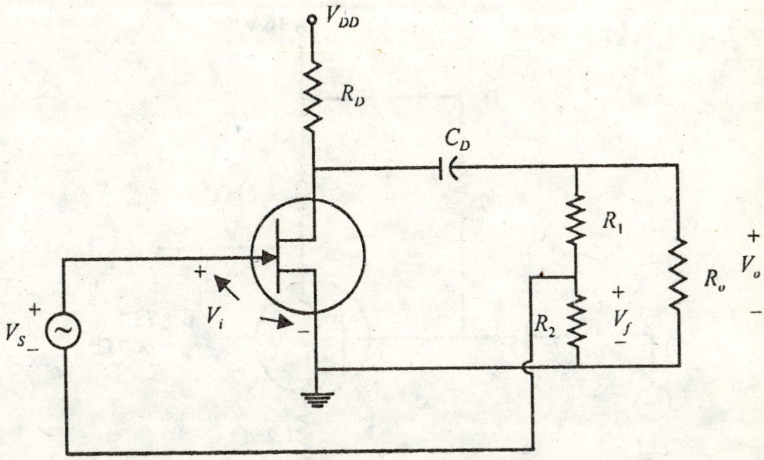

Fig. P. 8.3

Solution. The above figure is the FET amplifier stage with voltage series feedback. The load resistance R_L is given as

$$R_L \equiv R_o \parallel R_D = \frac{R_o R_D}{R_o + R_D}$$

$$\Rightarrow \quad R_L = \frac{40 \times 8}{40 + 8} = 6.67\,k\Omega$$

Neglecting the resistances R_1 and R_2 in series (1000, kΩ), the gain without feedback is given by

$$A = -g_m R_L = -(5000 \times 10^{-6}) \times 6.67 \times 10^3$$

$$= -33.3$$

The feedback factor $\beta = \dfrac{-R_2}{R_1 + R_2} = \dfrac{-200}{800 + 200} = -0.2$

Thus, the gain with feedback is

$$A_f = \frac{A}{1 + \beta A} = \frac{-33.3}{1 + 33.3 \times 0.2} = \frac{-33.3}{7.67}$$

$$\Rightarrow \quad A_f = -4.347$$

Problem 8.4. For the circuit shown with the given informations calculate the circuit gain with and without feedback $R_B = 600\ \Omega$, $R_E = 1.2\ k\Omega$, $R_C = 4.7\ k\Omega$, $\beta = 75$. Use $V_{CC} = 16$ V.

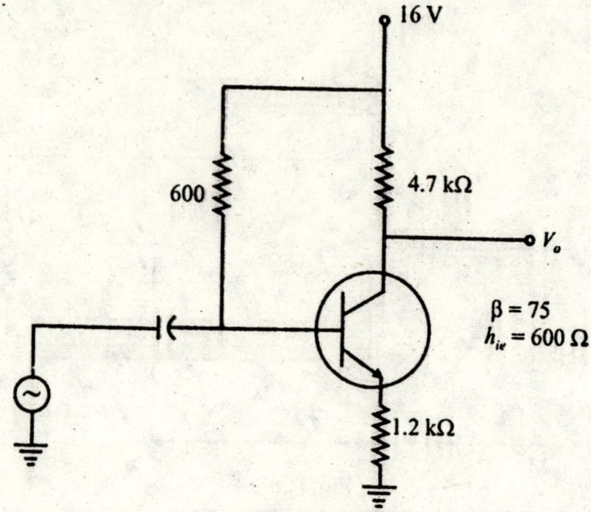

Fig. P. 8.4

Solution. The configuration given in the question is BJT amplifier with current-series feedback. The gain without feedback is

$$A = \frac{-hf_e}{hi_e + R_E} = \frac{-75}{600 + 1200} = -0.0417$$

The feedback factor $\beta = \dfrac{V_f}{I_o} = -R_E = -1.2 = -1200$

The factor $(1 + \beta A)$ is then

$$(1 + \beta A) = 1 + (-1200)(-0.0417) = 1 + 50 = 51$$

The gain with feedback is then $A_f = \dfrac{I_o}{V_s} = \dfrac{A}{1+\beta A}$

$$\Rightarrow \qquad A_f = \frac{-0.417}{51} = -8.176 \times 10^{-4}$$

And the voltage gain with feedback A_{Vf} is

$$A_{vf} = \frac{V_o}{V_s} = A_f R_C = -8.176 \times 10^{-4} \times 4.7 \times 10^3$$

$$\Rightarrow \qquad Av_f = -3.843$$

Without feedback ($R_E = 0$), the voltage gain is

$$A_v = \frac{-R_C}{re} = \frac{-4.7 \times 10^3}{15.5} = -303.2$$

Problem 8.5. An FET phase shift oscillator having $g_m = 6000\ \mu s$, $r_d =$ 36 kΩ and feedback resistor $R = 12$ kΩ is to operator at 2.5 kHz. Select C for specified oscillation operation and also calculate R_D, for $A > 2q$ to ensure oscillation action.

Solution. For a phase shift oscillator, the frequency of oscillation is given as

$$f = \frac{1}{2\pi R_C \sqrt{6}} \Rightarrow C = \frac{1}{\pi R_f \sqrt{6}}.$$

$$\Rightarrow \quad C = \frac{1}{2 \times 3.14 \times 12 \times 10^3 \times 2.5 \times 10^3 \times \sqrt{6}}$$

$$\Rightarrow \quad C = 2.167 \times 10^{-9}\ F$$

$$\Rightarrow \quad C = 2.167\ \text{nF}$$

The gain without feedback

$$|A| = g_m R_L, \quad \text{For } A > 29. \text{ Lets take } A = 40$$

$$R_L = \frac{|A|}{g_m} = \frac{40}{6000 \times 10^{-6}} = 6.67\ \text{k}\Omega$$

Now, the resistance R_L is

$$R_L = R_D \parallel r_d$$

$$\Rightarrow \quad R_L = \frac{R_D \times r_d}{R_D + r_d}$$

from here we calculates R_D as

$$\frac{1}{R_D} = \frac{1}{R_L} - \frac{1}{r_d}$$

$$\Rightarrow \quad \frac{1}{R_D} = \frac{1}{6.67} - \frac{1}{36}$$

$$\Rightarrow \quad \frac{1}{R_D} = 0.1499 - 0.0278$$

$$= 0.1221$$

$$\Rightarrow \quad R_D = 8.188\ \text{k}\Omega$$

Problem 8.6. Calculate the operating frequency of BJT phase shift oscillator for $R = 6$ kΩ, $C = 1500$ pF and $R_C = 18$ kΩ.

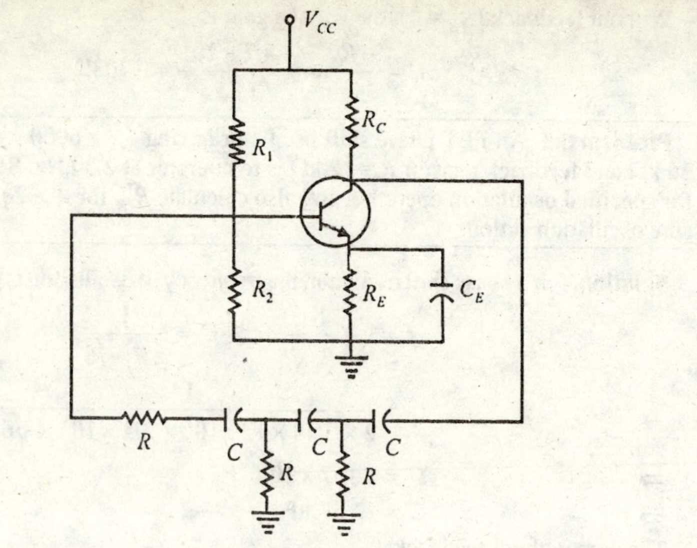

Fig. P. 8.6

Solution. The frequency of oscillation of a phase sift oscillation is:

$$f = \frac{1}{2\pi R_C \cdot \sqrt{6}} = \frac{1}{2\pi \times 6 \times 10^3 \times 1500 \times 10^{-12} \times \sqrt{6}}$$

$$\Rightarrow \qquad f = \frac{10^9}{2\pi \times 6 \times 1500 \times \sqrt{6}}$$

$$\Rightarrow \qquad f = 7.2194 \times 10^3 \text{ Hz}$$

$$\Rightarrow \qquad f = 7.22 \text{ kHz}$$

Problem 8.7. Calculate the frequency of a Wein Bridge oscillator circuit shown, when $R = 10$ kΩ and $C = 2400$ pF.

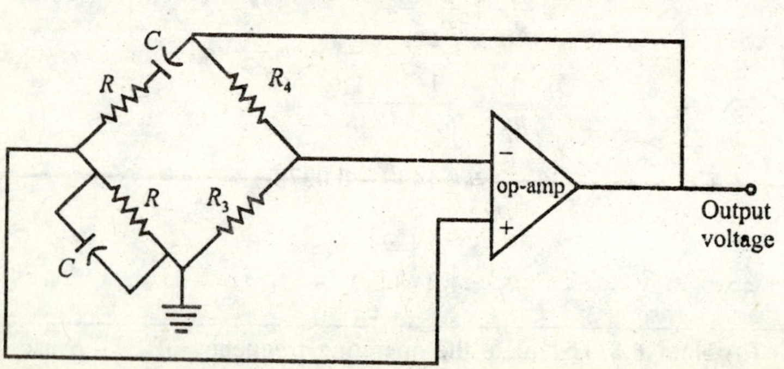

Fig. P. 8.7

Solution. The frequency of oscillation of the Wein Bridge oscillator is

$$f = \frac{1}{2\pi\sqrt{R_1 R_2 C_1 C_2}}$$

Here $R_1 = R_2 = R$ and $C_1 = C_2 = C$

Thus $$f = \frac{1}{2\pi RC}$$

Thus $$f = \frac{1}{2\pi \times 10 \times 10^3 \times 2400 \times 10^{-12}}$$

$$\Rightarrow \quad f = \frac{10^9}{2\pi \times 10 \times 2400} = 6.6314 \text{ kHz}$$

Problem 8.8. For an FET colpitts oscillator shown determine the circuit oscillating frequency, if $C_1 = 750$ pF, $C_2 = 2500$ pF, and $L = 40$ μH.

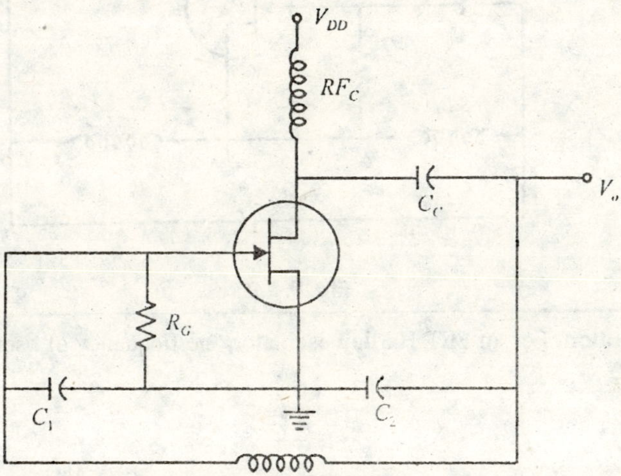

Fig. P. 8.8

Solution. The frequency of oscillation of a colpitts oscillator is given by

$$f = \frac{1}{2\pi\sqrt{LC_{eq}}} \quad \text{where, } C_{eq} = \frac{C_1 C_2}{C_1 + C_2}$$

$$\Rightarrow \quad C_{eq} = \frac{750 \times 2500}{750 + 2500}$$

$$C_{eq} = 576.92 \text{ pF}$$

$$\Rightarrow \qquad f = \frac{1}{2\pi\sqrt{40 \times 10^{-6} \times 576.92 \times 10^{-12}}}$$

$$\Rightarrow \qquad = 1.047 \times 10^{-3} \times 10^{9} = 1.047 \text{ MHz}$$

Problem 8.9. Calculate the oscillator frequency for an FET Hartley oscillator is $C = 250$ pF, $L_1 = 1.5$ mH, $2 = 1.5$ mH, and $m = 0.5$ mH.

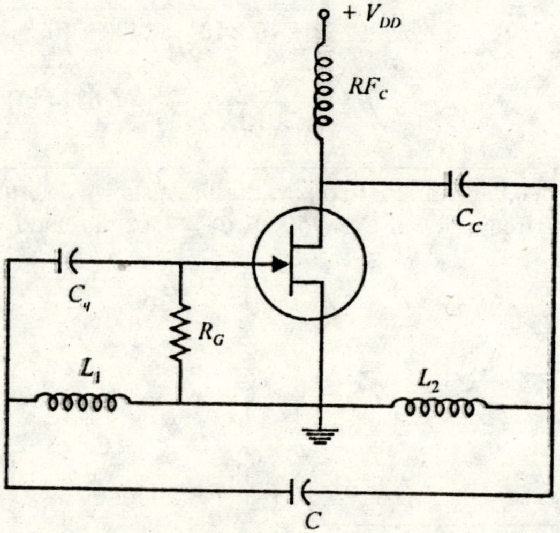

Fig. P. 8.9

Solution. For an FET Hartley oscillator, the frequency of oscillation is given by

$$f = \frac{1}{2\pi\sqrt{L_{eq}\, C}}$$

where, $$L_{eq} = L_1 + L_2 + 2M = 1.5 + 1.5 + 2 \times 0.5 = 4 \text{ mH}$$

$$\Rightarrow \qquad f = \frac{1}{2\pi\sqrt{4 \times 10^{-3} \times 250 \times 10^{-12}}}$$

$$\Rightarrow \qquad f = \frac{10^7}{2\pi\sqrt{0.4 \times 250}}$$

$$\Rightarrow \qquad f = \frac{10^7}{2\pi \times 10} = \frac{10^6}{2\pi} = 0.159 \text{ MHz}$$

$\Rightarrow \qquad f =$ [illegible]

Problem 8.10. Calculate the amplifier gain of the circuit shown for an op-amp gain $A = 10^5$ and resistances $R_1 = 2\ \text{k}\Omega$ and $R_2 = 200\ \Omega$.

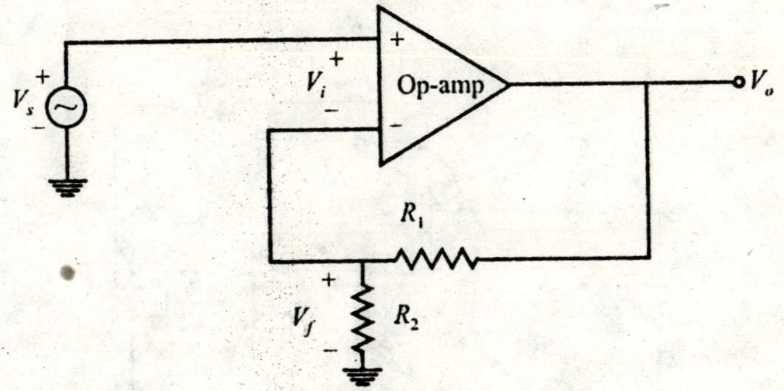

Fig. P. 8.10

Solution. Figure P. 8.10 is an example of voltage series feedback is an op-amp connection. The feedback factor:

$$\beta = \frac{R_2}{R_1 + R_2} = \frac{200}{200 + 2000} = \frac{200}{2200} = 0.091$$

The gain of the amplifier with feedback

$$A_f = \frac{A}{1 + \beta A} = \frac{10^5}{1 + (0.091) \times 10^5}$$

$$\Rightarrow \qquad A_f = \frac{10^5}{9091.91} = 10.999$$

Note that since $\beta A \gg 1$

$$A_f \cong \frac{1}{\beta} = \frac{1}{0.091} = 10.989$$

Problem 8.11. The circuit shown below shows a feedback amplifier. Let $R_s = R_e = 1\ \text{k}\Omega$, $R_C = 4\ \text{k}\Omega$. The transistor parameters are $b_{ie} = 2\ \text{k}\Omega$, $h_{fe} = 200$, and $h_{re} = b_{oe} = 0$. Assume that the inverting amplifier input resistance is infinite and gain $A_v = -2000$. Let the feedback factor $\beta = \frac{1}{150}$. Find the gain of the amplifiere $A_{vf} = \frac{V_o}{V_s}$.

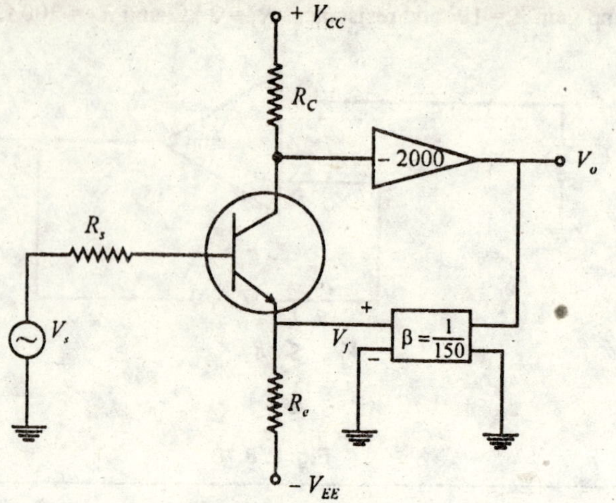

Fig. P. 8.11

Solution. Our steps will be as follows:

First, calculate the voltage gain of the transistor only bt this A_v.

Now with the voltage gain A_v and with inverting amplifier gain, we can have the block diagram as

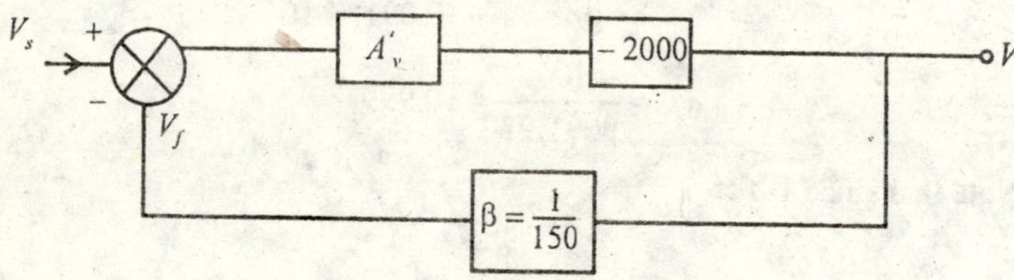

Fig. 8.11 (a)

and final voltage, gain with feedback will be A_{vf} as

$$A_{vf} = \frac{A_v \times (-2000)}{1 + \beta \times A_v \times (-2000)}$$

Now, to calculate A_v',

We simply proceed with the help of bybrid model, which is very similar to remodel except that β_{re} should replaced by β_{ie} and B should be replaced by h_{fe}.

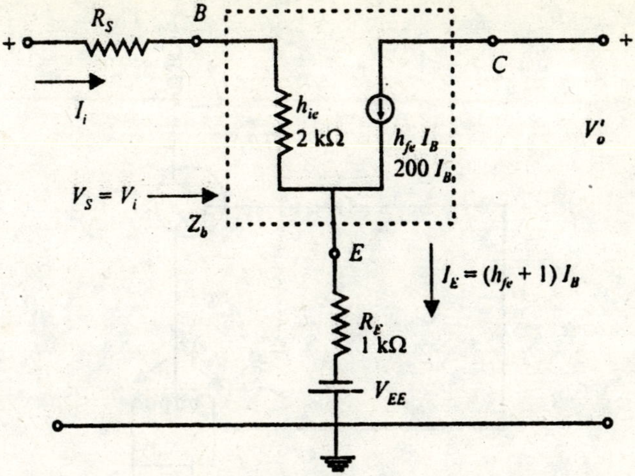

Fig. 8.11 (b)

Hence $$Z_b = h_{ie} + (h_{fe} + 1) R_E$$

{Similar to remodel $Z_b = \beta_{re} + (\beta + 1) R_E$}

or $$Z_b = 2\text{ k}\Omega + 201 \times 1\text{ k}\Omega = 203\text{ k}\Omega$$

Now $$A_v' = \frac{V_o'}{V_s} = \frac{h_{fe} R_C}{Z_b} \quad \{\text{Similar to remodel } A_v = -\frac{\beta R_C}{Z_b}\}$$

or $$A_v' = -\frac{200 \times 4\text{ k}\Omega}{203\text{ k}\Omega} = -\frac{800\text{ k}\Omega}{203\text{ k}\Omega} = -3.94$$

Now, from the blcok diagram analysis

$$A_{vf} = \frac{A_v' \times (-2000)}{1 + \beta \times A_v' \times (-2000)} = \frac{(-3.94) \times (-2000)}{1 + \frac{1}{150} \times (-3.94)(-2000)}$$

or $$A_{vf} = \frac{7880}{1 + \frac{7880}{150}} = \frac{7880}{1 + 52.534} = \frac{7880}{53.534}$$

or $$A_{vf} = 147.2$$

Problem 8.12. A Hartley oscillator has $L_1 = 2$ mH and $L_2 = 20$ mH. C is a variable capacitor to cover a range of frequencies from 950 kHz to 2050 KHz. Determine the extreme value of C and the minimum value of h_{fe} of the transistor.

Solution. A Hartley oscillator has the network as given below:

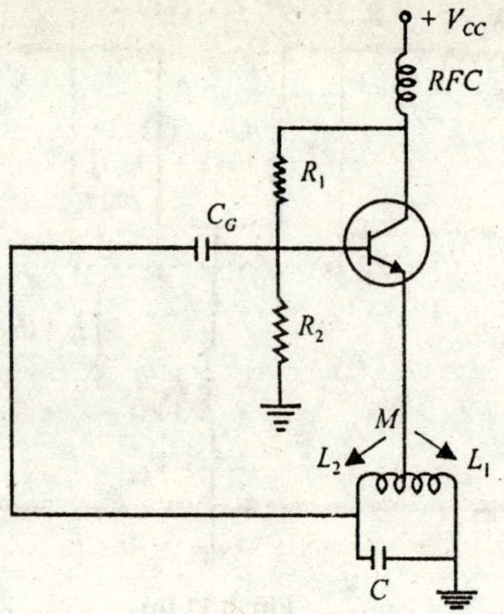

Fig. 8.12

Frequency of oscillation is given by $f = \dfrac{1}{2\pi\sqrt{L_{eq}\,C}}$

Nothing is said about the mutual inductance, so we will consider it negligible.
Two inductances L_1 and L_2 are in series

Hence $\quad L_{eq} = L_1 + L_2$

Therefore

$$f = \frac{1}{2\pi\sqrt{(L_1 + L_2)C}}$$

or
$$f^2 = \frac{1}{4\pi^2\,(L_1 + L_2)\,C} \Rightarrow C = \frac{1}{4\pi^2 f^2\,(L_1 + L_2)}$$

$$L_1 + L_2 = 2\text{ mH} + 20\text{ mH} = 2\text{ mH} + 0.02\text{ mH} = 2.02\text{ mH}.$$

or
$$C = \frac{1}{4 \times \pi^2\ f^2 \times 2.02 \times 10^{-3}} = \frac{12.54}{f^2}$$

For freq. $\quad f = 950$ kHz

$$C = \frac{12.54}{(950)^2 \times 10^6} = 13.89\text{ pF}$$

or $\quad C = 13.89$ pF, i.e. upper limit

For frequency $\quad f = 2050$ kHz

$$C = \frac{12.54}{(2050)^2 \times 10^6} = 2.984 \times 10^{-12} \text{ F}$$

or $$C = 2984 \text{ pF}$$

∴ Value of C is 13.89 pF to 2.984 pF

Required value of $h_{fe} = h_{fe\text{-min.}} = \frac{L_1}{L_2}$

or $$h_{fe\text{-min}} = \frac{2 \text{ mH}}{20 \text{ mH}} = 100$$

Problem 8.13. A three stage feedback amplifier has d.c. gain without feedback = – 10^4. The open loop poles are all at f = 2 MHz. Determine the maximum value of β, for which the amplifier is unstable and the corresponding frequency.

Solution. The two stage feedback amplifier is stable for all real positive values of β making $1 + A_0^2\ \beta > 1$.

or, a two stage feedback amplifier whose maximum phase change is less than 180° is stable, but a three stage amplifier with feedback with maxmimum. Phase change more than 180° can become unstable for large values of β.

Let B_C is the critical value when the amplifier oscillates then $B_C = \frac{8}{A_0^3}$, where A_0 = gain without feedback and frequency of oscillation is given by

$\omega_c = \sqrt{3} \times \omega_h$, ω_h = cut-off frequency of feedback amplifier

Here, given $|A_0| = 10^4$, and f_h = 2 MHz

Therefore $$\beta_{max} = \frac{8}{A_0^3} = \frac{8}{(10^4)^3} = 8 \times 10^{-12}$$

and from $f_c = f_h \times \sqrt{3}$, we get

$$f_c = (2 \text{ MHz}) \times \sqrt{3} = 2\sqrt{3} \text{ MHz}.$$

Problem 8.14. A single stage basic amplifier with gain A_m = 1000 has negative feedback $\beta = \frac{1}{10}$, applied to make it into a feedback amplifier, f_L = 20 Hz and f_H = 50 kHz for the basic amplifier. Find the values of f_L and f_H with feedback. At f_H, what is the gain of the basic amplifier?

Solution. Given feedback factor $\beta = \frac{1}{10}$, as we know that by applying feedback (negative). Bandwidth of the system increases. Hence, lower cut-off frequency becomes more low and higher cut-off frequency becomes more high, both by a factor $(1 + \beta A_m)$, given $A_m = 1000$, $f_L = 20$ Hz

after applying feedback

$$f'_L = \frac{f_L}{1 + \beta A_m}, \text{ and } f'_H = f_H (1 + \beta A_m)$$

$$(1 + \beta A_m) = 1 + \frac{1}{10} \times 1000 = 1 + 100 = 101$$

Hence $$f'_L = \frac{20 \text{ Hz}}{101} = \frac{20}{101} \text{ Hz} = 0.198 \text{ Hz}$$

and $$f'_H = (50 \text{ kHz}) \times 101 = 50 \times 101 \text{ kHz} = 5.05 \text{ MHz}$$

The gain with feedback is given by

$$A_{vf} = \frac{A_m}{1 + A_m \beta}$$

or $$A_{vf} = \frac{1000}{1 + 100} = \frac{1000}{101} = 9.9 \approx 10$$

Problem 8.15. For the feedback amplifier shown in Fig. P. 8.15 determine the nature of feedback, values of input and output resistances and gain $\frac{V_o}{I_s}$.

The circuit is simple for the purpose of evaluation.

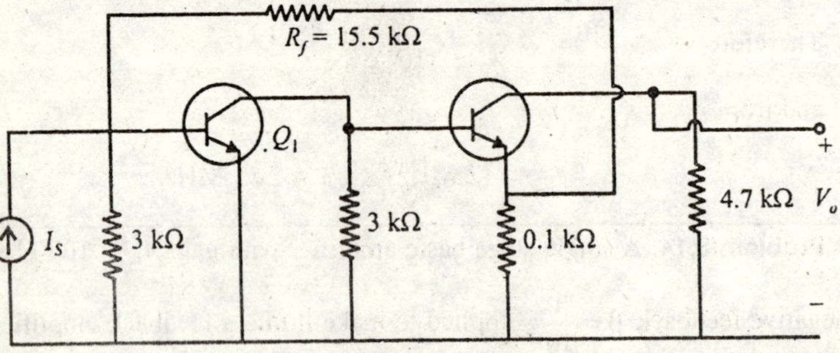

Fig. P. 8.15

Given h_{ie} = 1 kΩ and β = 50 for the transistor.

Solution. Figure P. 8.15 can be redrawn as follow:

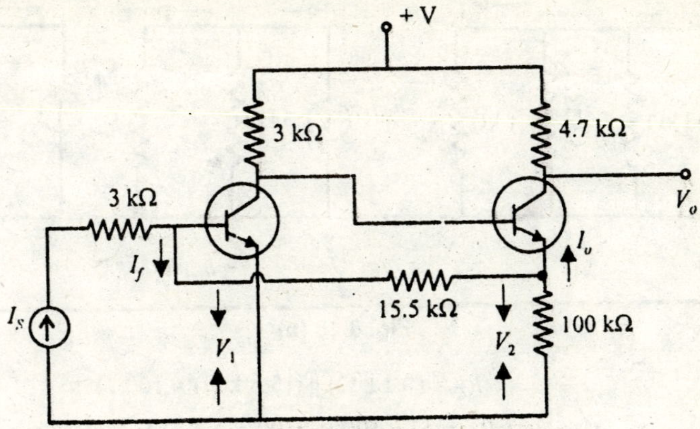

Fig. 8.15 (a)

Hence, the feedback is current shunt.

Now $$I_f = \frac{V_1 - V_2}{R_f} \approx -\frac{V_2}{R_f} \quad \text{... (1)}$$

Current through 100 Ω resistor = $I_f - I_o$

$$I_f = \frac{-(I_f - I_o)100\ \Omega}{R_f} \quad \text{[from (1)]}$$

or $$I_f = \frac{-(I_f - I_o)100\ \Omega}{15.5\ \text{k}\Omega}$$

or $$I_f\left(1 + \frac{100\Omega}{15.5\text{k}\Omega}\right) = I_o\ \frac{100\ \Omega}{15.5\ \text{k}\Omega}$$

or $$I_f\left(\frac{15.5\ \text{k}\Omega + 100\ \Omega}{15.5\ \text{k}\Omega}\right) = I_o \times \frac{0.1\ \text{k}\Omega}{15.5\ \text{k}\Omega}$$

or $$\frac{I_f}{I_o} = \frac{0.1\ \text{k}\Omega}{15.5\ \text{k}\Omega + 0.1\ \text{k}\Omega} = \frac{0.1}{15.6}$$

or feedback factor $$\beta = \frac{I_f}{I_o} = \frac{0.1}{15.6} = \frac{1}{156}$$

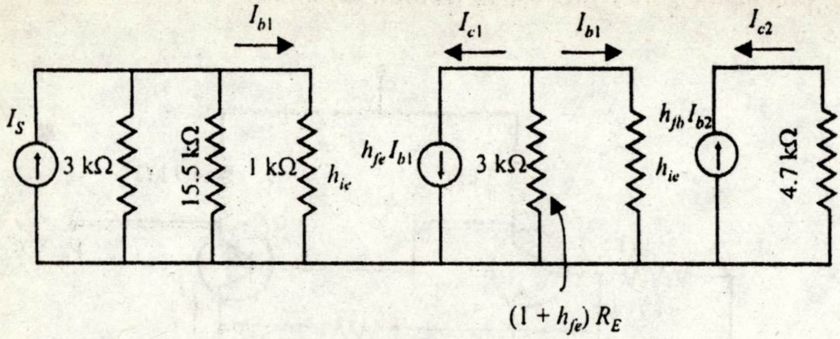

Fig. 8.15 (b)

$$R_E = (0.1\ \text{k}\Omega) \parallel (15.5\ \text{k}\Omega) \approx 100\ \Omega$$

$$(1 + h_{fe})\ R_E = 51 \times 100 = 5100 = 5.1\ \text{k}\Omega$$

$$A_i = \frac{I_o}{I_S} = -\frac{I_{C2}}{I_S}, \quad \frac{I_{c2}}{I_{b2}} = h_{fe} = 50$$

$$I_{b2} = -\frac{I_{c1}.3\ \text{k}\Omega}{(1\ \text{k}\Omega) + (5.1\ \text{k}\Omega) + 3\ \text{k}\Omega} = -\frac{3I_{c1}}{9.1}$$

Now $\dfrac{I_{b2}}{I_{c1}} = -\dfrac{3}{9.1}$

$\dfrac{I_{c1}}{I_{b1}} = h_{fe} = 50$, from current division rule, we have

$\dfrac{I_{b1}}{I_S} = \dfrac{2.516}{3.516}$, $\quad (3\ \text{K} \parallel 15.5\ \text{K} = 2.516)$

Hence, $A_i = \dfrac{I_{c2}}{I_s} = -\dfrac{I_{c2}}{I_{b2}} \times \dfrac{I_{b2}}{I_{c1}} \times \dfrac{I_{c1}}{I_{b1}} \times \dfrac{I_{b1}}{I_s}$

or $A_i = -50 \times \left(-\dfrac{3}{9.1}\right) \times 50 \times \left(\dfrac{2.516}{3.516}\right) = 589.76$

or $A_i \approx 590$

Now $\dfrac{V_o}{I_S} = -\dfrac{I_{c2}}{I_S} \times (4.7\ \text{k}\Omega) = 590 \times 4.7\ \text{k}\ \Omega$

$$= 2773\ \text{k}\Omega = 2.773 \times 10^6$$

$$R_i = (3\ \text{k}\Omega) \parallel (15.5\ \text{k}\Omega) \parallel (\text{k}\Omega) = (2.516\ \text{k}\Omega) \parallel (1\ \text{k}\Omega)$$

$$= \frac{2.516 \times 1}{3.516}\ \text{k}\Omega = 0.715\ \text{k}\Omega$$

Input resistanace $R_i = 715\ \Omega$

and $$R_{if} = \frac{715}{1 + 6.41 \times 10^{-3} \times 590} = 150\ \Omega$$

$$R_{of} = \frac{1}{0} = \infty. \qquad (\because\ D = 1 + A\beta = 1 - 1 = 0)$$

$$R_{of} = \frac{1}{h_{oe}} = \infty$$

$$R_{of} = R_{of} \parallel R_c = 4.7\ \text{k}\Omega.$$

Problem 8.16. Figure P. 8.16 shows an RC-phase shift oscillator. Solve the network to find the minimum value of h_{fe} for the transistor for oscillations to be possible. Also, determine the frequency of such oscillations. Take $C = 0.01\ \mu$F and $h_{ie} = 2$ kΩ.

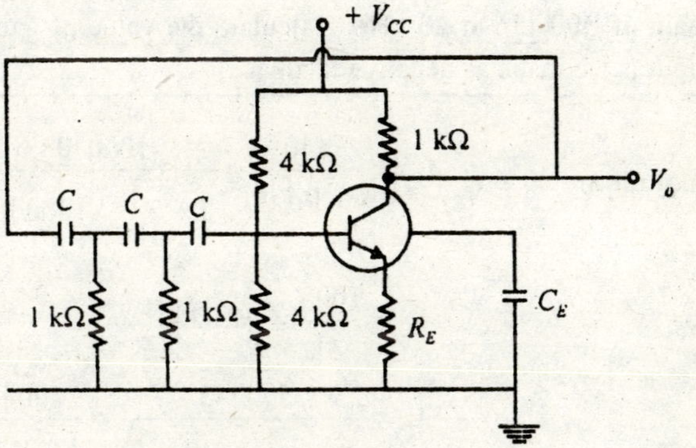

Fig. P. 8.16

Solution. In a BJT phase shift oscillator, for oscillations to take place, the minimum gain is

$$h_{fe} > \frac{4R_c}{R} + 23 + \frac{29R}{R_c}$$

Putting $R_c = 1\ \text{k}\Omega,\ R = 1\ \text{k}\Omega$

or $$h_{fe} > \frac{4 \times 1\ \text{k}\Omega}{1\ \text{k}\Omega} + 23 + \frac{29 \times 1\ \text{k}\Omega}{1\ \text{k}\Omega}$$

or $$h_{fe} > 4 + 23 + 29 = 56$$

i.e. $h_{fe} > 56$, therefore $h_{fe\text{-min}} = 56$

The input impedance h_{ie} of a BJT is low, and hence the output of the feedback network gets loaded.

So, a resistance is added in series with the feedback network and the base. This resistance

$$R' = R - (h_{ie} \parallel R_B) = 1\text{ K} - 1\text{ K} = 0$$

This means the input resistance of the BJT is fairly high, so as not to cause loading.

Frequency of oscillation $f_o = \dfrac{1}{2\pi RC\sqrt{6+4K}}$, where $K = \dfrac{R_C}{R}$

or
$$f_O = \frac{1}{2\pi \times 10^3 \times 0.01 \times 10^{-6}\sqrt{6+4}} = 0.05 \times 10^5$$

i.e. frequency of oscillation

$$f_O = 5\text{ kHz}$$

Problem 8.17. An amplifier hax a voltage gain of $1000\angle 0°$ at 100 Hz and a gain of $500\angle 45°$ at 20 kHz. Calculate the value of gain, if 2% of the output is feedback as negative feedback.

Solution.
$$A_f = \frac{A\angle\theta}{1+\beta A\angle\theta} = \frac{1000\angle 0°}{1+\dfrac{2}{100}\times 1000\angle 0°}$$

or
$$A_f = \frac{1000}{21} = 47.62\angle 0°$$

Also
$$A_f = \frac{500\angle 45°}{1+\dfrac{2}{100}\times 500\angle 45°} = \frac{500\angle 45°}{1+10\angle 45°}$$

or
$$A_f = \frac{500\angle 45°}{1+7.07+17.07} = \frac{500\angle 45°}{8.07+17.07}$$

$$= \frac{500\angle 45°}{10.725\angle 41.22°}$$

or
$$A_f = 46.62\angle 45° - 41.22° = 46.62\angle 3.78°$$

or
$$A_f = 46.62\angle 3.78°$$

Problem 8.18. For the two stage feedback amplifier shown Fig. P. 8.18 (*a*) identify the nature of feedback, (*b*) draw the equivalent circuit without overall feedback. Obtain A_v, R_i, and R_o, and (*c*) determine $D = 1 + A\beta$.

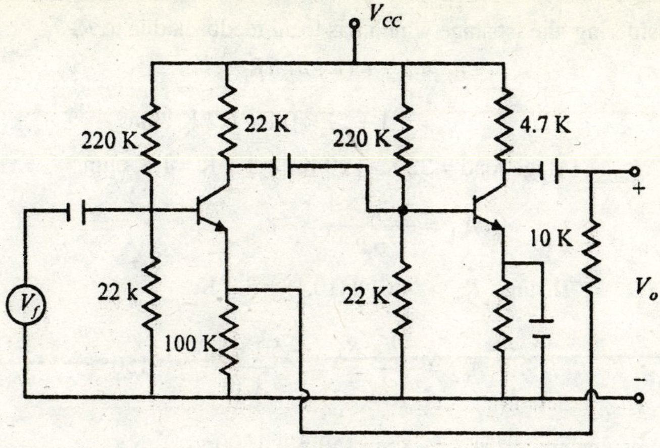

Fig. P. 8.18

Hence, write down A_{vf}, R_{if}, R_{of}. Given h_{ie} = 1.1 K and h_{fe} = 50.

Solution. (*a*) The first stage has a local feedback due to R_E = 100 ohm, which carries the collector current and its voltage is in series opposition to the signal. It is current series feedback.

Overall, a fraction of the output voltage V_o is fed in series opposition to the input

$$\beta = \frac{0.1}{10.1} \approx \frac{1}{100}. \text{ It is voltage series feedback}$$

(*b*) In removing the overall feedback β

R_E is modified to $R'_E = 0.1$ K parallel to 10 K

$= R_E$ approx.

The output is loaded by 10.1 K ohms due to the feedback network.

The bias resistors of 22 K and 220 K, give

$$R_B = \frac{22 \times 220}{242} = 20 \text{K}$$

The circuit without feedback is as shown in Fig. 8.18 (a)

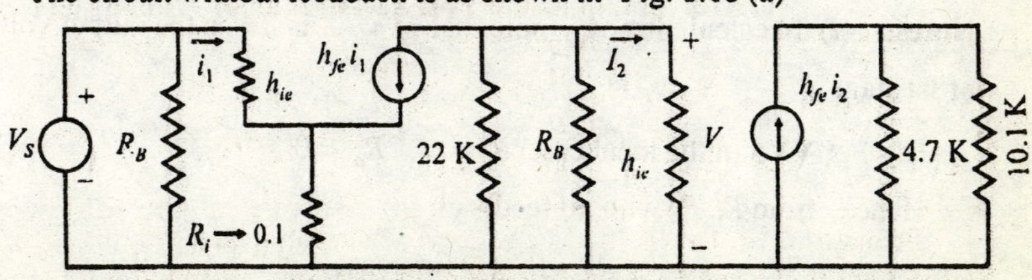

Fig. 8.18 (a)

Considering the Ist stage which has local feedback due to R_E

$$R_i = h_{ie} + (1 + h_{fe}) R_E$$

$$= 1.1 + 51 \times 0.1 = 6.2 \text{ K ohms}$$

$$\text{I stage load} = 22 \text{ K} \,||\, 20 \text{ K} \,||\, 1.1 \text{ K} = 1 \text{ K ohm}$$

$$A_{v1} = \frac{-h_{fe} \times 1}{6.2} = -8$$

$$\text{II stage } R_o = (4.6) \,||\, (10.1) = 3.3 \text{ K}$$

$$A_{v2} = \frac{-h_{fe} R_o}{h_{ie}} = \frac{-50 \times 3.3}{1.1} = -150$$

The results are $A_v = -8 \times -150 = +1200$

$$R_i = 6.2 \text{ K}, \quad R_o = 3.3 \text{ K}$$

(*c*) Feedback factor $D = 1 + A\beta$

$$A = 1200, \quad \beta = \frac{1}{100}$$

Hence $D = 13$

The effect of voltage series feedback is to make the amplifier a voltage controlled voltag esource (VVS). Hence, input resistance increases by factor *D*, output resistance decreases by factor $\frac{1}{D}$.

$$R_{if} = 6.2 \times = 80.6 \text{ K ohms}$$

$$R_{of} = \frac{3.3}{13} = 0.25 \text{ K ohms}$$

$$A_{Vf} = \frac{1200}{13} = 92$$

Notes: (*i*) In calculating A_{v2} note that $i_2 = \frac{V}{h_i}$ and V is the output voltage of Ist stage.

(*ii*) For VVS in the ideal case $R_i = \infty, \quad R_o = 0$

Hence, from R_i, R_o with no feedback

$$R_{if} = R_i \times D \text{ (increase)}$$

$$R_{of} = \frac{R_o}{D} \text{ (decrease)}$$

(*iii*) R_B is across V_S. Hence, the voltage across the input of the transistor V_i across R_i is also V_S. If there was a source resistance R_S, then the input to the transistor is only a fraction of V_S. By Norton equivalent

$$V_i = \frac{V_S}{R_S} \times R$$

Where $$R = R_S \parallel R_B \parallel R_i$$

Problem 8.19. A three stage feedback amplifier is made up of 3 identical FET common source amplifiers.

Each stage has a drain load R_d coupled ($C_c R_g$) to the next stage R_g being the gate to ground resistor. Overall 'voltage sereis' feedback is given from R_g of the last stage with $|\beta| = R_1/R_g$.

Determine overall gain with feedback and output resistance with feedback in midband.

Given: $g_m = 6 \times 10^{-3}$ mhos, $r_d = 10\ K$, $R_d = 50$ K

$R_g = 1$ Meg Ω, $R_1 = 40$ ohms

Solution. Single stage load $= r_d \parallel R_d \parallel R_g = R_L$

$$= 10\text{ K} \parallel 50\text{ K} \parallel 1\text{Meg }\Omega$$

$$= \frac{1000}{0.121} = 8.3\text{ K ohms}$$

$$\text{Single stage gain} = -g_m R_L = \frac{-6 \times 10^{-3} \times 10^3}{0.121} = -50$$

$$A_v = (-50)^3 = -125 \times 10^3$$

$$-\beta = \frac{40}{10^6} = 4 \times 10^{-5}$$

$$D = 1 + A\beta = 1 + 125 \times 10^3 \times 4 \times 10^{-5} = 6$$

$$A_{vf} = \frac{A_v}{D} = \frac{-125 \times 10^3}{6} = -20.8 \times 10^3$$

$$R_o = \frac{1000}{0.121} = 8.3\text{ K ohms taking } R_g \text{ as part of load.}$$

$$R_{of} = \frac{R_o}{D} = \frac{8.3}{6} = 1.39\text{ K ohms}$$

Note that the feedback amplifier is VVS and the output impedance is reduced by feedback.

Problem 8.20. (*a*) The gain of an amplifier varies due to parameters changes. Feedback is given to reduce this variation. Establish a relation between % change in gain and feedback.

(*b*) An amplifier has $A_v = 1000 \pm 100$. Determine the feedback needed to keep the gain within $\pm$ 0.1 percent. Find A_{vf}.

Solution. (*a*) Gain with feedback A_{vf} is related to A, thus

$$A_{vf} = \frac{A}{1 + A\beta}$$

Taking log and differentiating

$$\log A_{vf} = \log A - \log(1 + A\beta)$$

$$\frac{dA_{vf}}{A_{vf}} = \frac{dA}{A} - \frac{\beta dA}{1 + A\beta} = \frac{dA}{A} \quad \frac{1}{1 + A\beta}$$

$$\frac{dA_{vf}}{A_{vf}} = \frac{1}{D} \times \frac{dA}{A}$$

Thus, % change in $A_{vf} = \frac{1}{D} \times$ % change in A.

Where $D = 1 + A\beta$

(*b*) $$\frac{dA_{vf}}{A_{vf}} = 0.1\% = \frac{0.1}{100} = 10^{-3}$$

$$\frac{dA}{A} = \frac{100}{1000} = 10^{-1}$$

$$D = 1 + A\beta = \frac{10^{-1}}{10^{-3}} = 100$$

Since $A_v = 1000$

$$\beta = \frac{99}{1000} = \frac{1}{10} \text{ approx}$$

$$A_{vf} = \frac{A}{D} = \frac{1000}{100} = 10$$

Problem 8.21. For the circuit shown in Fig. P. 8.21, determin eht etype of feedback. Obtain the voltage gain, input and output impedances in the midband. The transistor parameters are

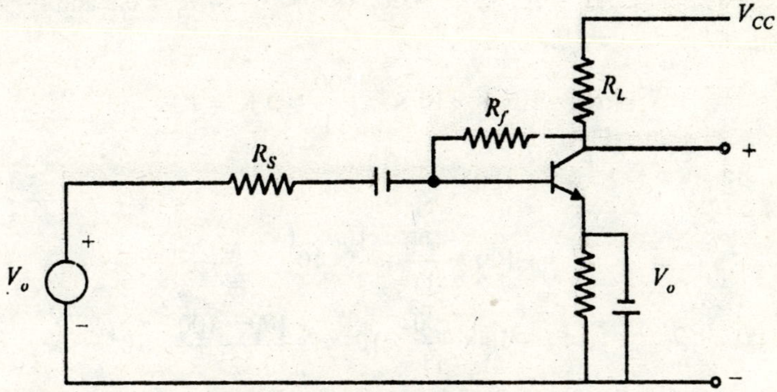

Fig. P. 8.21

$$h_{ie} = 1\text{ K},\ h_{fe} = 100$$

$$R_S = 1\text{ K},\ R_L = 10\text{ K},\ R_f = 100\text{ K}$$

Solution. R_f is across the output and samples the output voltage V_o. At the input it is in shunt resulting in current feedback. The feedback is 'voltage shunt'.

$$\beta = \frac{I_f}{V_o} = -\frac{1}{R_f} = -\frac{1}{100 \times 10^3}\text{ mho}$$

The feedback amplifier is CVS. Hence, both input and output impedances are reduced by feedback.

The amplifier gain to be considered is $\frac{V_o}{I_S} = A_z$, where I_S is the current source $= \frac{V_S}{R_S}$.

The equivalent circuit of the amplifier without feedback will have R_f across R_L, and R_f across input.

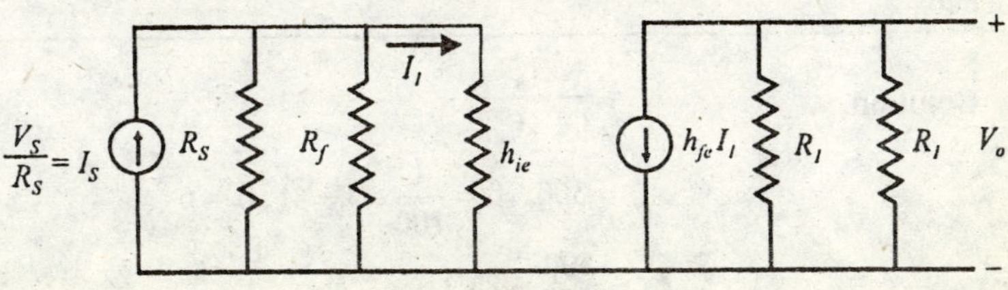

Fig. 8.21 (a)

$$A_z = \frac{V_o}{I_S} = \frac{V_o}{I_l} \times \frac{I_l}{I_S}$$

Since $R_S = h_{ie} = 1$ K and $R_f = 100$ K >> h_{ie}, $R_i = 0.5$ and $\frac{I_l}{I_S} = \frac{1}{2}$

$$R_f \parallel R_L = 100 \text{ K} \parallel 10 \text{ K} = \frac{100}{11} = 9 \text{ K} = R_o$$

$$\frac{V_o}{I_l} = -100 \times 9 \times 10^3$$

$$A_z = -100 \times \frac{100}{11} \times \frac{1}{2} \times 10^3$$

$$= -50 \times \frac{100}{11} \times 10^3 = \frac{-100 \times 100}{22} \times 10^3$$

$$R_i = 0.5 \text{ K}, \ R_o = 9 \text{ K}$$

$$D = 1 + A_z, \beta = 1 + \frac{100 \times 100}{22} \times \frac{1}{100} = 5.55$$

Hence
$$A_{zf} = \frac{-50}{5.55} \times \frac{100}{11} \times 10^3$$

$$A_{Vf} = \frac{A_{zf}}{R_S} = -82$$

$$R_{if} = \frac{0.5}{D} = 90 \text{ ohms}$$

$$R_{of} = \frac{9}{D} = 1.63 \text{ K ohms}$$

Problem 8.22. A single stage amplifier has a midband gain $A_m = 500$, lower and upper cut-off frequencies $f_l = 50$ H, $f_h = 1$ kHz. A negative feedback with $\beta = \frac{1}{100}$ is given. Determine gain and cut-off frequencies with feedback.

Solution.
$$A_f = \frac{A_m}{1 + A_m \beta} \quad \text{...(1)}$$

$$A_m = 500, \ \beta = \frac{1}{100}, \ 1 + A_m \beta = 6$$

$$A_f = \frac{500}{6} = 83.3$$

$$f_L = \frac{f_l}{(1+\beta A_m)} = \frac{50\,\text{Hz}}{6} = 8.3\text{ Hz}$$

and $$f_H = f_h(1+\beta A_m) = 1\times 10^3 \times 6 = 6\text{ kHz}$$

Thus, f_H has increase and f_L has decreased resulting in a increased bandwidth.

Problem 8.23. The basic amplifier has a gain function $A(s) = \dfrac{A_o}{1+\dfrac{s}{\omega_1}}$, where A_o is the gain at $s = 0$ and ω_1 is the upper cross-over. Two stages and three stages cascade amplifiers are built using such basic stages. The cascade amplifier is given an overall negative feedback β. Discuss the stability of the feedback amplifier of two stages and of three stages.

Solution. (*a*) Two stages amplifier without feedback

$$A_2(s) = \frac{A_0^2}{\left(1+\dfrac{s}{\omega_1}\right)\left(1+\dfrac{s}{\omega_2}\right)}$$

assuming that the upper cut-off frequencies ω_1 and ω_2 are different.
The feedback amplifier gain $A_{2f}(s)$ is given by

$$A_{2f}(s) = \frac{A_2}{1+\beta A_2} = \frac{A_0^2}{\left(1+\dfrac{s}{\omega_1}\right)\left(1+\dfrac{s}{\omega_2}\right) + A_0^2\beta}$$

$$= \frac{A_0^2}{(1+A_0^2\beta)}\cdot\frac{1}{D}$$

Where $$D = \left[1 + A_0^2\beta + s\left(\frac{1}{\omega_1}+\frac{1}{\omega_2}\right) + \frac{s^2}{\omega_1\,\omega_2}\right] + \left[1 + A_0^2\beta\right]$$

$$= 1 + \frac{s(\omega_1+\omega_2)}{\omega_1\,\omega_2(1+A_0^2\beta)} = \frac{s^2}{\omega_1\,\omega_2(1+A_0^2\beta)}$$

$$= 1 + \frac{s(\omega_1+\omega_2)}{\omega_0^2} + \frac{s^2}{\omega_0^2}$$

$$= 1 + \frac{s}{\omega_0} . \frac{\omega_1 + \omega_2}{\omega_0} + \frac{s^2}{\omega_0^2}$$

$$= 1 + \frac{1}{Q} \frac{s}{\omega_0} + \left(\frac{s}{\omega_0}\right)^2$$

Where $$Q = \frac{\omega_0}{\omega_1 + \omega_2}$$

and $$\omega_0^2 = \omega_1 \omega_2 (1 + A_0^2 \beta)$$

D is a quadratic with real coefficients. Its roots always have a negative real part given $-\omega_0/2Q$. Hence, the poles of A_{2f} are in the LHP of s.

The two stages feedback amplifier is stable for all real positive value of β making $1 + A_0^2 \beta > 1$.

(*b*) Three stages amplifier without feedback is given by

$$A_3(s) = \frac{A_0^3}{\left(1 + \frac{s}{\omega_1}\right)\left(1 + \frac{s}{\omega_2}\right)\left(1 + \frac{s}{\omega_3}\right)}$$

where the cross over frequencies are ω_1, ω_2, and ω_3. If they are taken equal and labelled ω_L.

$$A_3(s) = \frac{A_0^3}{\left(1 + \frac{s}{\omega_h}\right)^3}$$

$$\text{‘}1 + A\beta\text{’} = 1 + \frac{\beta(A_0^3)}{\left(1 + \frac{s}{\omega_h}\right)^3}$$

For $s = j\omega$ the phase changes result in $A\beta$ becomig negative. This happens when the total phase of

$$\left(1 + \frac{s}{\omega h}\right)^3 = \left(1 + \frac{j\omega}{\omega_h}\right)^3 \text{ is } 180^\circ$$

or phase of $1 + \frac{j\omega}{\omega_h}$ is 60°

$$\tan \frac{\omega_c}{\omega_h} = \tan 60° = \sqrt{3}$$

where ω_c is the particular frequency at which there is a sign change. The value of $A\beta$ at this frequency is given by

$$\beta A_0^3 /(1 + j\sqrt{3})^3 = \frac{\beta A_0^3}{8} \underline{\lfloor -180°} = \frac{\beta A_0^3}{8}$$

For $$'1 + A\beta' = 0,\ \beta_c A_0^3 / 8 = 1 \text{ or } \beta_c = \frac{8}{A_0^3}$$

where β_c is the critical value when the amplifier oscillates and the frequency of oscillation is $\omega_c = \sqrt{3}\,\omega_h$.

Thus, a two stages amplifier with feedback, whose maximum phase change is less than 180° is stable. A three stages amplifier with feedback with maximum phase change more than 180° can become unstable for large values of β.

Notes: The closed loop gain poles are given by

$$\left(1 + \frac{s}{\omega_h}\right)^3 + \beta A_0^3 = 0$$

For the critical value $\beta_c A_0^3 = 8$

$$\left(1 + \frac{s}{\omega_h}\right)^3 + 8 = 0$$

$$S^3 + 3S^2 + 3S + 9 = 0$$

Where $$S = \frac{s}{\omega_h},\ (S + 3)\ (S^2 + 3) = 0$$

The poles are at $$s = \sqrt{3}\ \omega_h \text{ and } \pm\sqrt{3}\ j.\omega_h.$$

We have already seen that $\omega_c = \sqrt{3}\ \omega_h$. Further, if β is greater than β_c, the poles are RHP of a plane and the feedback amplifier is unstable.

Reader's Notes

CHAPTER

9

Operational Amplifier

General Features of op-amp

Symbol

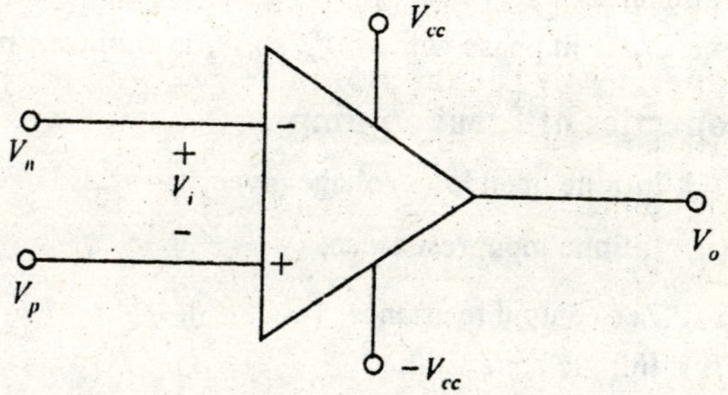

Fig. 9.1

Ckt. Representation

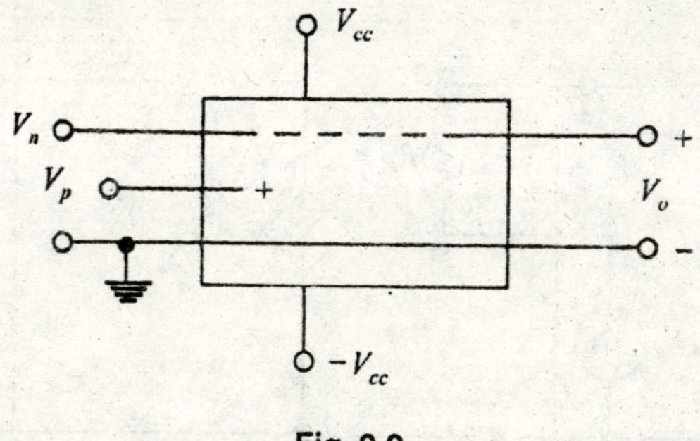

Fig. 9.2

An op-amp is a direct coupled, high gain voltage amplifier designed to amplify signals over a wide frequency range (from d.c. upwards), typically it has two *i/p* terminals and one *o/p* terminal and a gain of at least 10^5.

An op-amp is basically a differential amplifier responding to the difference in voltages applied at the – ve (inverting) and +ve noninverting) terminal.

Thus $\boxed{V_o = -A_v\left(V_n - V_p\right) = -A_v V_i}$

Where $V_i = V_n - V_p$

A_v = voltage gain

V_n = – ve (noninverting) terminal voltage

V_p = + ve (noninverting) terminal voltage

V_o = output voltage

N.B. If $V_p = 0, \; V_o = -A_v V_n$

i.e. *o/p* is inverted (180° out of phase) w.r.t. V_n

So, V_n is the inverting voltage

Similarly, if $V_n = 0, \; V_o = +A_v V_p$

i.e. *o/p* is in phase with *i/p* V_p, so, V_p is noninverting voltage.

Properties of Ideal op-amp

(i) Infinite open loop voltage given, $(\infty - A_v)$.

(ii) Infinite input resistance, $(\infty - A_v)$.

(iii) Zero output resistance, $(O - R_o)$.

(iv) Infinite bandwidth.

(v) Infinite common mode rejection ratio $(\infty - \text{CMRR})$.

Equivalent Ckt. of an op-amp

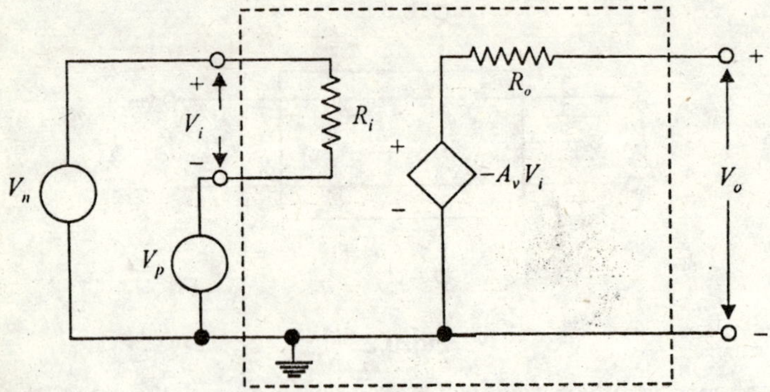

Fig. 9.3

$R_i = \infty$, means current into inverting and noninverting terminals is almost zero.

$A_v = \infty$, means for any finite V_o, $V_i = V_n - V_p = 0$.

$\therefore$ the inverting and noninverting terminals have the same potential. This property, i.e. $V_i = 0$ is known as *virtual grounding.*

N.B.: For linear operation the output voltage cannot exceed $\pm V_{cc}$.

$\Rightarrow$ Normaly op-amp is used with a feedback network that determines the f^{ns} performed.

Operational Amplifier Architecture

Four stage structure cascaded.

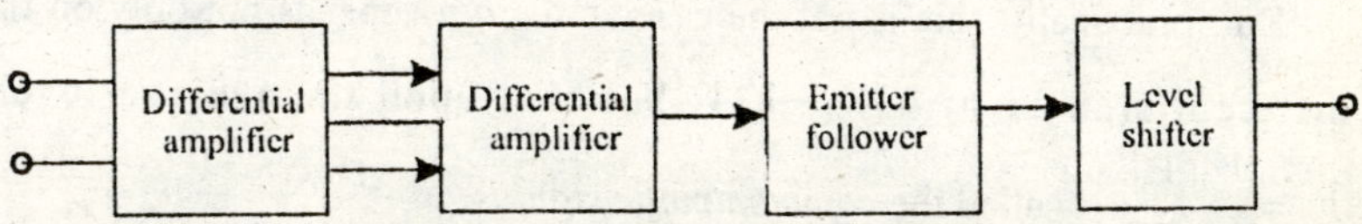

Fig. 9.4

First stage: Double-ended output, provides high input resistance, high gain for different signals and rejection of signals common to both terminals.

Second stage: Single output different amp, provides more gain and more discrimination.

Third stage: Is an emitter follower, a buffer to provide low output resistance and isolation of the amplifier from the load.

Fourth stage: Is a combination level shifter and drive stage. The level shifting correct for any d.c. offset that have been introduced by the bias n/w or by compound imbalances.

The driver is a power amplifier to provide large *o/p* current with a low R_o.

op-amp Parameters

(i) ***Open loop voltages gain*** (A_v): When the op-amp is used without any feedback, the differential voltage gain is open loop gain (voltage).

So $$\boxed{A_v = \frac{V_o}{V_n - V_p}}$$

Practically $\rightarrow 10^3$ + to 10^5.

Since, V_o is limited to $\pm V_{cc}$, the supply voltage for such a high gain, the differential *i/p* voltage should be very small, ideally *Zero*.

(ii) ***Input resistance (R_i):*** The open loop incremental resistance looking into the two i/p terminals and is typically 2 MΩ.

(iii) ***Output resistance (R_o):*** It's usually between 50 Ω and 500 Ω, e.g. 741 IC, being 75 Ω.

(iv) ***Common mode rejection ratio (CMRR):*** Ideally, if two equal voltages are applied to the i/p of an op-amp, the *o/p* is zero, i.e. the signal which is common to both inputs should not contribute anything to the *o/p* voltage.

But, practically, this is not the case as the *o/p* depends not only on the different signals $v_d = \left(V_p - V_n\right)$, but also upon the average levels $V_c = \dfrac{V_p + V_D}{2}$ called the *common mode signals.*

Thus $$V_o = A_d\left(V_p - V_n\right) + A_c\left(\frac{V_p + V_n}{2}\right)$$

$$V_o = A_d V_d + A_c V_c$$

$$\Rightarrow \quad V_o = A_d V_d\left(1 + \frac{V_c}{V_d}\cdot\frac{1}{\text{CMRR}}\right)$$

Where A_d = different mode gain

A_c = common mode gain

$$\text{CMRR} = \left|\frac{A_d}{A_c}\right|$$

expressed in dB as 20 log $|A_c / A_d|$

An op-amp is designed with high CMRR (approx. 90 dB), so that $V_o \approx A_d\ V_d$, thus, with high CMRR, op-amp rejects the common mode signals, and the *o/p* voltage is contributed only by the differential voltage signals V_d

$$\therefore \quad A_v = A_d$$

Application of op-amp

While the ideal op-amp is a linear device in that the output is directly proportional to *i/p* for all *i/p* voltages, the practical's characteristic is as shown.

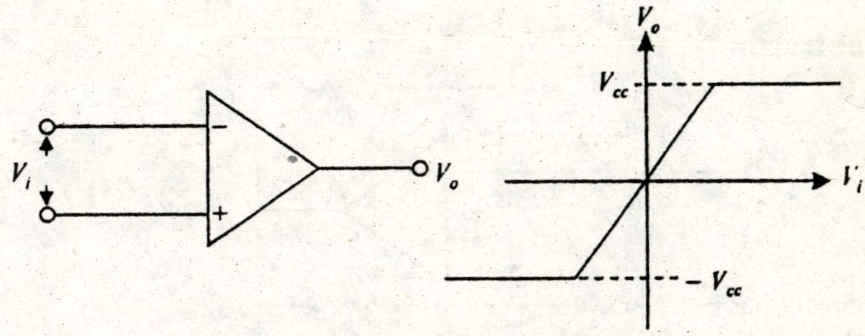

Fig. 9.5 op-amp transfer characteristic

Thus, the practical op-amp acts as a linear device over a certain range of the *i/p* and as a nonlinear (saturating) device otherwise.

Linear Applications

Adder or Summing Amplifier: Since, the inverting terminals acts as a virtual ground so, no current flows into op-amp.

So $$i_1 + i_2 + i_3 = I_f = -\frac{V_o}{R_F}$$

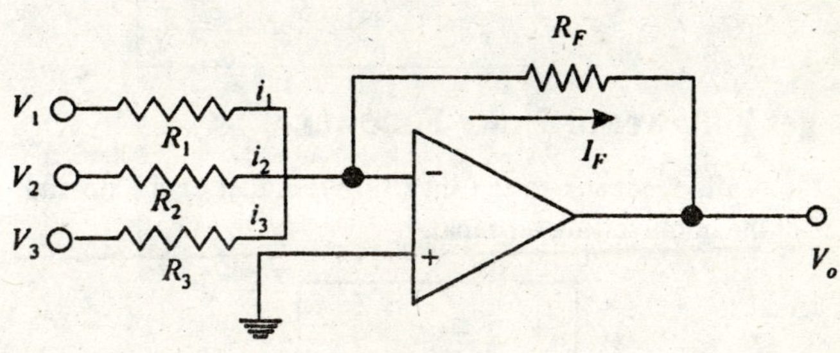

Fig. 9.6

$$\Rightarrow \quad V_o = -\left[(R_F/R_1)V_1 + (R_F/R_2)V_2 + (R_F/R_3)V_3\right]$$

If $$R_F = R_1 = R_2 = R_3$$

$$\boxed{V_o = -(V_1 + V_2 + V_3)}$$, thus acts as an inverting adder.

If $R_1 = R_2 = R_3 = 3R_F$

$$\boxed{V_o = -\frac{1}{3}(V_1 + V_2 + V_3)} = \text{average of 3 } o/p.$$

Subtractor

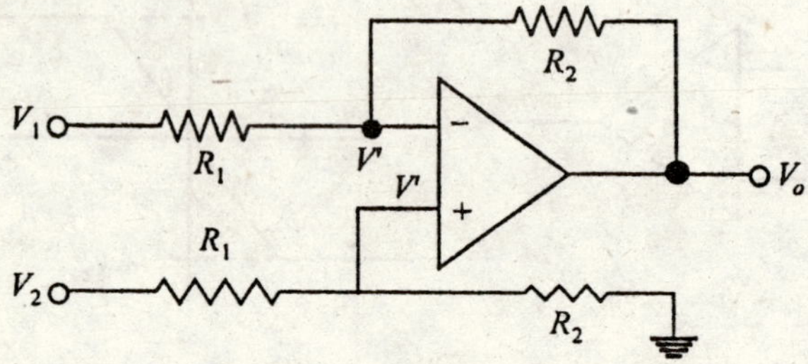

Fig. 9.7

Assuming ideal

$$(V_1 - V')/R_1 = (V' - V_o)/R_2$$

and $$(V_2 - V')/R_1 = V'/R_2$$

or $$\boxed{V_o = \left(\frac{R_2}{R_1}\right)(V_2 - V_1)}$$

$\Rightarrow$ $$\boxed{V_o = V_2 - V_1 \text{ for } R_2 = R_1}$$

Voltages Follower or Unity Follower

Here, the o/p is feedback to the inverting terminal of the op-amp with *i/p* applied at the noninverting terminal.

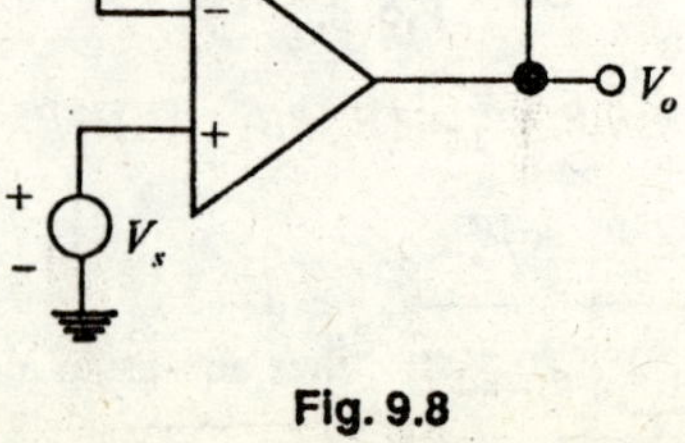

Fig. 9.8

It's in fact a noniverting amplifier with $R_F = 0$ and R_I as open ckt.

$$\frac{V_o}{V_s} = 1 + \frac{R_F}{R_1} = 1 + 0 = 1$$

Thus, the *o/p* followers the *i/p*.

For Practical

If open loop gain—A, input impedance R_i and *o/p* impedance of R_o then in the voltage follower configuration, *input impedance becomes* AR_i and *o/p impedance becomes* R_o/A.

This ckt. is used in isolating a high impedance source from a low impendance load. Such isolating ckt. is called *a buffer* and this particular type of buffer is called *unit gain buffer*.

Current to Voltage Converter (Transistance Amplifier)

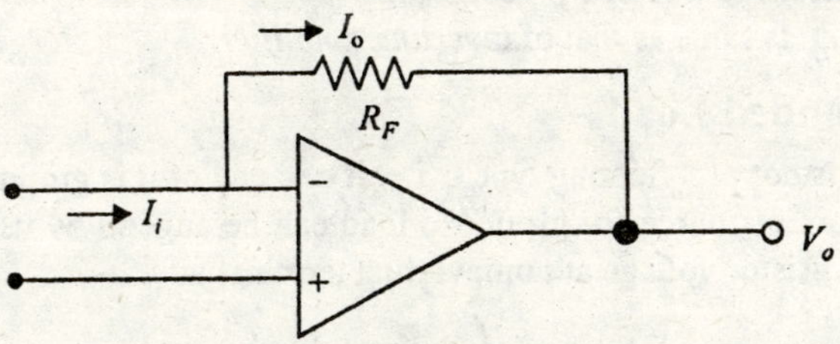

Fig. 9.9

The *i/p* current I_i is applied directly to the inverting *i/p* of the op-amp.

$$I_F = I_i \quad \text{and} \quad \boxed{V_o = T_i R_F}$$

Disadvantage and Care

The disadvantages is that the *i/p* bias current gets added to the *i/p* current.

Hence $\qquad V_o = -(I_B + I_i)\, R_F$

∴ Care should be taken to keep I_B as small as possible.

Often a capacitor is used in ||el with R_F to reduce noise at high frequencies.

Voltages to Current Converter (Transconductance Amplifier)

Using the concept of virtual ground.

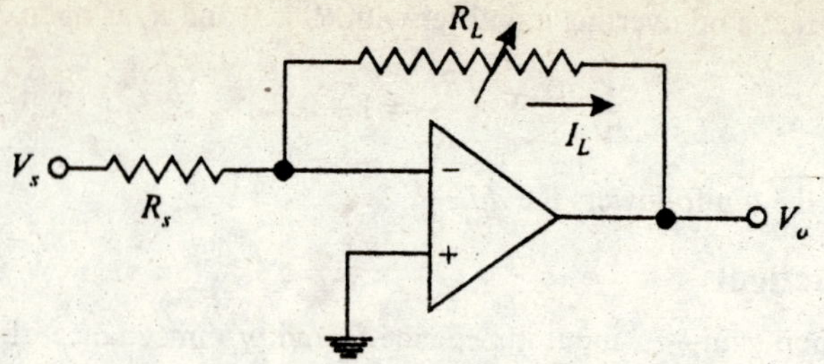

Fig. 9.10

$$\frac{V_s}{R_s} = -\frac{V_o}{R_L} = -I_L$$

$$\Rightarrow \qquad \boxed{I_L = -\frac{V_s}{R_s}}$$

i.e. *o/p* current is impedance of the load resistance R_L and ckt. works as a constant current source of V_s = constant.

Here, ckt. is same as that of *inverting amplifier.*

For Grounded Load

If the load is not of "floating type", i.e. if one end of it is grounded, robust constant current source for grounded load can be augend by using also an external transistor voltage at noninverting terminal is

$$V_1 = V_s\left(\frac{R_2}{R_1 + R_2}\right)$$

at inverting terminal is nearly V_E. Due to feedback these two i/p voltages become equal, so that $V_1 = V_E$.

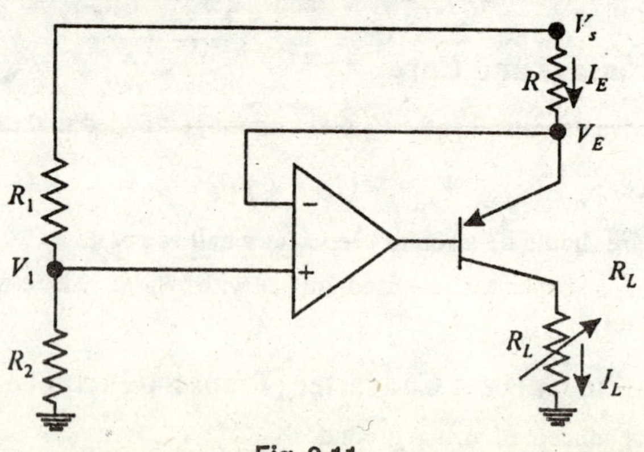

Fig. 9.11

A current I_E is set-up in the emitter

$$I_E = \frac{(V_s - V_E)}{R} = \frac{V_s - V_E}{R}$$

$$\Rightarrow \qquad I_E = \left(\frac{V_s}{R}\right) \frac{R_1}{R_1 + R_2}$$

and $\qquad I_L = I_C \approx I_E$ impedance of R_L

Constant Voltage Source

$$V_o = -\left(\frac{R_F}{R_s}\right) V_s$$

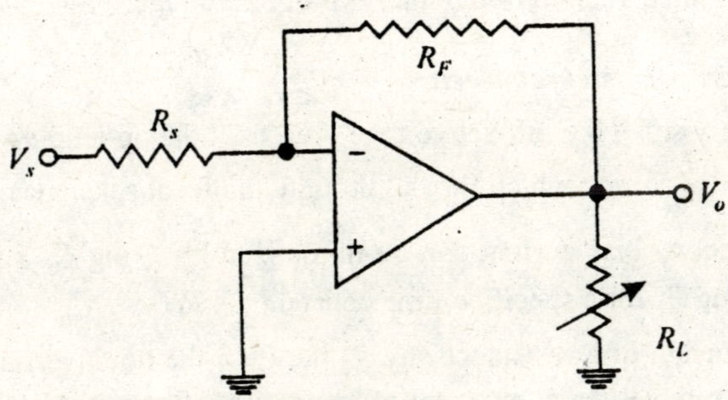

Fig. 9.12

independent of the load R_L. Here, ckt, has low *o/p* impedance, and hence, act as a voltage reference source, of v_s is a reference voltage such as a cell or a zener device.

Integrator

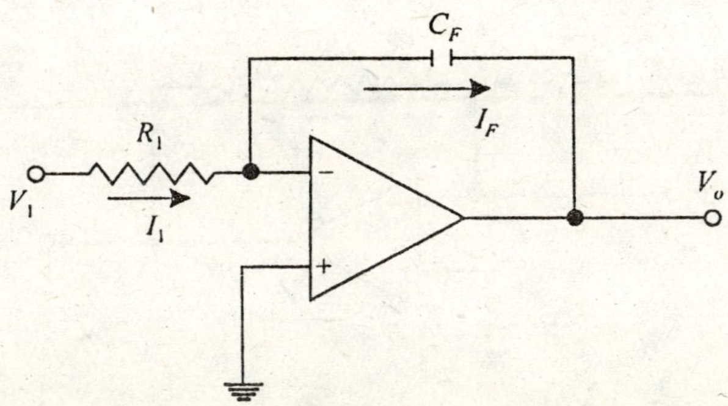

Fig. 9.13

Because of virtual grounding at the *i/p*

$$I_1 = \frac{V_1}{R_1} = I_F$$

$$V_o = -\left(\frac{1}{C_F}\right)\int I_F \, dt$$

$$\Rightarrow \qquad \boxed{V_o = -\left(\frac{1}{C_F \, R_1}\right)\int V_1 \, dt}$$

i.e. *o/p* is a scaled version of the intergral of *i/p* voltage.

Sources of error is integrator here

(i) A very small d.c. off-set voltage present at the op-amp *i/p*.

(ii) i/p bias current which flows 'through' the feedback capacitor.

The effect of bias current can be minimized by ↑ ing C_F and simultaneously ↓ ing R_1 for a specified time constant $C_F R_1$.

⇒ Inclusion of a resistance say R_2 between the noninverting terminal and ground reduces the component of bias current flowing towards *C*. This effect is minimized when $R_2 = R_1$ (*i/p* resistance).

Differentiator

Just interchange the role of *R* and *C* of integrator. Here, the *o/p* voltage is proportional to the time rate of change of *i/p* voltage.

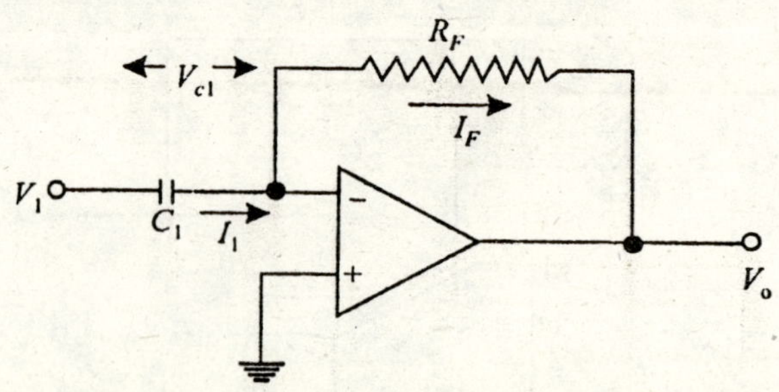

Fig. 9.14

$$I_1 = C_1 \frac{dV_{c1}}{dt}, \quad \text{for ideal, } V_{c1} = V_1$$

$$\therefore \quad -\frac{V_o}{R_F} = C_1 \frac{D_{v1}}{dt}$$

$$\boxed{V_o = -(R_F C_1)\frac{D_{v1}}{dt}}$$

Waveshaping and Waveform Generation

op-amp can also generate rectangular, triangular and square root wave-forms.

It can also be used in timing ckts. in the form of *multivibrators* based on alternative charging and discharging of a capacitor, these are also known as *relaxation oscillator.*

op-amp Comparator or Analog Comparator

It compares two analog *i/p* voltages and indicates, which one is larger by quickly generating an appropriate one of the two possible *o/p* signal levels.

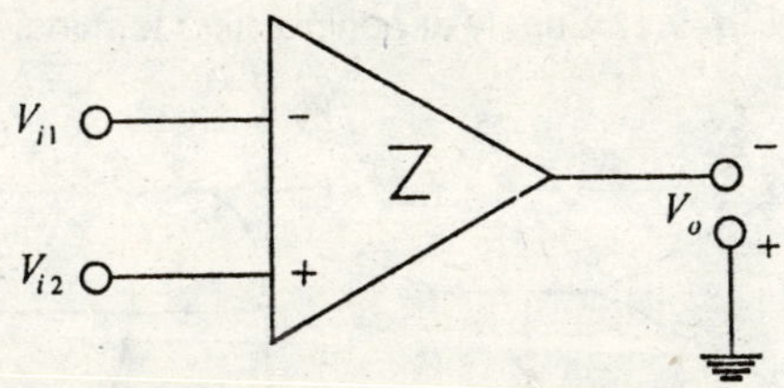

Fig. 9.15 Symbol of comparator

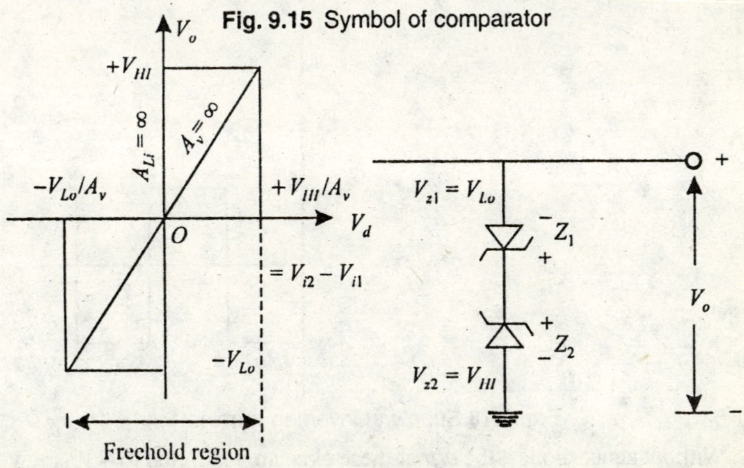

(a) *i/p* V_d – *o/p* V_o characteristic (b) Limiting ckt. connected in o/p terminal

Fig. 9.16

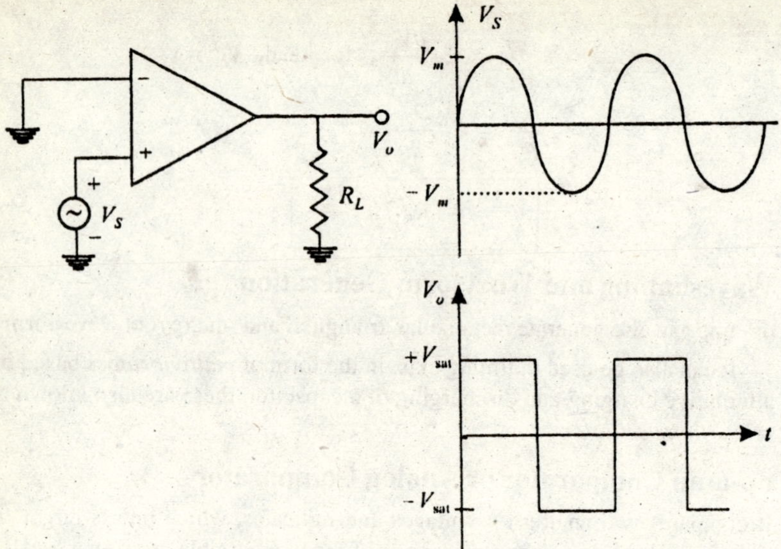

Fig. 9.17 Supply at noninverting terminal

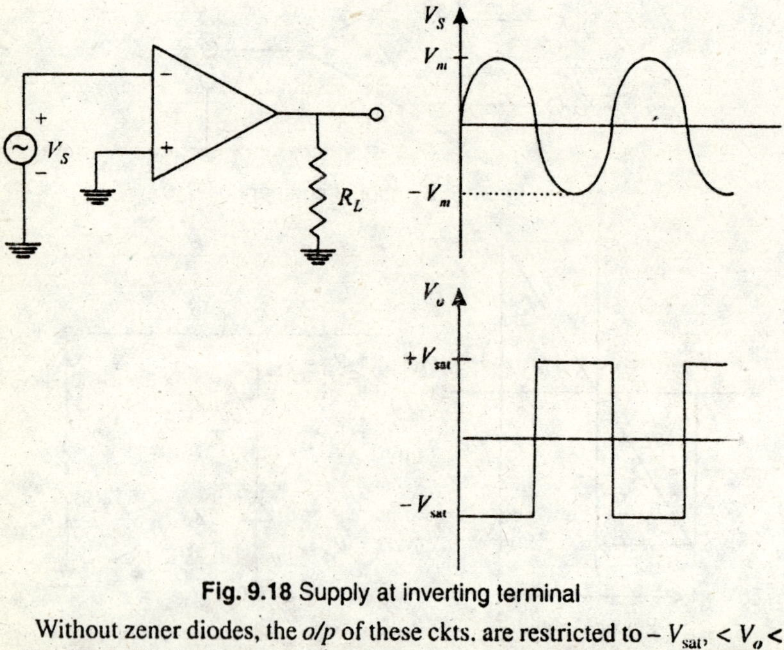

Fig. 9.18 Supply at inverting terminal

Without zener diodes, the *o/p* of these ckts. are restricted to $-V_{sat}$, $< V_o <$ $+V_{sat}$. These limitations can be regulated by using zener diodes, and we can have $-V_{io} < V_o < +V_{HI}$.

Mathematically, *Comparator's operation*

$$V_o = +V_{HI} = V\ (1), \quad \text{if} \quad V_d > 0, \ \text{i.e.}\ V_{i2} > V_{i1}$$

$$V_o = -V_{Lo} = V\ (0), \quad \text{if} \quad V_d < 0, \ \text{i.e.}\ V_{i2,} < V_{i1}$$

Thus, $V_d = 0$ is i/p transition point for change in the comparators *o/p*.

Main applications of the comparator are as a *voltage-level detector, as a square-wave generator.*

Schmitt Trigger or Regenerative Comparator

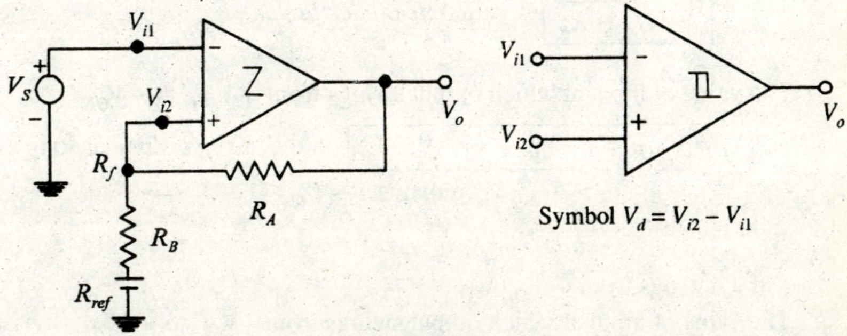

Fig. 9.19

V_{OTP} and V_{LTP} are *input voltages* at which triggering (transition) takes place.

V_{OH} and V_{OL} are *output voltages* about which the *o/p* swings. $|V_{OH}| = |V_{OL}|$.

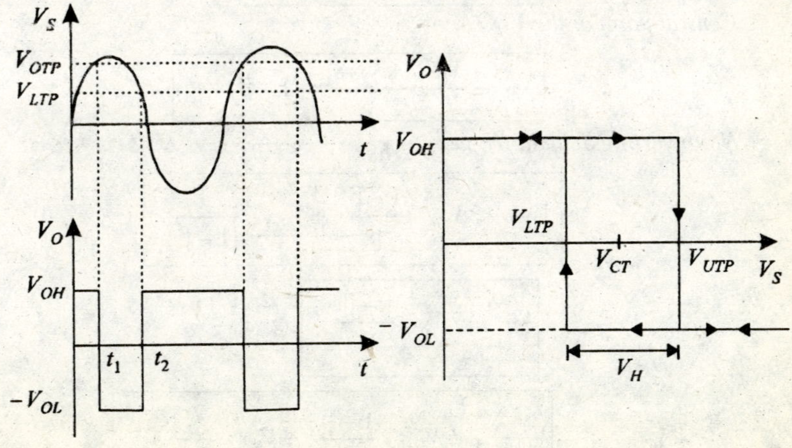

Fig. 9.20

Let initially $V_s = 0, V_o = +V_{OH}$

As long as $V_s < V_f$, V_o remains at $+V_{OH}$

i.e. $Vs < V_{UTP}$, $V_f = V_{OH} \dfrac{R_B}{R_A + R_B} + V_{\text{Ref}} \times \dfrac{R_A}{R_A + R_B}$

at $t = t_1$, $v_s = V_f = V_{UTP}$, $V_d = 0$, i.e. $v_{i2} = v_{i1}$ and v_o swings from $-V_{OH}$ to $-V_{OL}$.

$$\boxed{\beta = \frac{R_B}{R_A + R_B}}$$ called *feedback factor.*

The value of input at which output swings from $+V_{OH}$ to $-V_{OL}$

$V_f = \boxed{V_{UTP} = \beta V_{OH} + (1-\beta)\ V_{Ref}}$ in the *positive slope* as long as

$v_s > V_{LTP}$, V_o remains at $-V_{oL}$

at $t = t_2$, $v_s = v_f = V_{LTP}$, $v_d = 0$

and v_o swings from $-V_{OL}$ to $+V_{OH}$

The value of input at which output swings from $-V_{OL}$ to $+V_{OH}$

$V_{LTP} = \boxed{-\beta V_{OL} + (1-\beta) V_{\text{Ref}}}$ in the *negative slope*.

The circuit *exhibits hysteresis, dead zone or backlash.*

Hysteresis voltage (V_H)

$$|V_{OH}| = |V_{OL}| = V_o$$

$$\boxed{V_H = V_{UTP} - V_{LTP} = 2\beta V_o}$$

Central point of the V_H curve

$$\boxed{V_{ot} = \frac{V_{UTP} + V_{LTP}}{2} = (1-\beta) V_{\text{Ref}}}$$

Noninverting Schmitt trigger

$$\boxed{V_{UTP} = V_{OL} \frac{1-\beta}{\beta} + V_{\text{Ref}} \cdot \frac{1}{(1-\beta)}}$$

$$\boxed{V_H = 2 V_o \frac{1-\beta}{\beta}}$$

$$\boxed{V_{LTP} = -V_{OH} \frac{1-\beta}{\beta} + V_{\text{Ref}} \cdot \frac{1}{(1-\beta)}}$$

Multivibrators using op-amps

Astable or free-running multivibrators: o/p of Schmitt trigger is integrated by a *low-pass* $R_F C_1$. It has two quasistable state.

$$t_{p1} = R_F C_1 \ln \ 1 + \frac{2R_B}{R_A}$$

$t_{p2} = t_{p1}$ for *symmetrical wave* $T = t_{p1} + t_{p2}$

$$\boxed{T = 2R_F\ C_1 \ln \ 1 + \frac{2R_B}{R_A}} \qquad R_F\ C_1 = \tau$$

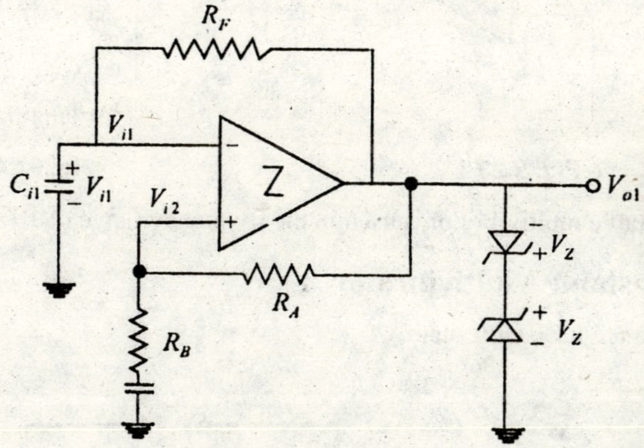

Fig. 9.21

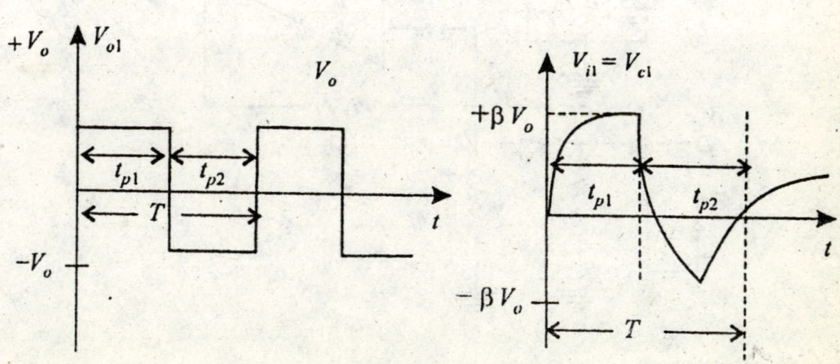

Fig. 9.22

or $$\boxed{T = 2\tau \ln \frac{1+\beta}{1-\beta}}, \text{ for } \textit{unsymmetrical}$$

$$t_{p1} \neq t_{p2},\ \tau_1 = R_{F1} C_1,\ \tau_2 = R_{F2} C_2,\ f = \frac{1}{T}$$

In order to get a *triangular wave o/p*, and integrator at V_{o1}.

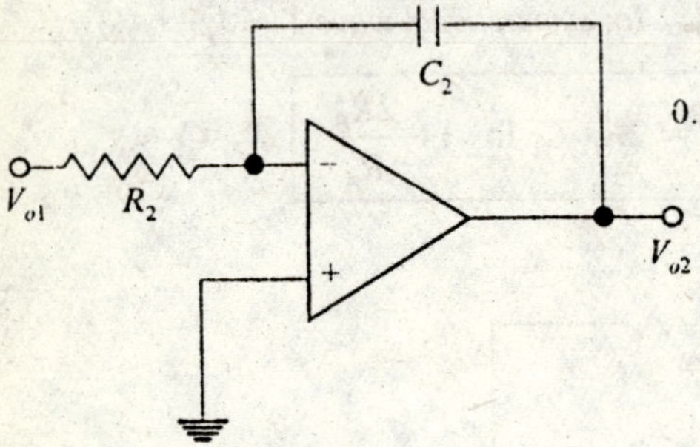

Fig. 9.23

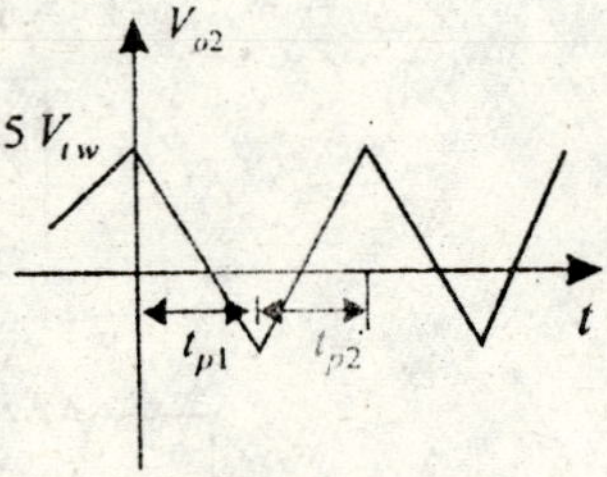

t_w- triangular wave

Fig. 9.24

Astable multivibrator is used in the frequency range of 10 Hz to 10 kHz.

Monostable Multivibrator

It has one quasistable state.

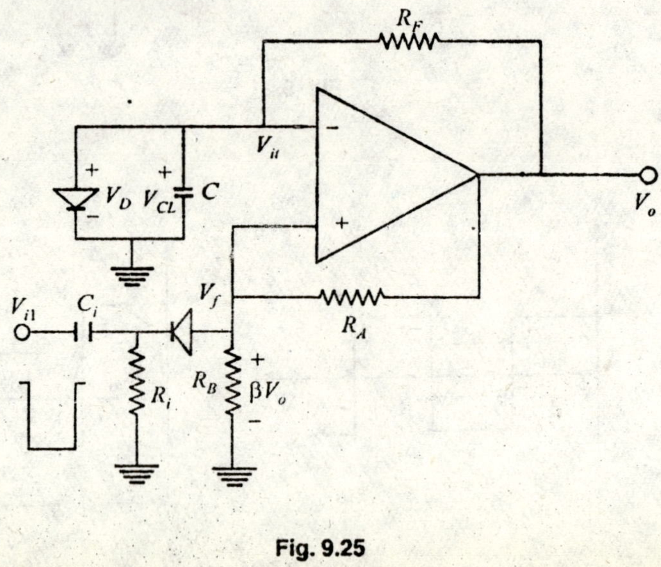

Fig. 9.25

Due to D_1, voltage V_{c1} (of astable) is clamped at V_D during +ve excursion.

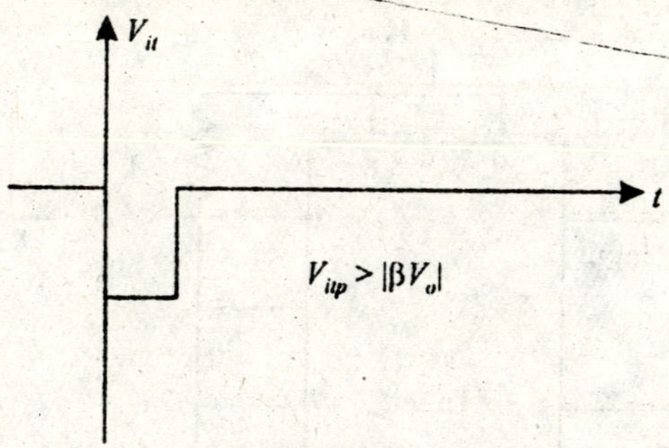

Fig. 9.26

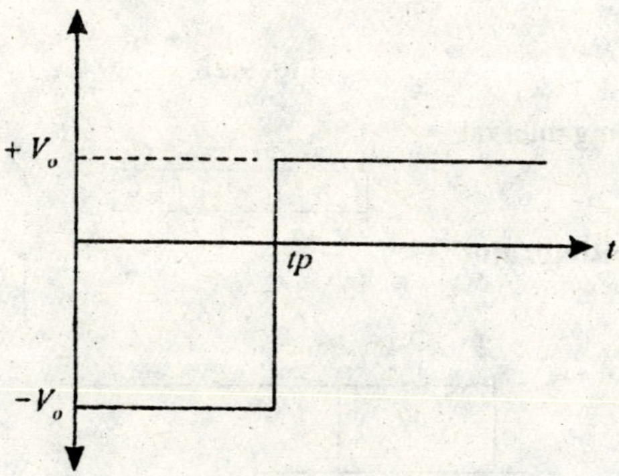

Fig. 9.27

At $t = t_p$, $V_{c2} = \beta V_o$ and comparator swings back to $i\beta$ value $+V_o$

$$\boxed{t_p = R_F C \ln \frac{V_o + V_D}{V - \beta V_o}} \qquad V_D = 0.7\text{ V.}$$

or $$\boxed{t_p = \tau \ln \frac{1 + V_D / V_o}{1 - \beta}} \qquad \text{For } R_A = R_B, \quad \beta = 0.5$$

$$\boxed{t_p = 0.693\,\tau} \qquad \text{For } V_o >> V_D$$

Multivibrators using 555 times

Monostable Multivibrator

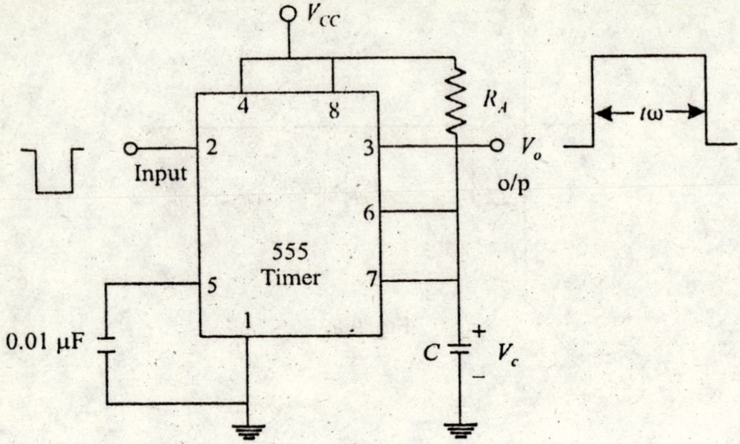

Fig. 9.28

o/p timing interval

$$\boxed{T = t_{\omega} = 1.1\, R_A\, C}$$

Stable Multivibrator

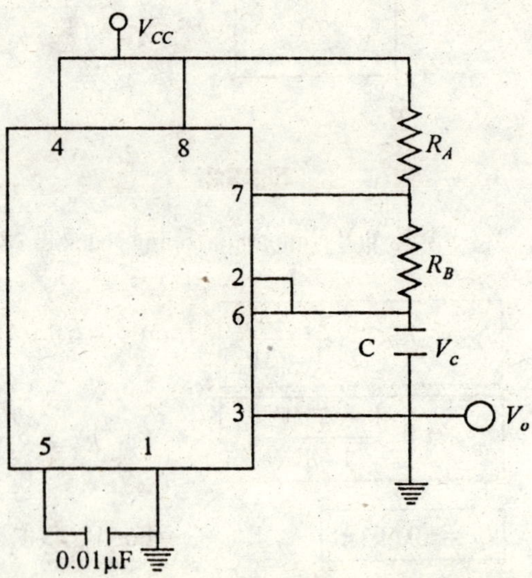

Fig. 9.29

Charging time

$$\boxed{T_1 \approx 0.7\, C\,(R_A + R_E) = 0.7\,\tau_1}$$

$$\tau_1 = (R_A + R_B)\,C$$

Discharging time

$$\boxed{T_2 \approx 0.7\,C\,R_B = 0.7\,\tau_2}$$

$$\tau_2 = R_B\,C$$

Frequency $f = \frac{1}{T} = \frac{1}{T_1 + T_2}$ $\boxed{f = \frac{1.4}{(R_A + 2R_B)\,C}}$

$$D = \text{Duty cycle} = \frac{T_{on}}{T_{on} + T_{off}}$$

$$\boxed{D = \frac{R_A + R_B}{R_A + 2R_B} \times 100\%}$$

Free-running multivibrator using logic gates

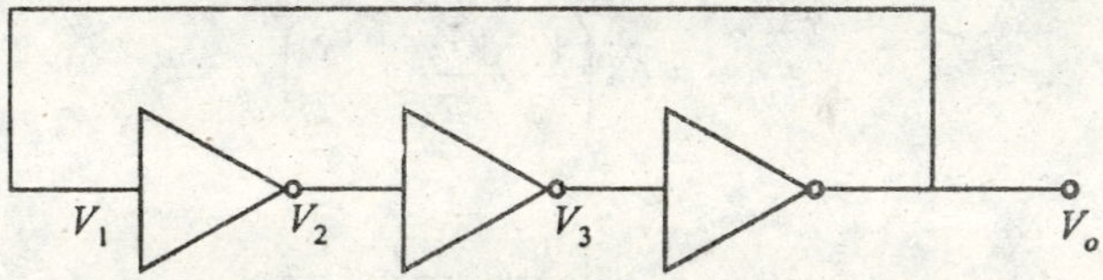

Fig. 9.30

Let delay time of each inverter, then = Δt

Time period of o/p square wave, $\boxed{T = 6.\,\Delta t}$

in general,

i.e. $\boxed{T = 2n\,\Delta t,}$ n = no. of inverters.

Problem 9.1. Calculate the output voltage of the circuit shown in Fig. P. 9.1:

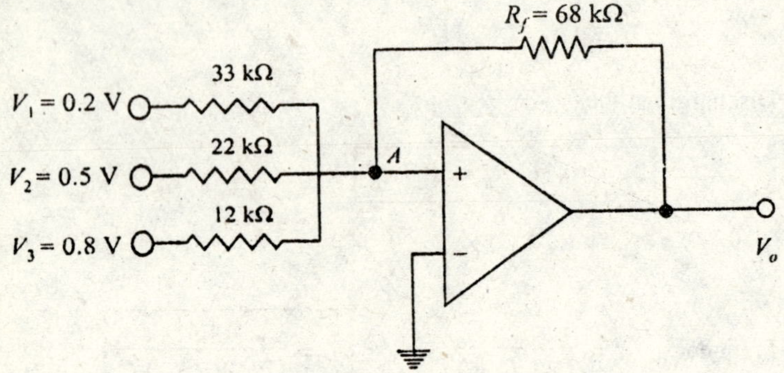

Fig. P. 9.1

Solution. Applying KCL at mode A

$$\frac{V_1 - V_A}{33\text{K}} + \frac{V_2 - V_A}{22\text{K}} + \frac{V_A - V_A}{12\text{K}} = \frac{V_A - V_o}{68\text{K}}$$

$$\Rightarrow \quad \frac{V_1}{33} + \frac{V_2}{22} + \frac{V_3}{12} + \frac{V_o}{68} \qquad \{\because V_A = 0\}$$

$$\Rightarrow \quad V_o = -68\left(\frac{V_1}{33} + \frac{V_2}{22} + \frac{V_3}{12}\right)$$

$$\Rightarrow \quad V_o = -68\left(\frac{0.2}{33} + \frac{0.5}{22} + \frac{0.8}{12}\right)$$

$$\Rightarrow \quad V_o = -68\,(0.00606 + 0.02273 + 0.0667)$$

$$\Rightarrow \quad V_o = -68\,(0.04999)$$

$$\Rightarrow \quad V_o = -3.399$$

$$\Rightarrow \quad V_o \simeq -3.4\text{ V}$$

Problem 9.2. What input voltage results in an output of 2 V in the circuit shown below?

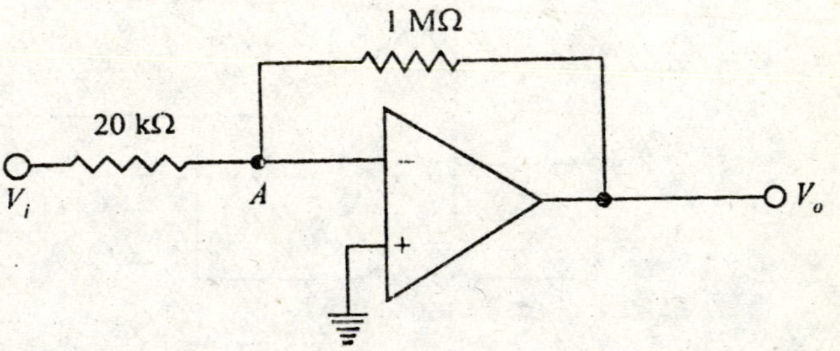

Fig. P. 9.2

Solution. Apply KCL law at point *A*, voltage at point *A* is zero, since the other point is grounded.

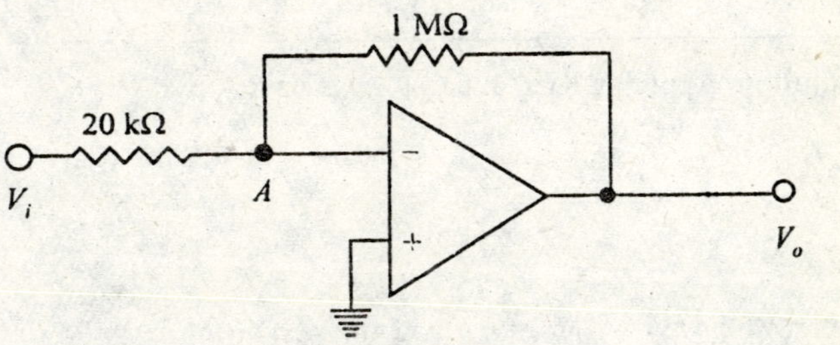

Fig. 9.2 (a)

$$\Rightarrow \quad \frac{V_i - V_A}{20 \text{ K}} = \frac{V_A - V_o}{10^3 \text{ K}}$$

$$\Rightarrow \quad \frac{V_i - V_o}{20} = \frac{0 - 2}{1000} \qquad \{\text{Since } V_o = 2 \text{ V}\}$$

$$\Rightarrow \quad V_i = \frac{-2 \times 20}{1000}$$

$$\Rightarrow \quad V_i = \frac{-40}{1000}$$

$$\Rightarrow \quad V_i = -0.04 \text{ V}$$

Problem 9.3. What input must be applied to the input terminal of the Fig. P. 9.3 to result an output of 2.4 V?

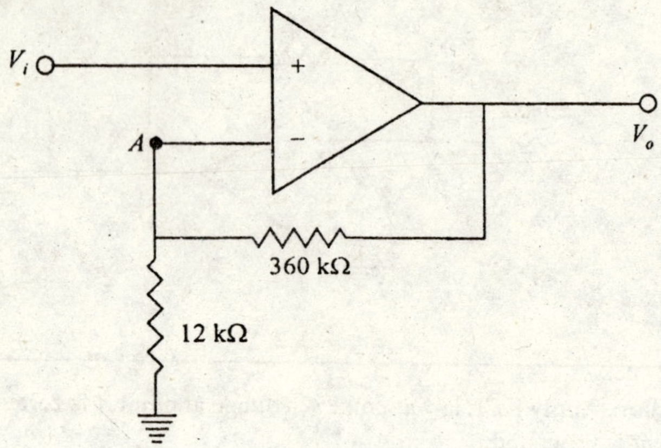

Fig. P. 9.3

Solution. Applying KCL at point *A*, we have

$$\frac{0 - V_A}{12\text{ K}} = \frac{V_A - V_o}{360\text{ K}}$$

Here, we have $V_A = V_i$ and $V_o = 2.4$ V

$$\Rightarrow \quad \frac{0 - V_i}{12\text{ K}} = \frac{V_i - 2.4}{360\text{ K}}$$

$$\Rightarrow \quad \frac{-V_i}{12} = \frac{V_i - 2.4}{360}$$

$$\Rightarrow \quad -360\,V_i = 12\,V_i - 28.8$$

$$\Rightarrow \quad (360 + 12)\,V_i = 28.8$$

$$\Rightarrow \quad V_i = \frac{28.8}{372} = 0.077419\text{ V}$$

Problem 9.4. What range of output voltage is developed in circuit shown in Fig. P. 9.4?

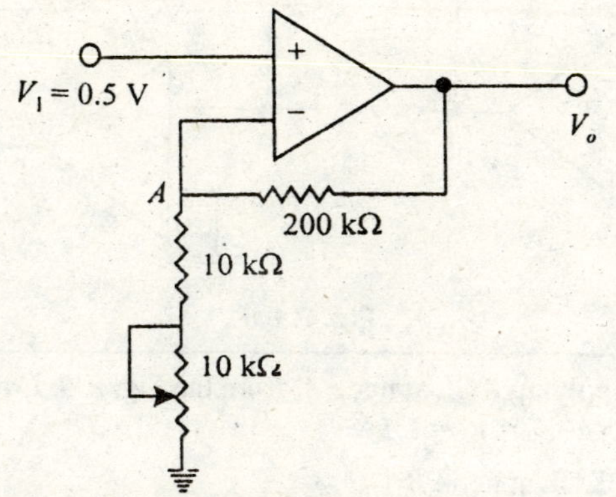

Fig. P. 9.4

Solution. Applying KCL at node V_A, for minimum output voltage take the whole resistance 10 + 10 = 20 kΩ

$$\frac{O - V_A}{20\text{ K}} = \frac{V_A - V_o}{200\text{ K}}$$

$$\Rightarrow \quad \frac{0 - 0.5}{20} = \frac{0.5 - V_o}{200}$$

$$\Rightarrow \quad -5 = 0.5 - V_o$$

$$\Rightarrow \quad V_o = 5.5\text{ V}$$

For maximum output voltage reduce the variable resistance (10 kW) to zero, we have

$$\frac{0 - V_A}{10\text{ K}} = \frac{V_A - V_o}{200\text{ K}}$$

$$\Rightarrow \quad \frac{-0.5}{10} = \frac{0.5 - V_o}{200}$$

$$\Rightarrow \quad -10 = 0.5 - V_o$$

$$\Rightarrow \quad V_o = 10.5\text{ V}$$

Problem 9.5. Calculate the voltage output for the given circuit.

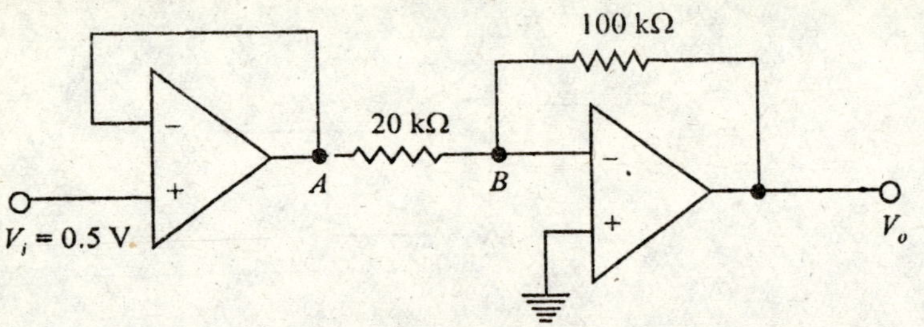

Fig. P. 9.5

Solution. Applying KCL at node A, from the Fig. P. 9.5 we have the voltage of node A, i.e. $V_A = V_1 = 1.5$ V

Now, apply KCL at node B

$$\frac{V_A - V_B}{20\text{ K}} = \frac{V_B - V_o}{100\text{ K}} \quad \{V_B = 0\}$$

$$\Rightarrow \quad \frac{1.5 - 0}{20\text{ K}} = \frac{0 - V_o}{100\text{ K}}$$

$$\Rightarrow \quad \frac{1.5}{20} = \frac{-V_o}{100} \qquad \Rightarrow V_o = -\frac{1.5 \times 100}{20} = -7.5\text{ V}$$

Problem 9.6. Calculate the output voltage V_2 and V_3 in the given circuit.

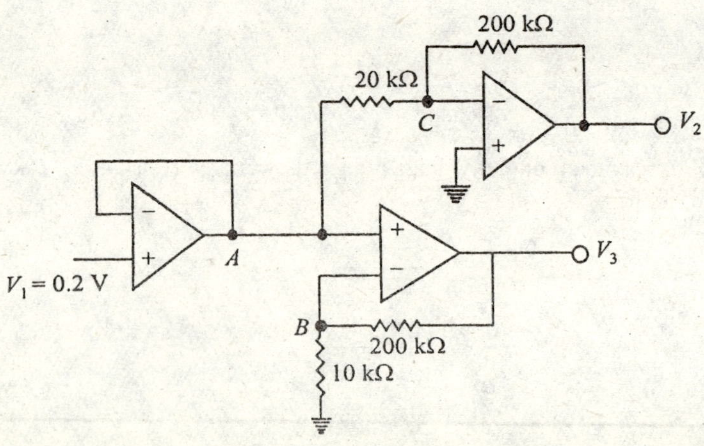

Fig. P. 9.6

Solution. Figure P. 9.6 we have that the voltage of node A is $V_A = 0.2$ V.

Applying KCL at node B to get the output voltages V_3

$$\frac{0 - V_B}{10\text{K}} = \frac{V_B - V_3}{200\text{ K}} \quad \{\text{Now, } V_B, = V_A = 0.2 \text{ V}\}$$

$$\Rightarrow \quad \frac{0 - 0.2}{10\text{ K}} = \frac{0.2 - V_3}{200\text{ K}}$$

$$\Rightarrow \quad V_3 = 4 + 0.2 = 4.2\text{V}$$

Now, again apply KCL at mode C, to get the output voltage V_2

$$\frac{V_A - V_C}{20\text{ K}} = \frac{V_C - V_2}{200\text{ K}} \quad \{\because V_C = 0\}$$

$$\Rightarrow \quad \frac{0.2 - 0}{20\text{ K}} = \frac{0 - V_2}{200\text{ K}}$$

$$\Rightarrow \quad \frac{0.2}{20} = \frac{-V_2}{200} \quad \Rightarrow \quad V_2 = -2\text{ V}$$

Problem 9.7. Calculate the output voltage V_o of the given circuit.

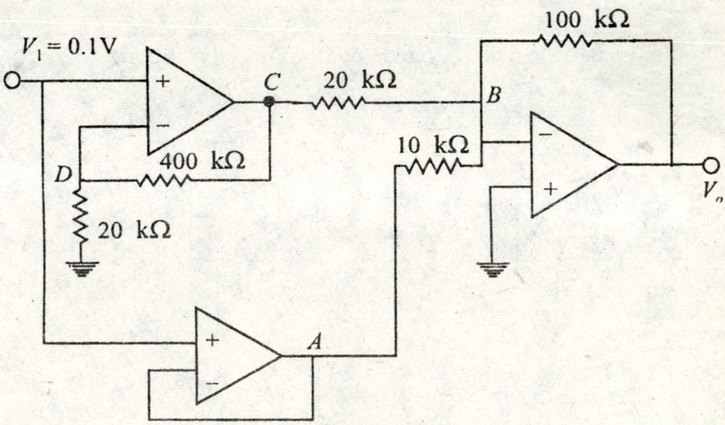

Fig. P. 9.7

Solution. Now, from the above figure, we have

$$V_A = V_i = 0.1 \text{ V}$$

Apply nodal analysis (KCL) at node D to get voltage at point C, V_c

$$\Rightarrow \quad \frac{0 - V_D}{20\text{ K}} = \frac{V_D - V_C}{400\text{ K}} \qquad (\text{Now, } V_D = V_1 = 0.1\text{ V})$$

$$\Rightarrow \quad \frac{-0.1}{20\text{ K}} = \frac{0.1 - V_c}{400\text{ K}}$$

$$\Rightarrow \quad V_C = 0.1 + 2$$

$$\Rightarrow \quad V_C = 2.1\text{ V}$$

Now, again apply KCL at node B to get the output final voltage V_o

$$\frac{V_C - V_B}{20\text{ K}} = \frac{V_A - V_B}{10\text{ K}} = \frac{V_B - V_D}{10\text{ K}} \qquad (V_B = 0\text{ V, since grounded})$$

$$\Rightarrow \quad \frac{2.1 - 0}{20\text{ K}} + \frac{0.1 - 0}{10\text{ K}} = \frac{0 - V_o}{100\text{ K}}$$

$$\Rightarrow \quad \frac{2.1}{20} = \frac{0.1}{10} = \frac{-V_o}{100}$$

$$\Rightarrow \quad (1.05 + 0.1)\,10 = -V_o$$

$$\Rightarrow \quad 1.15 \times 10 = -V_o$$

$$\Rightarrow \quad V_o = -11.5\text{ V}$$

Problem 9.8. Calculate V_o, the output for the given circuit.

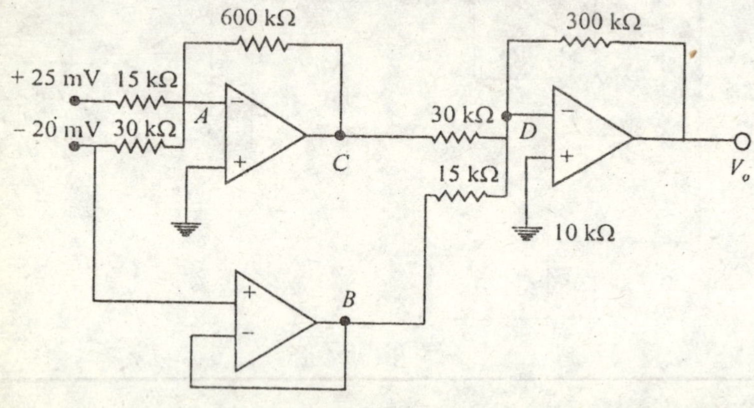

Fig. P. 9.8

Solution. From the Fig. P. 9.8 we find that voltage at node B is $V_B = -20\,\text{mV}$.

$$V_A = 0 \; \{\because \text{ since it is grounded}\}$$

To get the voltage at node C, V_c we apply KCL at node A

$$\frac{25 - V_A}{15\text{ K}} + \frac{(-20) - V_A}{30\text{ K}} = \frac{V_A - V_C}{600\text{ K}}$$

$$\Rightarrow \qquad \frac{25}{15} - \frac{-20}{30} = \frac{-V_c}{600}$$

$$\Rightarrow \qquad 25 - 10 = \frac{-V_c}{40}$$

$$V_C = -600\text{ mV}$$

To determine output voltage we apply KCL nodc D

$$\frac{V_C - V_D}{30\text{ K}} + \frac{V_B - V_D}{15\text{ K}} = \frac{V_D - V_O}{300\text{ K}} \qquad \begin{Bmatrix} V_D = 0\text{V} \\ V_B = 20\text{ mV} \end{Bmatrix}$$

$$\Rightarrow \qquad \left(\frac{-600 - 0}{30\text{ K}}\right) + \left(\frac{-20 - 0}{15\text{ K}}\right) = \left(\frac{0 - V_o}{300\text{ K}}\right)$$

$$\Rightarrow \qquad \frac{-600}{30} - \frac{20}{15} = \frac{-V_o}{300}$$

$$\Rightarrow \qquad -20 - \frac{4}{3} = \frac{-V_o}{300}$$

$$\Rightarrow \qquad V_o = \left(20 + \frac{4}{3}\right) \times 300$$

$$\Rightarrow \qquad V_o = \frac{64}{3} \times 300 = 6400\text{ mV}$$

$$\Rightarrow \qquad V_o = 6.4\text{ V}$$

Problem 9.9. Find the cut-off frequency of a first order low-pass filter in the given circuit.

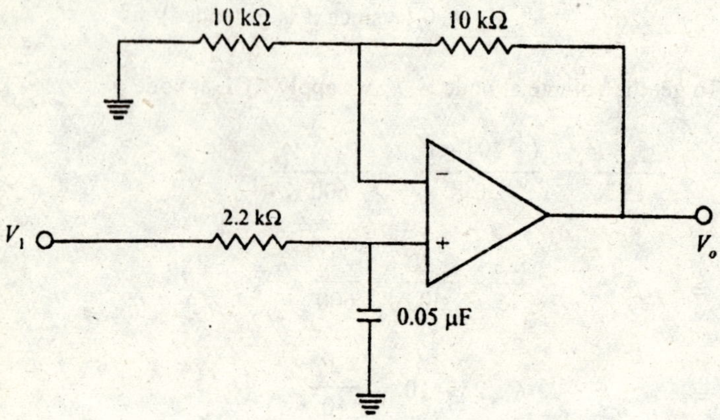

Fig. P. 9.9

Solution. The high cut-off of the circuit as the low-pass filter is given by the formula.

$$f_{OH} = \frac{1}{2\pi R_1 C_1}$$

Where $R_1 = 2.2 \text{ k}\Omega = 2.2 \times 10^3 \ \Omega$

and $C_1 = 0.05 \ \mu F = 5 \times 10^{-8} F$

$$\therefore \quad f_{OH} = \frac{1}{2\pi \times 2.2 \times 10^3 \times 5 \times 10^{-8}}$$

$$\Rightarrow \quad f_{OH} = \frac{100000}{2\pi \times 2.2 \times 5}$$

$$\Rightarrow \quad f_{OH} = 1447 \text{ Hz}$$

$$\Rightarrow \quad f_{OH} = 1.447 \text{ k}\Omega$$

$$\Rightarrow \quad f_{OH} = 1.45 \text{ k}\Omega$$

Problem 9.10. Calculate the lower cut-off frequency and the upper cut-off frequency of the band pass filter in the given circuit.

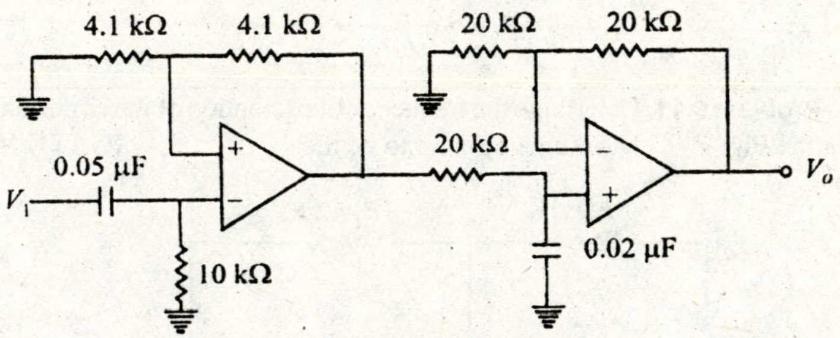

Fig. P. 9.10

Solution. From in the Fig. P. 9.10, we find that it is a *B* and pass filter, which has both lower cut-off frequency f_{OL} and upper cut-off frequency f_{OH} both.

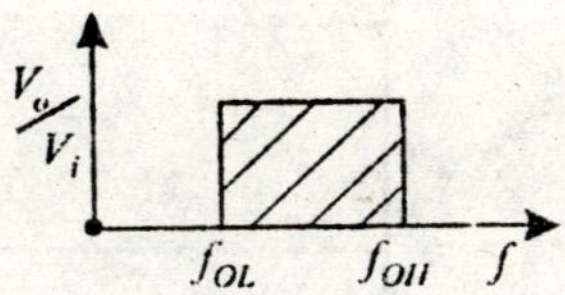

Fig. 9.10 (a)

The lower cut-off frequency is given by

$$f_{OL} = \frac{1}{2\pi R_1 C_1}$$

where, $R_1 = 10\text{ k}\Omega$ and $C_1 = 0.05\ \mu F$

$$f_{OL} = \frac{1}{2\pi \times 10\text{K} \times 0.05 \times 10^{-6}}$$

$$f_{OL} = \frac{10^5}{2\pi \times 50} = 318.3\text{ Hz}$$

The upper cut-off frequency is given by

$$f_{OH} = \frac{1}{2\pi R_2 C_2}$$

where, $R_2 = 20\text{ k}\Omega$ and $C_2 = 0.02\ \mu F$

$$f_{OH} = \frac{1}{2\pi \times 20 \times 10^3 \times 0.02 \times 10^{-6}}$$

$$f_{OH} = \frac{10^5}{2\pi \times 20 \times 2}$$

$$f_{OH} = 397.9 \text{ Hz}$$

Problem 9.11. Determine the frequency of oscillation of the circuit shown in the Fig. P. 9.11. Assume op-amp to be ideal. **(GATE 1997)**

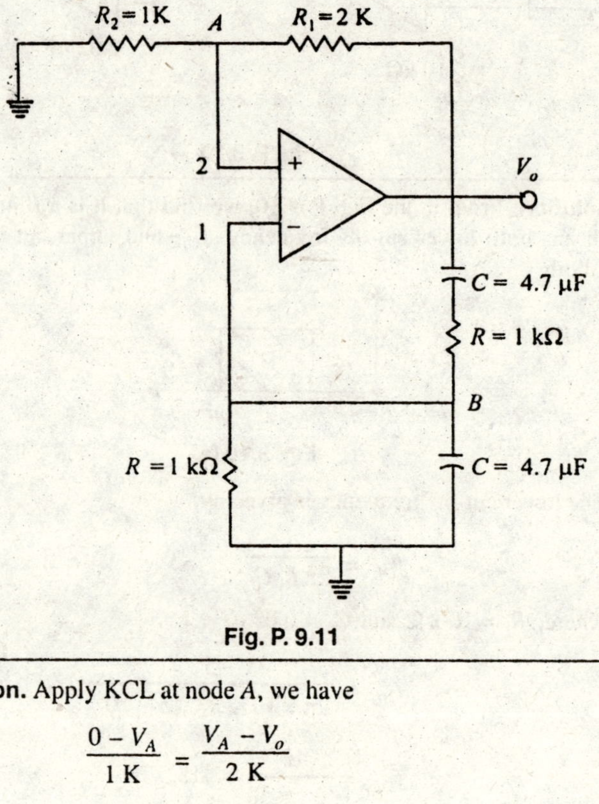

Fig. P. 9.11

Solution. Apply KCL at node A, we have

$$\frac{0 - V_A}{1\text{ K}} = \frac{V_A - V_o}{2\text{ K}}$$

$$-2V_A = V_A - V_o$$

$$V_o = 3\,V_A$$

Now, apply KCL at node B, we have

$$\frac{V_B - V_o}{R + \frac{1}{CS}} + \frac{V_B - 0}{R \| \frac{1}{CS}} = 0 \qquad \{V_A = V_B\}$$

$$\Rightarrow \quad \frac{(V_A - V_o)\,CS}{(RCS+1)} + \frac{V_A\,(RCS+1)}{R\times 1} = 0$$

$$\Rightarrow \quad V_A\left[\frac{CS}{(1+RCS)} + \frac{(1+RCS)}{R}\right] = \frac{V_o\,CS}{(1+RCS)}$$

$$\Rightarrow \quad V_A\left[\frac{RCS + (1+RCS)^2}{R\,(1+RCS)}\right] = \frac{V_o\,CS}{(1+RCS)}$$

$$\Rightarrow \quad V_A\left[RCS + 1 + R^2C^2S^2 + 2RCS\right] = 3\,V_A\,RCS$$

$$\Rightarrow \quad R^2C^2S^2 + 3\,RCS + 1 = 3\,RCS$$

$$\Rightarrow \quad R^2C^2S^2 + 1 = 0 \qquad \text{(put } s = j\omega\text{)}$$

$$\Rightarrow \quad -R^2C^2S^2 + 1 = 0$$

$$\Rightarrow \quad \omega^2 = \frac{1}{R^2C^2}$$

$$\omega = \frac{1}{RC}$$

$$\Rightarrow \quad 2\pi f = \frac{1}{RC}$$

$$\Rightarrow \quad f = \frac{1}{2\pi RC}$$

where, $R_1 = 1\ \text{k}\Omega$ and $C = 4.7\ \mu F$

$$\Rightarrow \quad f = \frac{1}{2\pi\times 1\times 10^3\times 4.7\times 10^{-6}}$$

$$\Rightarrow \quad f = \frac{1000}{2\pi\times 4.7}$$

$$\Rightarrow \quad f = \frac{1000}{29.516} = 33.8799\ \text{Hz} \quad \Rightarrow \ \simeq 33.88\ \text{Hz}$$

Alternatively, we can solve the problem directly since it a Wein Bridge, the frequency of oscillation is given by

$$\Rightarrow \quad f = \frac{1}{2\pi\sqrt{R_1R_2C_1C_2}}$$

Here $\quad R_1 = R_2 = 1\ k\Omega$ and $C_1 = C_2 = 4.7\ \mu F$

$$\Rightarrow \quad f = \frac{1}{2\pi\sqrt{R^2C^2}} \Rightarrow f = \frac{1}{2\pi R_C}$$

$$\Rightarrow \quad f = 33.88 \text{ Hz}$$

Problem 9.12. Show that the circuit given in the Fig. P. 9.12 will work as an oscillator frequency. **(GATE 1998)**

$$f = \frac{1}{2\pi\ RC}, \text{ if } R_1 = 2R_2$$

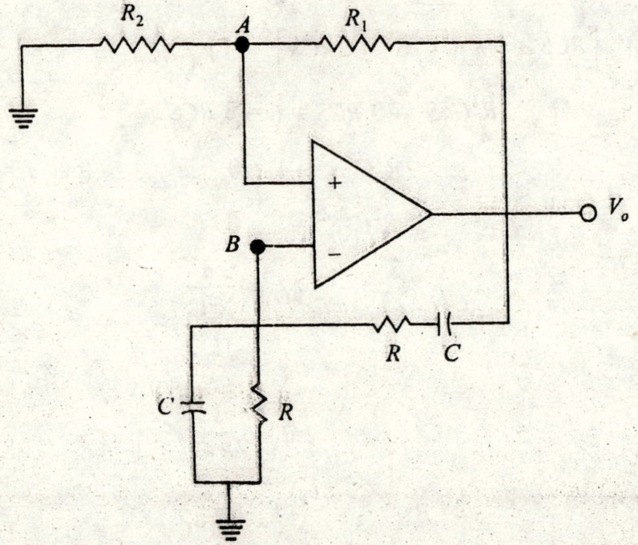

Fig. P. 9.12

Solution. Apply KCL at node *A*

$$\frac{0 - V_A}{R_2} = \frac{V_A - V_o}{R_1} \Rightarrow V_A\left(\frac{1}{R_1} + \frac{1}{R_2}\right) = V_o$$

$$\Rightarrow \quad V_o = V_A\left(\frac{R_1 + R_2}{R_2}\right) \quad \text{... (1)}$$

Again apply KCL at node *B*

$$\frac{0 - V_B}{R \parallel \frac{1}{CS}} = \frac{V_B - V_o}{R + 1/CS}$$

$$\frac{-V_B(1+RCS)}{R}=(V_B-V_o)\frac{C_s}{1+RCS}$$

$$V_B\ \frac{C_S}{1+RCS}+\frac{1+RCS}{R}=\frac{V_oCS}{1+RCS}$$

$$V_B\ \frac{RCS+1+2RCS\ R^2C^2S^2}{R\,(1+RCS)}=\frac{V_oCS}{(1+RCS)}$$

$$V_B\left(R^2C^2S^2+3RCS+1\right)=RCS\,V_o$$

Now, from equation (1), put the value of V_o and $V_A = V_B$

$$V_B\ \left(R^2C^2S^2+3RCS+1\right)=RCS\ V_B\ \frac{R_1+R_2}{R_2}$$

$$\left(R^2C^2S^2+3RCS+1\right)=RCS\ \frac{R_1+R_2}{R_2}$$

$$R^2C^2S^2+\ 3RCS-RCS\ \frac{(R_1+R_2)}{R_2}\ S+1=0$$

Now, equation the imaginary part equal to zero.

$$3RC-RC\left[\frac{R_1+R_2}{R_2}\right]=0$$

$$3-\frac{R_1+R_2}{R_2}=0$$

$$3R_2-R_1-R_2=0$$

$$2R_2-R_1=0$$

$$\boxed{R_1=2R_2}$$

Problem 9.13. In the circuit given in Fig. P. 9.13, R_S=2 kΩ, R_L=5 kΩ for the op-amp $A = 10^5$, R_i = 100 kΩ, R_o = 50 Ω. For V_o = 10 V. Calculate V_S and V_o/V_S and estimate the input resistance of the circuit. **(GATE 1998)**

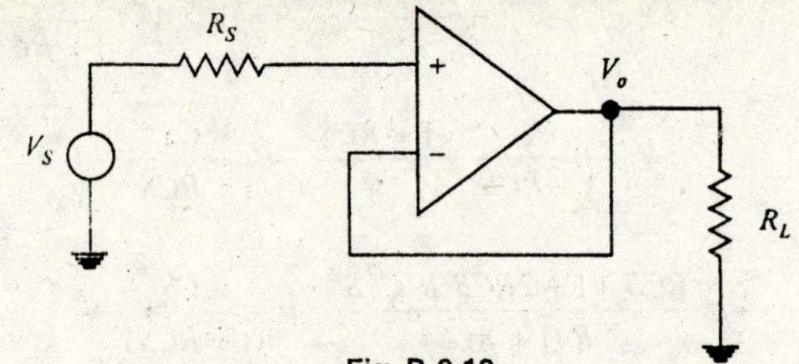

Fig. P. 9.13

Solution. We can resolve the circuit in the following way:

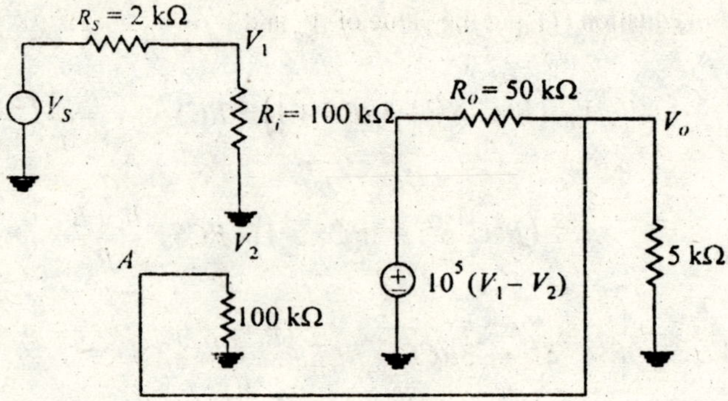

Fig. 9.13 (a)

Now, we have

$$\Rightarrow \qquad V_1 = \frac{100\text{K}}{102\text{K}} V_S$$

and
$$\frac{V_2}{5\text{ K}} + \frac{V_2}{100\text{ K}} + \frac{V_2 - 10^5 (V_1 - V_2)}{50} = 0$$

$$\Rightarrow \qquad 2V_2 + V_2 + 2 \times 10^3 V_2 = 10^{10} (V_1 - V_2)$$

$$\Rightarrow \qquad V_2 + (2 + 1 + 2 \times 10^3 + 10^{10})\, 10^{10} V_1$$

$$\Rightarrow \qquad V_2 \approx V_1$$

(a) For

$$V_o = 10\,\text{V}$$

$$V_2 = V_1 = 10\,\text{V}$$

$$V_s = 10.2\,\text{V}$$

(b)

$$\frac{V_o}{V_s} = \frac{10}{10.2} = 0.98$$

(c) $R_{in} = 100\ \text{k}\Omega$ = input resistance of the circuit.

Problem 9.14. For the op-amp shown in Fig. P. 9.14, determine the output voltage V_o. Assume that the op-amps are ideal. **(GATE 2001)**

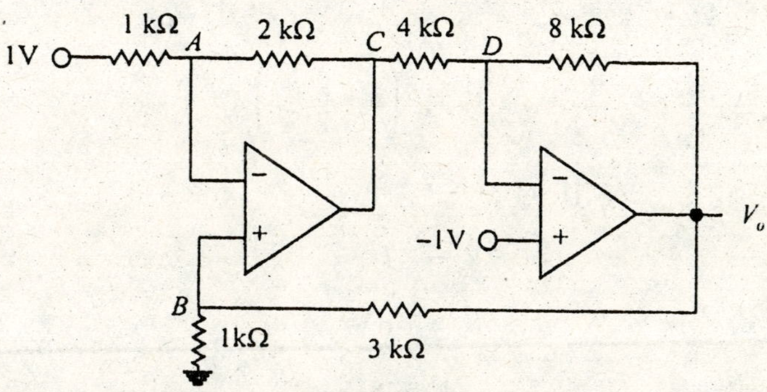

Fig. P. 9.14

Solution. Apply KCL at node *A*. We have

$$\Rightarrow \quad \frac{1 - V_A}{1\,\text{K}} = \frac{V_A - V_c}{2\,\text{K}}$$

$$\Rightarrow \quad V_C = 3\,V_A \qquad \text{... (1)}$$

Again KCL is applied at node *B*

$$\frac{0 - V_B}{1\,\text{K}} = \frac{V_B - V_o}{3\,\text{K}}$$

$$\Rightarrow \quad -3\,V_B = V_B - V_o$$

$$\Rightarrow \quad V_o = 4\,V_B \qquad \{\because V_A = V_B\}$$

$$\Rightarrow \quad V_o = 4V_A \qquad \text{... (2)}$$

Again apply KCL at node D, to get the output voltage V_o

$$\Rightarrow \quad \frac{V_C - V_D}{4\text{ K}} = \frac{V_D - V_o}{8\text{ K}} \qquad \{\because V_D = -1\text{V}\}$$

$$\Rightarrow \quad \frac{3V_A + 1}{4} = \frac{-1 - V_o}{8}$$

$$\Rightarrow \quad 6V_A + 2 = -1 - V_o$$

$$V_o = -(3 + 6V_A)$$

From equation (2), we have

$$\Rightarrow \quad V_o = -\left(3 + 6 \times \frac{V_o}{4}\right)$$

$$\Rightarrow \quad V_o = -\left(3 + \frac{3}{2}V_o\right)$$

$$\Rightarrow \quad (1 - 1.5)V_o = -3$$

$$\Rightarrow \quad -0.5V_o = -3$$

$$\Rightarrow \quad V_o = +6\text{V}$$

$$\Rightarrow \quad V_o = 6\text{V}$$

Problem 9.15. Figure P. 9.15, find the output voltage in steady state condition, where (i) Switch S is open, (ii) Switch S is closed.

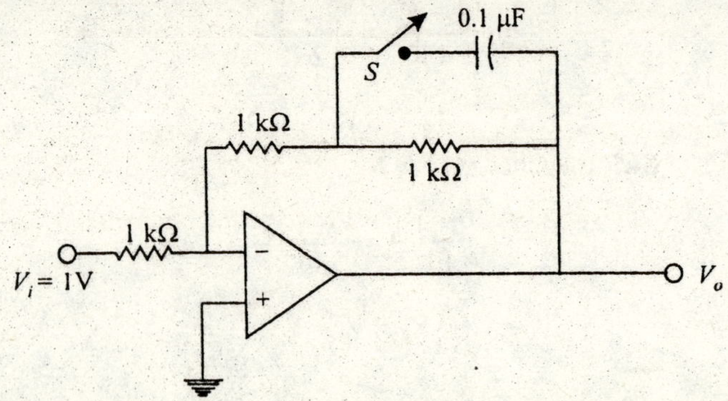

Fig. P. 9.15

Solution. (i) When the switch S is open.

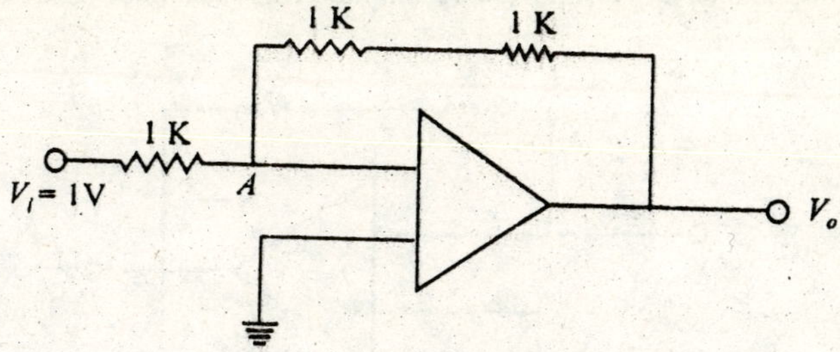

Fig. 9.15 (a)

Apply KCL at node A, we have

$$\Rightarrow \qquad \frac{V_i - V_A}{1\text{ K}} = \frac{V_A - V_o}{2\text{ K}} \qquad \{\because V_A = 0\text{ V}\}$$

$$\Rightarrow \qquad \frac{1-0}{1} = \frac{0-V_o}{2}$$

$$\Rightarrow \qquad V_o = 2\text{ V}$$

(ii) When the switch S is closed (under steady state capacitor is open circuit for $\omega = 0$).

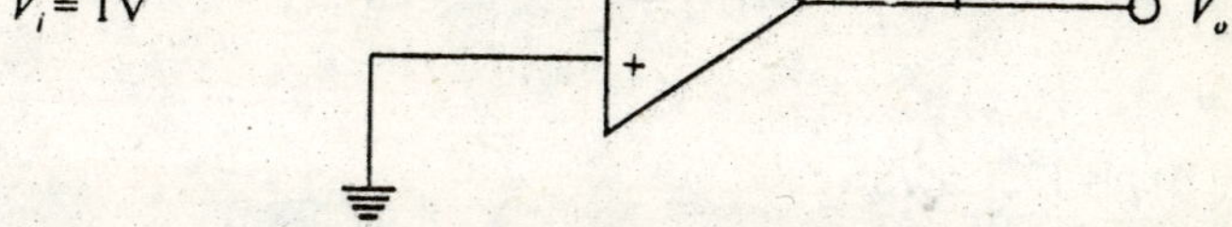

Fig. 9.15 (b)

Under steady state condition the capacitor is short-circuit, and hence apply KCL at node A.

$$\Rightarrow \qquad \frac{V_i - V_A}{1\text{ K}} = \frac{V_A - V_o}{1\text{ K}}$$

$$\Rightarrow \qquad V_i - V_A = V_A - V_o$$

$$\Rightarrow \qquad V_o = 2\ V_A - V_i$$

Now $\qquad V_A = 0\text{ V}, V_i = 1\text{ V}, \ V_o = -1\text{ V}$

Problem 9.16. For the circuit shown in the Fig. P. 9.16, determine the input impedance Z_i. Assume the op-amp to be an ideal one. **(GATE 1995)**

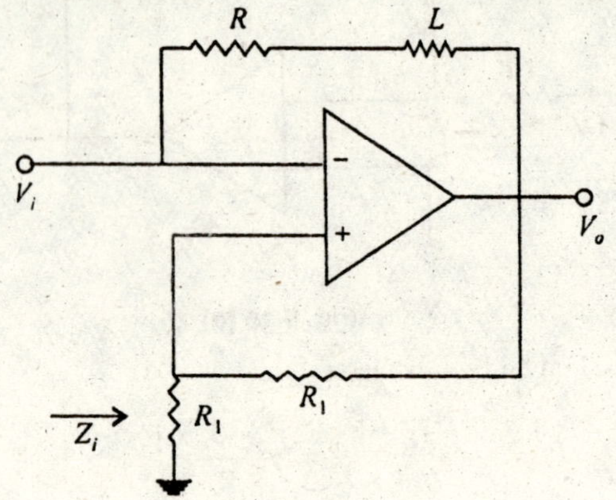

Fig. 9.16

Solution. Apply KCL at V_o, we have

$$\frac{V_i - V_o}{R + LS} = \frac{-V_o - V_i}{R_1}$$

$$\frac{(V_i - V_o)}{R + LS} = \frac{-V_i - V_o}{R_1}$$

$$R_1 V_i - R_1 V_o = -V_i (R + LS) + (R + LS) V_o$$

$$(R_1 + R + LS) V_i = (R_1 + R + LS) V_o$$

$$V_i = V_o \qquad \text{... (1)}$$

Apply KCL at A

$$\frac{0 - V_i}{R_1} = \frac{V_i - V_o}{R_1}$$

$$V_o = 2 V_i$$

Problem 9.17. For the circuit shown determine $V_o(s) / V_i(s)$, and hence write the equation for the magnitude and the phase response of V_o / V_i. If the value of $R_1 = 100$ k Ω and of $R = 10$ kΩ, determine the value of C to obtain a phase shift of 270° between V_o and V_i for an input frequency of 1000 rad/sec. **(GATE 1995)**

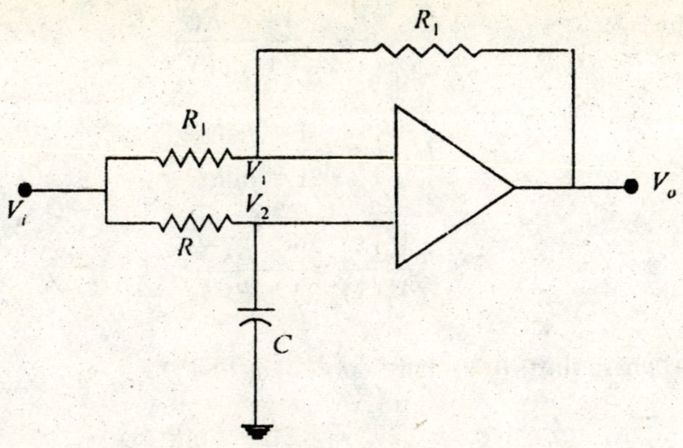

Fig. P. 9.17

Solution. From Fig. P. 9.17, we apply KVL .

$$V_1 = V_2 = \frac{V_i \dfrac{1}{C_s}}{R + \dfrac{1}{CS}} = \frac{V_i}{1 + RCS}$$

$$\Rightarrow \qquad V_1 = V_2 = \frac{V_i}{1 + j\omega RC} \qquad \ldots (1)$$

Apply KCL at node V_1

$$\frac{V_i - V_1}{R_1} = \frac{V_1 - V_o}{R_1}$$

$$\Rightarrow \qquad V_i - V_1 = V_1 - V_o$$

$$\Rightarrow \qquad 2\,V_1 = V_i + V_o$$

$$\Rightarrow \qquad \frac{2\,V_i}{1 + j\omega RC} = V_i + V_o$$

$$\Rightarrow \qquad V_i \left[\frac{2}{(1 + j\omega RC)} - 1\right] = V_o$$

$$\Rightarrow \qquad V_i \left[\frac{2 - 1 - j\omega RC}{1 + j\omega RC}\right] = V_o$$

$$\frac{V_o}{V_i} = \frac{1 - j\omega RC}{1 + j\omega RC}$$

$$H\,(j\omega) = \frac{1 - j\omega RC}{1 + j\omega RC}$$

$$\frac{V_o(S)}{V_i(S)} = \frac{1 - RCS}{1 + RCS}$$

Now, phase shift, $\theta = -\tan^{-1}\omega RC - \tan^{-1}\omega RC$

$$270 = -2\tan^{-1}\omega RC$$

$$-\tan^{-1}\omega RC = 135°$$

$$\omega RC = -\tan 135°$$

$$\omega RC = 1$$

$$10^3 \times 10^4 \times C = 1$$

$$C = \frac{1}{10^7} = 10^{-7} = 0.1\mu F.$$

Problem 9.18. (a) Obtain the transfer function for the circuit shown in Fig. P. 9.18. Assume the op-amp to be ideal. *Yi* represent the admittance of branch *i*. (b) If $Y_1 = C_1\ S$, $Y_4 = C_4^{\cdot}$, $Y_2 = G_2$, $Y_3 = G_3$, what is the nature of the filter that will be realized by the circuit ? **(GATE 1996)**

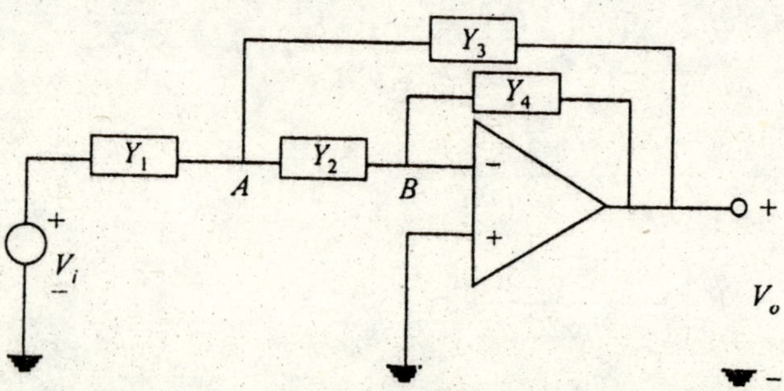

Fig. P. 9.18

Solution. (a) Apply KCL at node A

$$(V_A - V_i) Y_1 + (V_A - V_o) Y_3 + (V_A - V_B) Y_2 = 0$$

$$V_A (Y_1 + Y_3 + Y_2) = V_i Y_1 + V_o Y_3 + V_B Y_2$$

$$V_A (Y_1 + Y_2 + Y_3) = V_i Y_1 + V_o Y_3 \quad \text{... (1)} \quad \{\because V_B = 0\}$$

Now, again apply KCL at node B

$$(V_B - V_A) Y_2 + (V_B - V_o) Y_4 = 0$$

$$-V_A Y_2 = V_o Y_4 \qquad \{\because Y_B = 0\}$$

$$V_A = -\frac{Y_4}{Y_2} V_o$$

Put the value of V_A in equation (1)

$$\Rightarrow \qquad \frac{-Y_4}{Y_2}(Y_1 + Y_2 + Y_3) V_o = V_i Y_1 + V_o Y_3$$

$$V_o (Y_4 (Y_1 + Y_2 + Y_3) + Y_3 Y_2) = -Y_1 Y_2 V_i$$

$$\frac{V_o}{V_i} = \frac{-Y_1 Y_2}{[Y_4 (Y_1 + Y_2 + Y_3) + Y_3 Y_2]}$$

(b) Substituting the values of Yi, the transfer function is given by

$$\frac{V_o}{V_i} = \frac{-GS \times G_2}{[C_4 S (GS + G_2 + G_3) + G_2 G_3]}$$

$$\frac{V_o}{V_i} = \frac{-C_1 G_2 S}{C_1 C_4 S^2 + (G_2 + G_3) C_4 S + G_2 G_3}$$

This is of the form of B and pass filter transfer function

$$H(S) = \frac{AS}{S^2 + 2\xi \omega_o S + \omega_o^2}$$

Problem 9.19. A simple active is shown in Fig. P. 9.19. Assume the op-amp to be ideal. Derive the transfer function V_o/V_i of the circuit, and state the type of the filter, i.e. high pass, low pass, Band pass or Band reject. **(GATE 2001)**

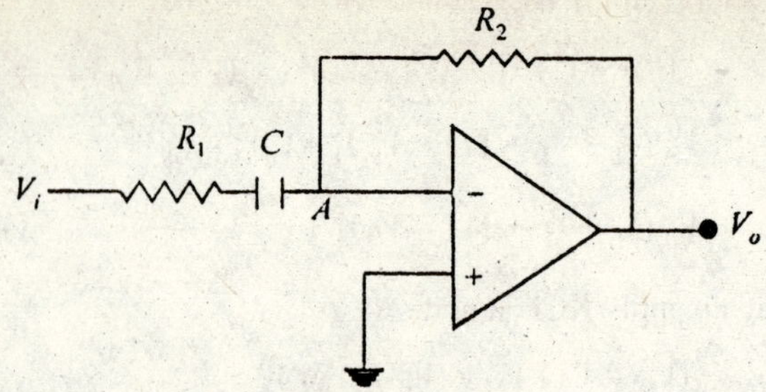

Fig. P. 9.19

Solution. Apply KCL at node A

$$\frac{(V_A - V_i)}{\left[R_1 + \frac{1}{CS}\right]} + \frac{(V_A - V_o)}{R_2} = 0$$

$$\Rightarrow \quad (V_A - V_i)\frac{CS}{[1 + R_1CS]} + \frac{(V_A - V_o)}{R_2} = 0$$

$$\Rightarrow \quad \frac{-CSV_i}{1 + R_1CS} + \frac{-1\,V_o}{R_2} = 0$$

$$\Rightarrow \quad \frac{V_o}{R_2} = \frac{-CSV_i}{1 + R_1CS}$$

$$\Rightarrow \quad \frac{V_o}{V_i} = \frac{-R_2CS}{R_1CS + 1}$$

$$\Rightarrow \quad \frac{V_o}{V_i} = \frac{-R_2CS}{R_1C(S + 1/R_1C)}$$

$$\Rightarrow \quad \frac{V_o}{V_i} = \frac{-(R_2/R_1)S}{(S + 1/R_1C)}$$

This is of the form of the first order high pass the filter.

Problem 9.20. Calculate the CMRR (in dB) for the circuit measurements of $V_d = 1\,\text{mV}$, $V_o = 120\,\text{mV}$ and $V_C = 1\,\text{mV}$, $V_o = 20\,\mu\text{V}$.

Solution.
$$A_d = \frac{V_o}{V_d} = \frac{120}{1} = 120$$

$$\Rightarrow \qquad A_c = \frac{V_o}{V_c} = \frac{20 \times 10^{-6}}{1 \times 10^{-3}} = 20 \times 10^{-3}$$

$$\Rightarrow \qquad \text{CMRR} = \frac{A_d}{A_c} = \frac{120}{20 \times 10^{-3}} = \frac{120 \times 10^{-3}}{20}$$

$$\Rightarrow \qquad \text{CMRR} = 6 \times 10^{-3}$$

in dB, $\text{CMRR} = 20 \log_{10} \frac{A_d}{A_c} = 20 \log_{10} 6 \times 10^3$

$$\text{CMRR} = 75.56 \text{ dB}.$$

Problem 9.21. Determine the output voltage of an op-amp for input voltage of $V_{i1} = 200\ \mu\text{V}$ and $V_{i2} = 140\ \mu\text{V}$. The amplifier has a differential gain of $A_d = 6000$ and the value of CMRR is (*a*) 200 (*b*) 10^5.

Solution.
$$V_d = V_{i1} - V_{i2} = (200 - 140)\ \mu\text{V}$$

$$\Rightarrow \qquad V_d = 60\ \mu\text{V}$$

$$\Rightarrow \qquad V_c = \frac{1}{2}(V_{i1} + V_{i2}) = \frac{1}{2}(200 + 140) = \frac{340}{2}$$

$$\Rightarrow \qquad V_c = 170\ \mu\text{V}$$

(a) The output voltage is given by

$$V_o = A_d V_d \left[1 + \frac{1}{\text{CMRR}} \frac{V_c}{V_d}\right]$$

$$= 6000 \times 60 \times 10^{-6} \left[1 + \frac{1}{200} + \frac{170}{60}\right]$$

$$= 365.1 \times 10^{-3} = 0.365 \text{ V}$$

(*b*) $\text{CMRR} = 10^5$

$$V_o = V_d A_d \left(1 + \frac{1}{\text{CMRR}} \frac{V_c}{V_d}\right)$$

$$= 60 \times 10^{-6} \times 6000 \left(1 + \frac{170}{60} \times \frac{1}{10^5}\right)$$

$$= 360 \times 10^{-3} = 0.36 \text{ V}$$

Problem 9.22. What is the output voltage of the op-amp circuit shown below:

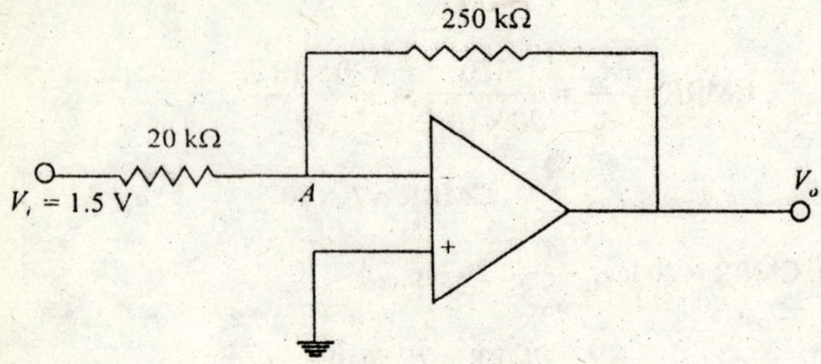

Fig. P. 9.22

Solution. Apply Kirchoffs current law at node *A*.

$$\frac{V_i - V_A}{20\text{K}} = \frac{V_A - V_o}{250\text{K}}$$

$$\Rightarrow \quad \frac{1.5 - 0}{20} = \frac{0 - V_o}{250} \quad \{\because V_A = 0\,\text{V (grounded)}\}$$

$$\Rightarrow \quad V_o = -\left[\frac{250}{20}\right] \times 1.5$$

$$V_o = -18.75\ \text{V}$$

Problem 9.23. What is the range of the voltage gain adjustment in the op-amp circuit shown in Fig. P. 9.23?

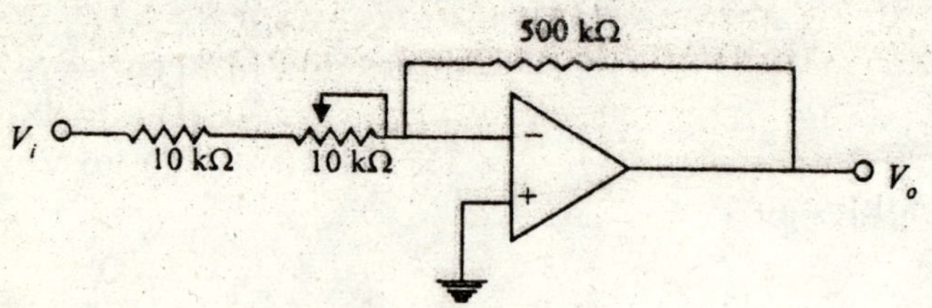

Fig. P. 9.23

Solution. The range in voltage gain can be obtained when the 10 kΩ is taken completely once and shorted completely in the second case.

Case I: 10 kΩ is taken completely.

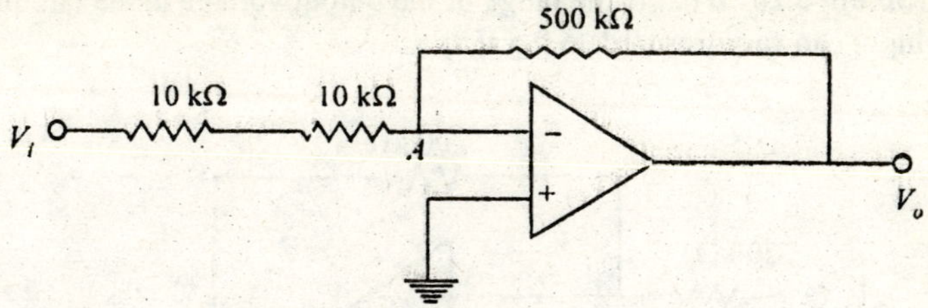

Fig. P. 9.23 (a)

Apply KCL at node A

$$\frac{V_i - V_A}{20\text{ K}} = \frac{V_A - V_o}{500\text{ K}} \quad \{V_A = 0\}$$

$$\Rightarrow \quad \frac{V_i}{20} = \frac{-V_o}{500}$$

$$\Rightarrow \quad \frac{V_o}{V_i} = -\frac{500}{20} = -25$$

Case II: Neglect (short) 10 k Ω

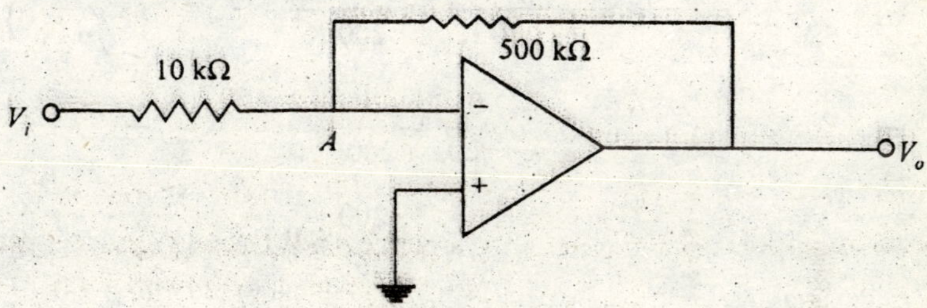

Fig. P. 9.23 (b)

Apply KCL at node A

$$\frac{V_i - V_A}{10\text{ K}} = \frac{V_A - V_o}{500\text{ K}} \quad \{\because V_A = 0\}$$

$$\Rightarrow \quad \frac{V_i}{10} = \frac{-V_o}{500}$$

$$\Rightarrow \quad \frac{V_o}{V_i} = \frac{-500}{10} = -50$$

Thus, the range of voltage gain is given as

$$-25 \le A_v \le -50$$

Problem 9.24. What is the range of the output voltage in the circuit, if the input can vary from 0.1 to 0.5 V ?

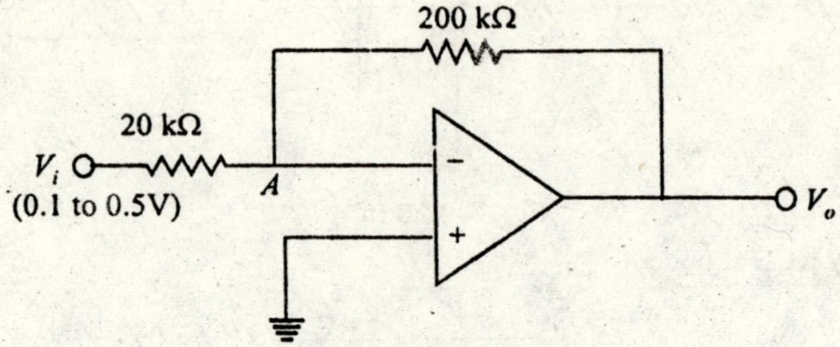

Fig. P. 9.24

Solution. To get the range of V_o, we take the two case as

Case I: When V_i = 0.1 V

Apply KCL at node *A*

$$\frac{V_i - V_A}{20\text{ K}} = \frac{V_A - V_o}{200\text{ K}} \qquad \{\because V_A = 0\}$$

$$\Rightarrow \qquad \frac{0.1}{20} = \frac{-V_o}{200}$$

$$\Rightarrow \qquad V_o = \frac{-200}{20} \times 0.1 = -1\text{V}$$

Case II: When V_i = 0.5 V

Apply KCL aat node *A* again

$$\Rightarrow \qquad \frac{V_i - V_o}{20\text{ K}} = \frac{0 - V_o}{200\text{ K}}$$

$$\Rightarrow \qquad V_o = -0.5 \times \frac{200}{20}$$

$$\Rightarrow \qquad V_o = -5\text{ V}$$

Thus, the range of output voltage is

$$-5 \le V_o \le -1$$

Problem 9.25. What output voltage result in the circuit for an input voltage of $V_1 = -0.3$ V?

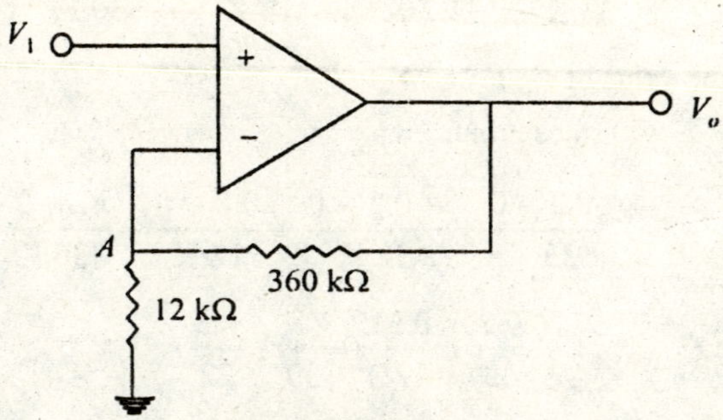

Fig. P. 9.25

Solution. The voltage $V_A = V_1 = 0.3$ V

Now, apply KCL at node A

$$\frac{0 - V_A}{12\text{ K}} = \frac{V_A - V_o}{360\text{ K}}$$

$$\Rightarrow \quad \frac{+0.3}{12} = \frac{-0.3 - V_o}{360}$$

$$\Rightarrow \quad 9 = -0.3 - V_o$$

$$\Rightarrow \quad V_o = -0.3 - 9 = -9.3\text{ V}$$

Thus, the output voltage for the input $V_i = 0.03$ V is −9.3 V. It is a noninverting op-amp ckt.

Problem 9.26. Calculate the output voltage developed by the circuit for $R_f = 330$ k Ω.

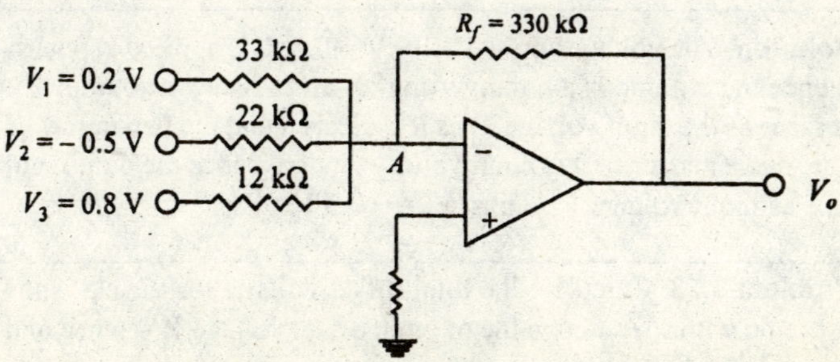

Fig. P. 9.26

Solution. Apply KCL at node *A*

$$\frac{V_1 - V_A}{33\text{ K}} + \frac{V_2 - V_A}{22\text{ K}} + \frac{V_3 - V_A}{12\text{ K}} + \frac{V_o - V_A}{330\text{ K}} = 0$$

Put the value of V_1, V_2, V_3, V_A

Here, $V_A = 0$ since grounded.

$$\Rightarrow \quad \frac{0.2 - 0}{33\text{ K}} + \frac{-0.5 - 0}{22\text{ K}} + \frac{0.8 - 0}{12\text{ K}} - \frac{V_o - 0}{330\text{ K}}$$

$$\Rightarrow \quad \frac{0.2}{33} + \frac{0.8}{12} + \frac{V_o}{330} - \frac{0.5}{22} = 0$$

$$\Rightarrow \quad \frac{V_o}{330} = -\frac{0.2}{33} - \frac{0.8}{12} + \frac{0.5}{22}$$

$$\Rightarrow \quad \frac{V_o}{330} = -6.06 \times 10^{-3} - 0.067 + 0.0227$$

$$\Rightarrow \quad V_o = -16.6098 \quad \Rightarrow \quad V_o \simeq -16.61\text{ V}$$

Problem 9.27. What output voltage result in the circuit for $V_1 = +0.5$ V?

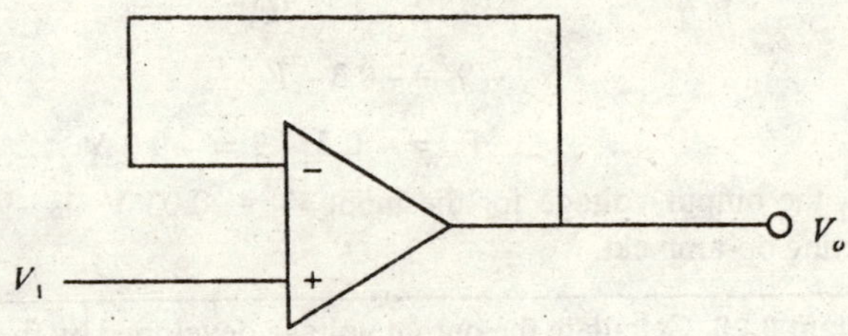

Fig. P. 9.27

Solution. The voltage applied is $V_1 = +0.5$. It is applied at positive input and hence the op-amp is a noninverting op-amp. Now, the output voltage V_o is the same as the input voltage V_1 as it is cleared that the two inputs to the op-amp are same and since it is noninverting device, hence the output voltage V_o is same as input voltage V_1. Thus, $V_o = +0.5$ V.

Problem 9.28. Calculate the total offset voltage for the ckt. shown for an op-amp with specified value of input offset voltage $Y_o = 6$ mV and input offset current $I_{io} = 120$ mA.

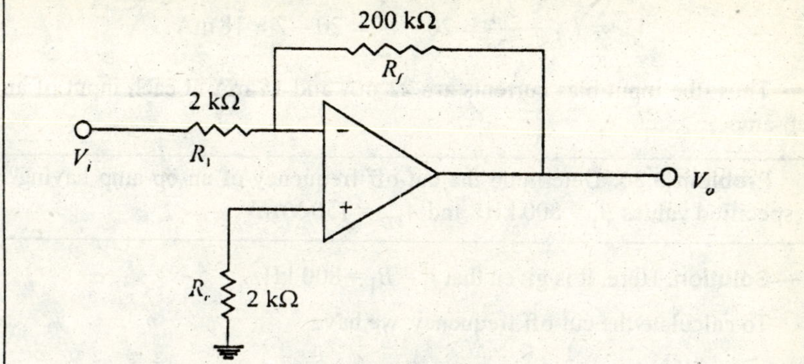

Fig. P. 9.28

Solution. The offset due to V_{io} is given by

$$V_o \text{ (offset due to } V_{io}) = V_{io} \frac{R_1 + R_f}{R_1}$$

$$= 6 \times 10^{-3} \left(\frac{2 + 200}{2}\right)$$

$$= 6 \times 10^{-3} \times \frac{202}{2}$$

$$= 606 \times 10^{-3} \text{ V} = 606 \text{ mV}$$

$$V_o \text{ (offset due to } I_{io}) = I_{io} R_f = 120 \times 10^{-9} \times 200 \text{ K}$$

$$= 120 \times 10^{-9} \times 200 \times 10^{+3}$$

$$= 24 \times 10^{-3} = 24 \text{ mV}$$

Therefore, the total offset voltage is given as

V_o (total offset) = V_o (offset due to V_{io}) + V_0 (offset due to I_{io})

$$= (606 + 24) \text{ mV} = 630 \text{ mV}$$

Problem 9.29. Calculate the input bias current at each input of an op-amp having specified values of $I_{io} = 4$ mA and $I_{iB} = 20$ mA.

Solution. The input bias current at each input of an op-amp having specified values of I_{io} and I_{iB} is given by

$$I_{iB}^{+} = I_{iB} + \frac{I_{io}}{2} = 20 + \frac{4}{2} = 20 + 2 = 22 \text{ mA}.$$

$$I_{iB}^{-} = I_{iB} + \frac{I_{io}}{2} = 20 - \frac{4}{2} = 20 - 2 = 18\,\text{mA}.$$

Thus, the input bias currents are 22 mA and 18 mA at each input of an op-amp.

Problem 9.30. Determine the cut-off frequency of an op-amp having specified values B_1 = 800 kHz and A_{VD} = 150 V/mV.

Solution. Here, it is given that $f_1 = B_1$ = 800 kHz.

To calculate the cut-off frequency, we have

$$f_c = \frac{f_1}{A_{VD}} = \frac{800 \times 10^3}{200 \times 10^3} = 4\,\text{Hz}.$$

Thus, the cut-off frequency of the op-amp is 4 Hz.

Problem 9.31. For an op-amp having a slow rate of SR = 2.4 $V/\mu S$. What is the maximum closed loop voltage gain that can be used when the input signal varies by 0.3 V in 10 μS?

Solution. Since the output voltage, $V_o = A_{CL} V_i$, where A_{CL} = closed loop voltage gain.

We have
$$\frac{\Delta V_o}{\Delta t} = A_{CL} \frac{\Delta V_i}{\Delta t}$$

Thus
$$A_{CL} = \frac{\Delta V_o / \Delta_t}{\Delta V_i / \Delta_t} = \frac{SR}{\Delta V_i / V_t}$$

Substituting the desired values, we have

$$A_{CL} = \frac{2.4\ \text{V}/\mu S}{0.3\ \text{V}/\ 10\ \mu S} = \frac{2.4 \times 10}{0.3} = 80$$

Thus, any closed loop voltage gain of magnitude greater than 80 would drive the output at a rate greater then the slow rate, so the maximum closed loop gain is 80.

Problem 9.32. For an input of V_1 = 50 mV in the circuit shown determine the maximum frequency that may be used. The op-amp slew rate SR = 0.4 V/μS.

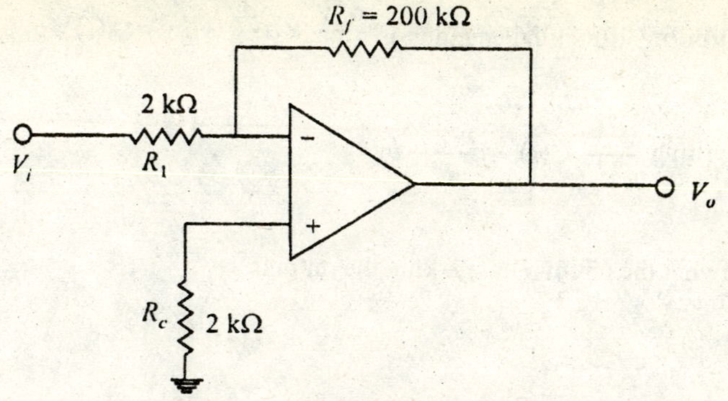

Fig. P. 9.32

Solution. The closed loop gain = $A_{CL} = \left|\frac{R_f}{R_1}\right| = \frac{200\text{ K}}{2\text{ K}} = 100$

The output voltage provides

$$K = A_{CL}V_i = 100 \times 50 \times 10^{-3}$$
$$K = 5\text{ V}$$

The op-amp frequency is given by

$$\omega \leq \frac{SR}{K} = \frac{0.4\,\text{V}/\mu S}{5} = 0.08 \times 10^6 = 80\text{ kHz}$$

Thus, the maximum value of frequency that no output distortion will result is kHz. Above this value there will be distortion in the output voltage.

Problem 9.33. In the circuit Fig. P. 9.33, the output V follows an equation of the form.

$$\frac{d^2v}{dt^2} + a\frac{dv}{dt} + bv = f\ (t)$$

Find a, b and $f(t)$.

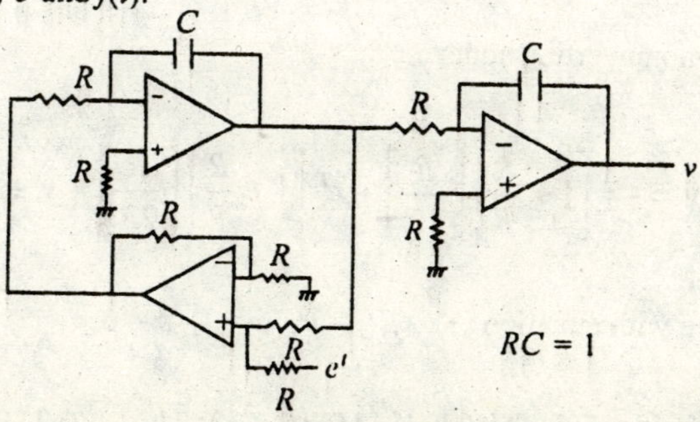

Fig. P. 9.33

Solution. Differential equation is $\dfrac{d^2v}{dt^2} + a\dfrac{dv}{dt} + bv = f(t)$

Rewriting $\dfrac{d^2v}{dt^2}\ f\ (t) - a\dfrac{dv}{dt} - bv$ ... (1)

Drawing the circuit and marking the outputs.

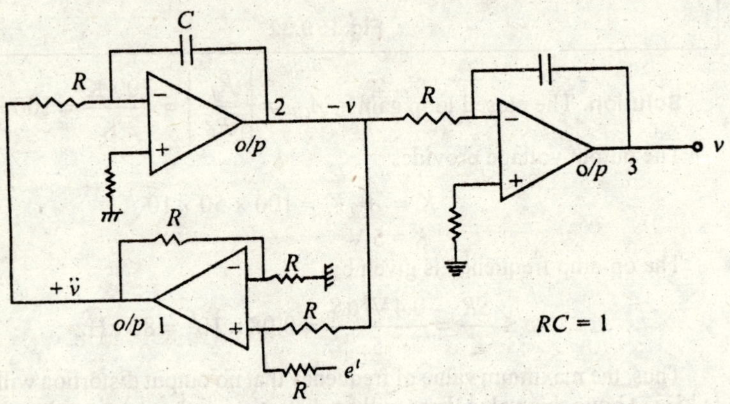

Fig. 9.33 (a)

$$(o/p)_1 = \ddot{v} = \frac{d^2v}{dt^2},\ (o/p)_2 = -v = -\frac{dv}{dt},\ (o/p)_3 = v$$

$(o/p)_1$ is a noninverting adder

$$\ddot{v} = -\dot{v}\left[1+\frac{R}{R}\right]\left[\frac{R}{2R}\right] + e^t\left[1+\frac{R}{R}\right]\left[\frac{R}{2R}\right] - \dot{v} + e^t$$

Comparing with equation (1), $f(t) = e^t,\ a = 1,\ b = 0$

The resistance R connected to $v(+)$ of $(o/p)_2$ and $(o/p)_3$ is a compensating resistor, which should be equal to resistance connected to $v(-)$.

Problem 9.34. Analyse the circuit.

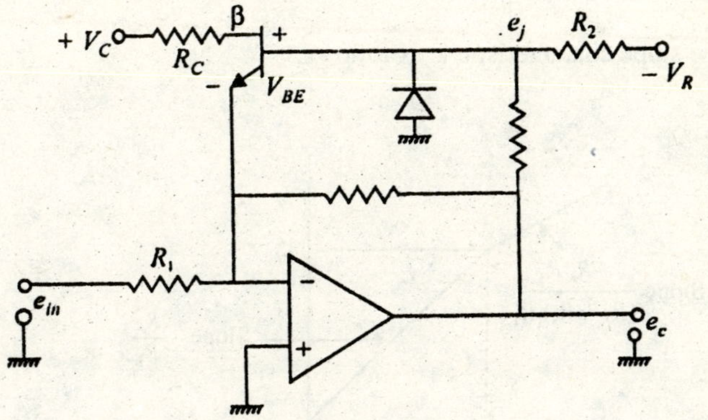

Fig. P. 9.34

Solution. This is feedback limiter circuit. It is of this form.

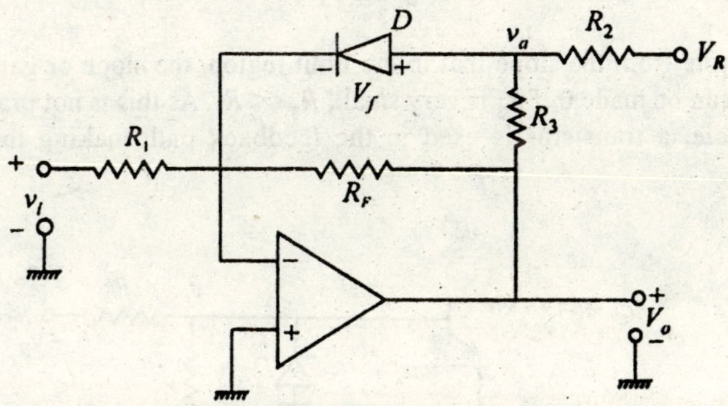

Fig. 9.34 (a)

Before diode D's conduction, i.e. when cut-off,

$$\text{Gain} \qquad A_{CL} = -\frac{R_F}{R_1}$$

When v_a exceeds the diode forward voltage V_f the diode conducts. The output voltage is then

$$v_o = V_R \frac{R_3}{R_2} + V_f + \frac{R_3}{R_2} V_f, \; A_{CL} = -\frac{R_F R_3}{(R_F + R_3)\, R_1}$$

The slope characteristic is as follows:

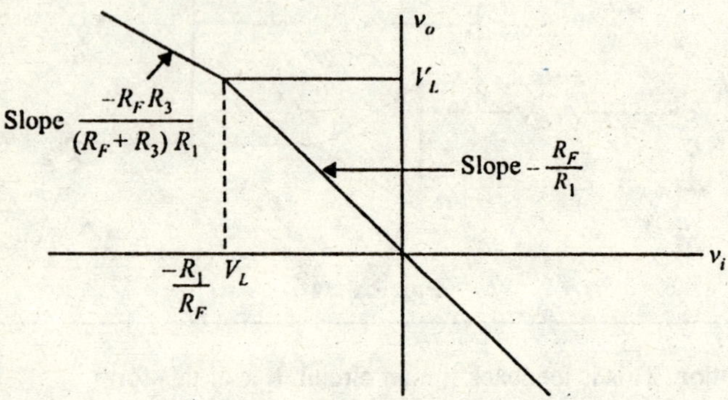

Fig. 9.34 (b)

We note from the slope that in the limit region, the slope or gain is not zero. It can be made 0 if R_3 is very small, $R_3 << R_1$. As this is not practically achievable, a transistor is used in the feedback path making the slope nearly 0.

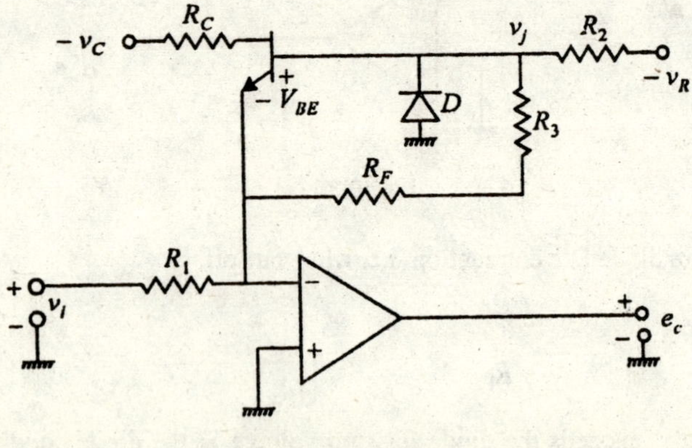

Fig. 9.34 (c)

When v_j exceeds V_{BE}, transistor is on and provides almost 0 resistance in the shunt path of R_F. Diode D protects the transistor from the reverse breakdown of V_{BE}.

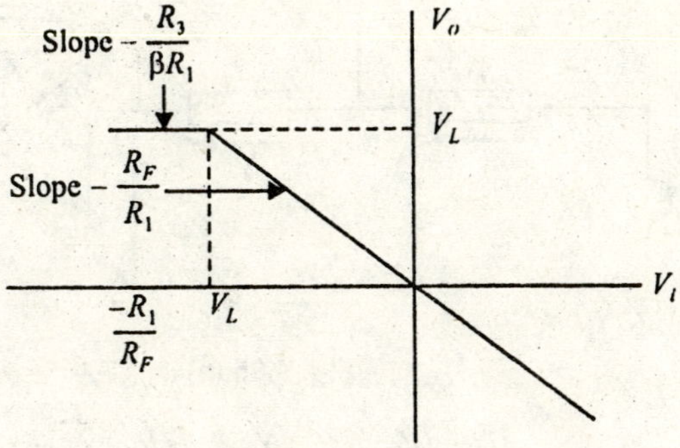

Fig. P. 9.34 (d)

The first circuit is called 'soft' limiting and the second 'hard' limiting. Limiting is not possible below V_{BE} or V_F in the 1st case.

Problem 9.35. Find the *TF*, $\frac{V_o(s)}{V_{in}(s)}$ when Y_i is the admittance of branch *i*. If $Y_1 = sC_1$, $Y_2 = G_2$, $Y_3 = G_3$, $Y_4 = sC_4$. What is the nature of the filter ?

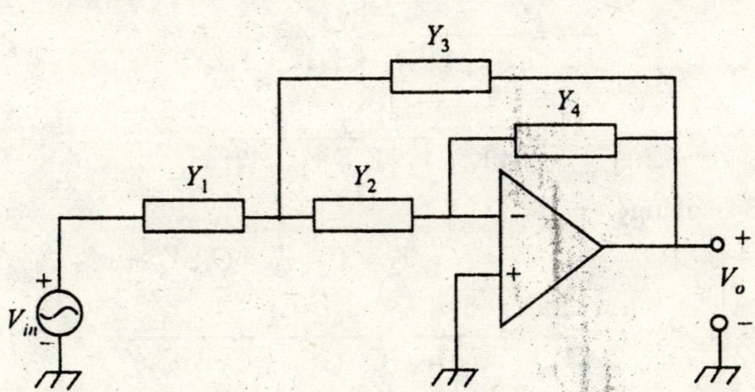

Fig. P. 9.35

Solution. (*a*)

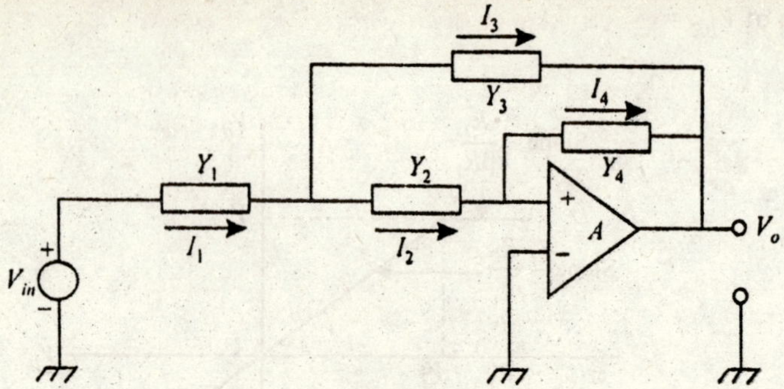

Fig. 9.35 (a)

$$I_1 = (V_{in} - V_A)\, Y_{1,}\ I_2 = V_A,\ Y_2 = I_4 = -V_o Y_4 \quad \text{... (1)}$$

$$I_3 = (V_A - V_o) Y_3$$

$$I_1 = I_2 + I_3$$

$$(V_{\text{in}} - V_A)\, Y_1 = -V_o Y_4 + (V_A - V_o)\, Y_3$$

$$V_{\text{in}}\, Y_1 = -V_o Y_4 + V_A Y_3 + V_o Y_3 + V_A Y_1$$

From (1) $$V_A = -\frac{V_o Y_4}{Y_2}$$

$\therefore$ $$V_{\text{in}} Y_1 - V_o Y_4 - \frac{V_o Y_4 Y_3}{Y_2} - V_o Y_3 - \frac{V_o Y_4 Y_1}{Y_2}$$

$$\frac{V_o}{V_{\text{in}}} = \frac{-Y_1 Y_2}{Y_2 Y_4 + Y_1 Y_4 + Y_2 Y_3 + Y_3 Y_4}$$

$$= \frac{-Y_1 Y_2}{Y_2 Y_3 + Y_4\,(Y_1 + Y_2 + Y_3)}$$

(b) Substituting

$$Y_1 = sC_1,\ Y_2 = G_2,\ Y_3 = G_3,\ Y_4 = sC_4$$

$$\frac{V_o}{V_{\text{in}}} = \frac{-sC_1 G_2}{G_2 G_3 + sC_4\,(sC_1 + G_2 + G_3)}$$

$$= \frac{-sC_1 G_2}{s^2 C_1 C_4 + sC_4 (G_2 + G_3) + G_2 G_3}$$

$$= \frac{-sG_2/C_4}{s^2 + s\dfrac{(G_2+G_3)}{C_1} + \dfrac{G_2G_3}{C_1C_4}}$$

This filter has a zero and two poles. It is of the general form.

$$H(s) = \frac{As}{s^2 + 2(\omega_o s + \omega_0^2)}$$, which is the TF of a band pass filter.

Problem 9.36. Deduce the input-output relation in the op-amp circuit shown in Fig. P. 9.36, and determine the output waveform for a symmetrical square wave input voltage of amplitude V_p and frequency f.

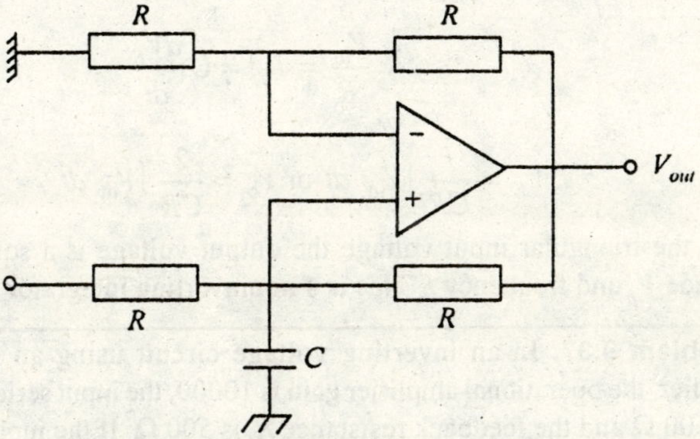

Fig. P. 9.36

Solution. The output V_o is divided equally by the voltage divider, so that

$$V_1 = V_o \times \frac{R}{2R} = \frac{V_o}{2}$$

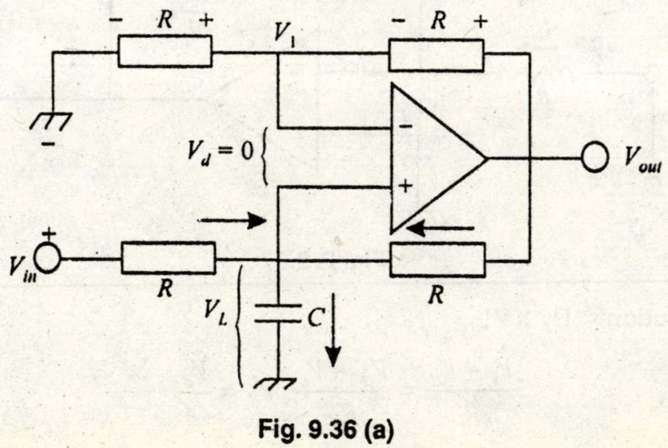

Fig. 9.36 (a)

The differential voltage across – and + input terminals of an ideal op-amp is 0. Hence, the + and – terminals are at $V_L = V_L$.

$$\therefore \qquad V_L = \frac{V_o}{2} \text{ or } V_o = 2\,V_L$$

Taking KCL at the positive input

$$\frac{V_L - V_{in}}{R} + \frac{V_L - V_o}{R} + I_L = 0$$

$$\frac{V_L + V_L - 2V_L - V_{in}}{R} = -\,I_L$$

$$\frac{V_{in}}{R} = I_L = C\frac{dV_L}{dt}$$

$$V_L = \frac{I}{CR}\int V_{in}\,dt \text{ or } V_o = \frac{2}{CR}\int V_{in}\,dt$$

For the triangular input voltage the output voltage is a square wave of amplitude V_p and frequency f. This is a noninverting integrator circuit.

Problem 9.37. In an inverting voltage circuit using an operational amplifier, the operational amplifier gain is 10000, the input series resistance R_i is 100 Ω and the feedback resistance R_f is 500 Ω. If the input voltage is 1 V, find the exact output voltage and the percentage error that would be committed by considering an infinite gain for the operational amplifier.

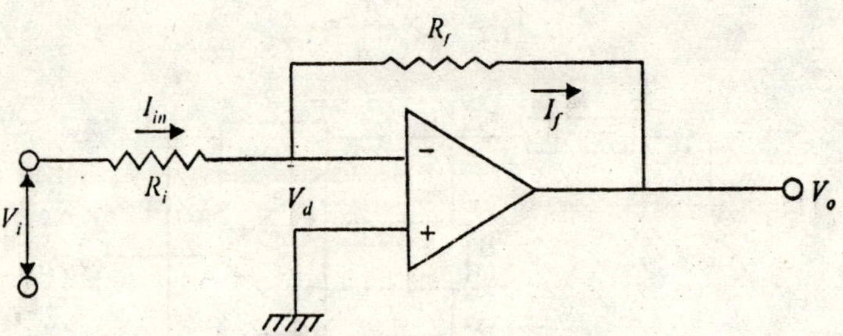

Fig. P. 9.37

Solution. By KVL

$$\frac{V_i - V_d}{R_i} = \frac{V_d - V_o}{R_f},\ V_d = \frac{V_o}{A}$$

$$\frac{V_i - \dfrac{V_o}{A}}{R_i} = \frac{\dfrac{V_o}{A} - V_o}{R_f}$$

$$\frac{V_i}{R_i} = \frac{\dfrac{V_o}{A} - V_o}{R_f} + \frac{V_o}{A} \times \frac{1}{R_i} = \frac{V_o\,(1 - A)}{AR_f} + \frac{V_o}{AR_i}$$

$$= V_o\left[\frac{(1 - A)}{AR_f} + \frac{1}{AR_i}\right]$$

$$\frac{V_o}{V_i} = \frac{1}{R_i} \times \frac{1}{\left[\dfrac{(1 - A)}{AR_f} + \dfrac{1}{AR_1}\right]}$$

Substituting $R_i = 100\ \Omega$, $R_f = 500\ \Omega$, and $A = 10000$

$$\frac{V_o}{V_i} = \frac{1}{100} \times \frac{1}{\dfrac{(1 - 10000)}{10000 \times 500} + \dfrac{1}{10000 \times 100}}$$

Simplifying II term's Dr

$$\frac{-9999 \times 100 + 500}{10000 \times 100 \times 500} = -\frac{999400}{10^4 \times 10^2 \times 500} = -\frac{9994}{500 \times 10^4}$$

$$\therefore \qquad \frac{V_o}{V_i} = -\frac{500 \times 10^4}{9994 \times 100} = -5.003$$

If we consider A to be infinite, then from equation (1)

$$\frac{V_o}{V_i} = \frac{1}{R_i} \times \frac{1}{\left(-\dfrac{1}{R_f}\right)} = -\frac{R_f}{R_i} = \frac{-500}{100} = -5$$

$$\text{Error} = \frac{5.003 - 5}{5.003} = 0.00059 \text{ or } 0.06\%$$

$\therefore$ Exact output voltage = 5.003 V, and error = 0.06%.

Reader's Notes

CHAPTER

10

Frequency Response of Transistors

Gain Versus Frequency Curve

(*i*) For RC coupled amplifiers

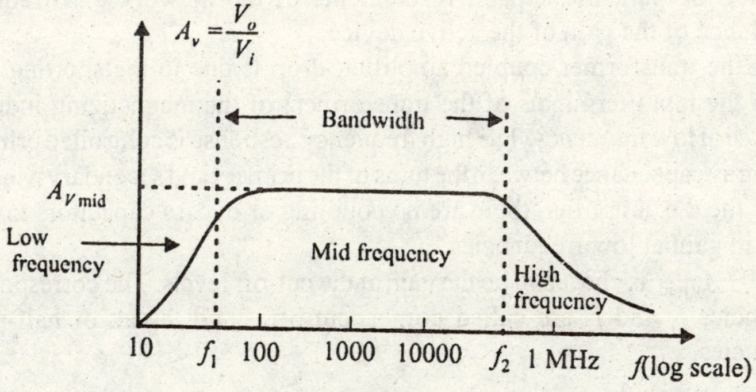

Fig. 10.1

(*ii*) For transformer coupled amplifier

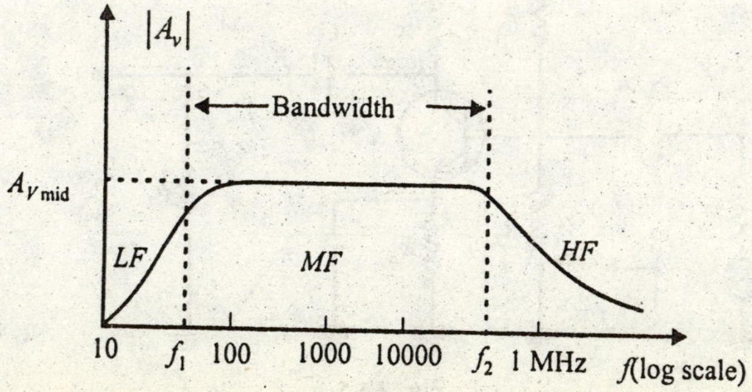

Fig. 10.2

(*iii*) For direct-coupled (d.c.) amplifier

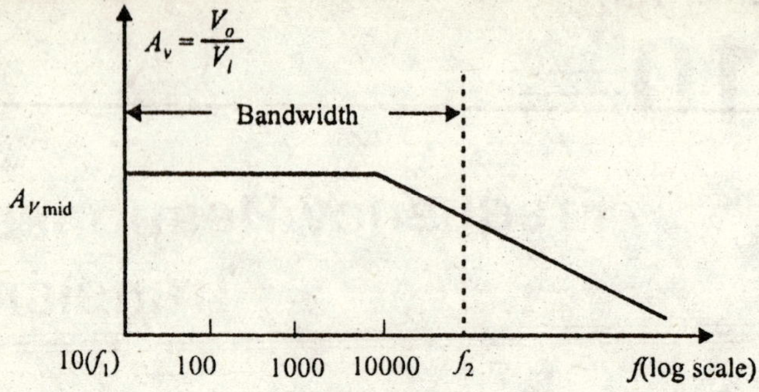

Fig. 10.3

For the *RC* coupled amplifier, the drop at low frequency is due to the increasing capacitance of C_C, C_S, and C_E, while its upper frequency limit is determined by either the parasitic capacitive elements of the network and frequency dependence of the gain of the active device.

For the transformer coupled amplifier, drop is due to the shorting effect (across the input terminals of the transformer) of the magnetizing inductive reactance at low frequency. The high frequency response is controlled primarily by the stray capacitance between the turns of the primary and secondary windings.

For the d.c. amplifier, there are no coupling or bypass capacitors to cause a drop in gain at low frequencies.

0.707 $A_{v\,mid}$ is chosen to be the gain at the cut-off levels. The corresponding frequencies f_1 and f_2 are called corner, cut-off, band, break or half-power frequencies.

Low Frequency Respose BJT Amplifier

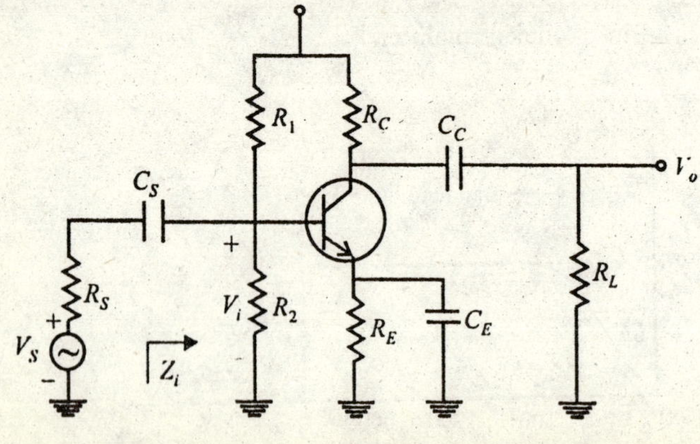

Fig. 10.4

Though the analysis is employed for voltage divider BJT bias configuration, but the result can be applied to any BJT.

The capacitors C_c, C_E, and C_S, i.e. coupling capacitor, bypass capacitor and capacitor connected to source will determine the low frequency resource.

Effect of C_S

The total resistance is $(R_S + R_i)$, and it is general form of the R–C configuration.

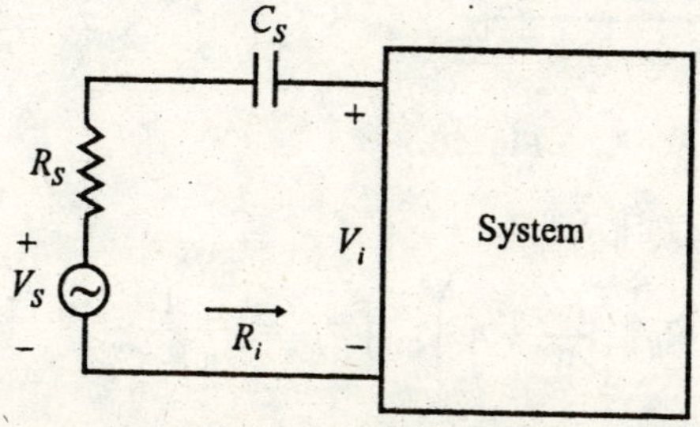

Fig. 10.5

Hence, the cut-off (lower) frequency is given by

$$f_{LS} = \frac{1}{2\pi(R_S + R_i)C_S}$$

and $$V_{i\ \text{mid}} = \frac{V_S R_i}{R_i + R_S}$$

$$R_i = R_B \,||\, \beta_{re} = R_1 \,||\, R_2 \,||\, \beta_{re}$$

and $$A_{v\ \text{mid}} = \frac{-R_C \,||\, R_L}{r_e}$$

Effect of C_C

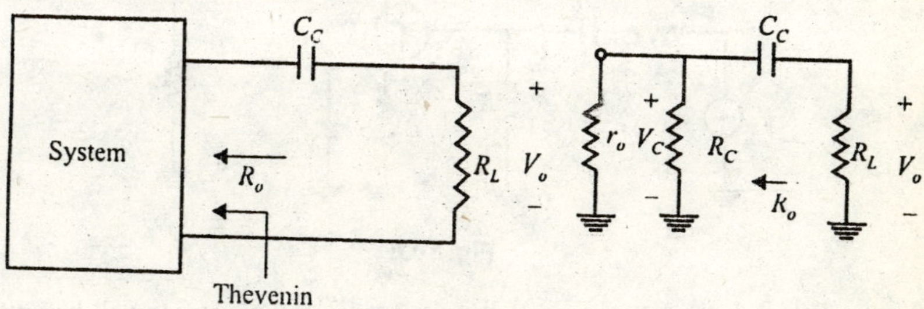

Fig. 10.6

$$f_{LC} = \frac{1}{2\pi(R_0 + R_L)C_C}, \quad R_o = R_C \,||\, r_o$$

If $\quad r_o = \infty,\ R_o = R_C$

Effect of C_E

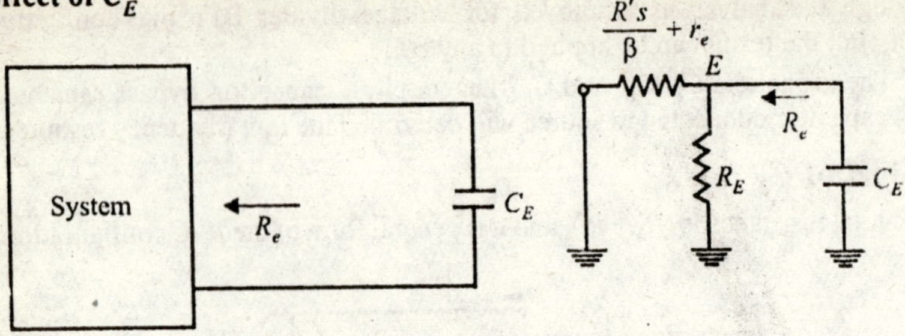

Fig. 10.7

Let $$R'_S = R_S \parallel R_1 \parallel R_2$$

$$R_e = R_E \parallel \left(\frac{R'_S}{\beta} + r_e \right) \text{ and } f_{LE} = \frac{1}{2\pi R_e C_E}$$

After evaluating the three frequencies, f_{LC}, f_{LS}, and f_{LE}, the resultant lower cut-off frequency will of the value which is highest of the three frequencies.

Including the effect of r_o, mid-band gain is given by

$$A_{v\,\text{mid}} = -\frac{R_C \parallel R_L \parallel r_o}{r_e}$$

Low Frequency Response FET Amplifier

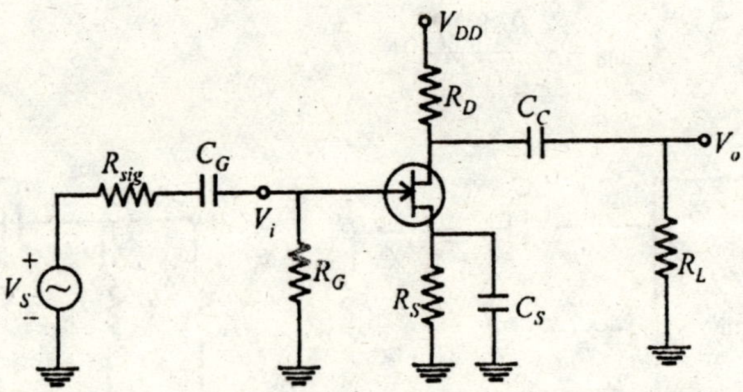

Fig. 10.8

The analysis of the FET amplifier in the low frequency region will be quite similar to that of the BJT amplifier.

Three concerned capacitors are C_G, C_S, and C_C.

C_G: The cut-off frequency determined by C_G will be

$$f_{LG} = \frac{1}{2\pi(R_{sig} + R_i)C_G}, \text{ where } R_i = R_G$$

$$C_c\text{: } f_{LC} = \frac{1}{2\pi(R_o + R_L)C_C}, \; R_o = R_D \parallel r_d$$

$$C_S\text{: } f_{LS} = \frac{1}{2\pi_{Req.} C_S}$$

$$\text{Req.} = \frac{R_S}{1 + R_S \dfrac{(1 + g_m r_d)}{(r_d + R_D \parallel R_L)}}$$

When $r_d = \infty$, Req.$= R_S \parallel \dfrac{1}{g_m}$

$A_{v\,\text{mid}}$ is given by $A_{v\,\text{mid}} = -g_m(R_D \parallel R_L)$

Again the frequency having the largest value will the lower cut-off frequency.

Miller Effect Capacitance

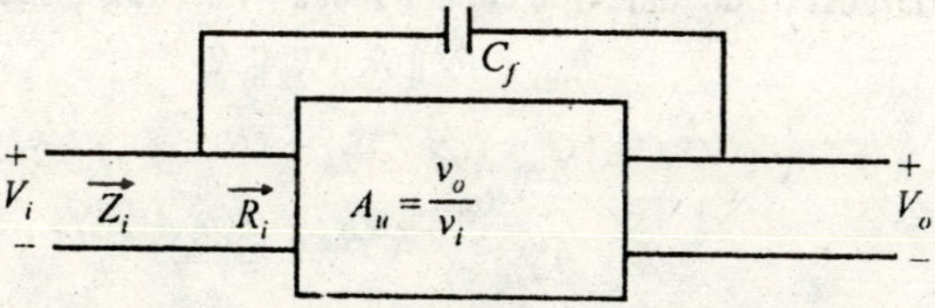

Fig. 10.9

The feedback capacitor C_f can be converted into equivalent Miller input and Miller output capacitances.

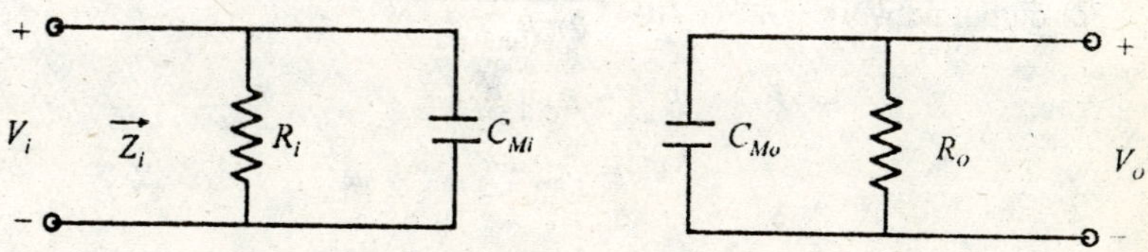

Fig. 10.10

Where $C_{M_i} = (1 - A_v)\, C_f$

and $C_{M_o} = \left(1 - \dfrac{1}{A_v}\right) C_f$

High Frequency Response BJT Amplifier

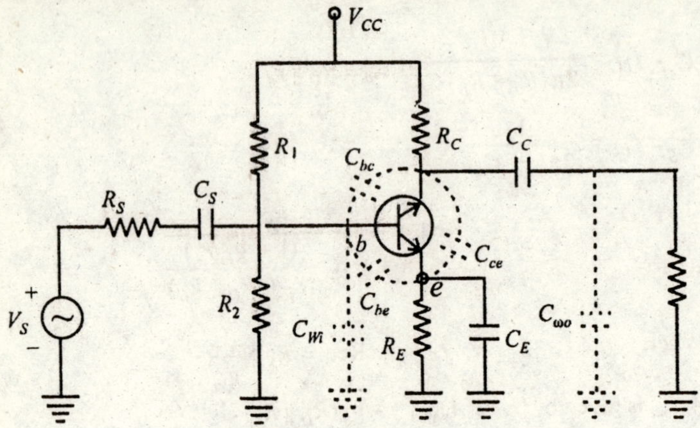

Fig. 10.11

$C_{\omega i}$ and $C_{\omega o}$ are called wiring capacitances and C_{ce}, C_{be}, and C_{bc} are called parasitic capacitances.

C_{be} is the largest of the parasitic capacitances with C_{ce} the smallest.

Let $$R_{th1} = R_S \parallel R_1 \parallel R_2 \parallel R_i$$

$$C_i = C_{\omega i} + C_{be} + C_{Mi}$$

or $$C_i = C_{\omega i} + C_{be} + (1 - A_v)C_{bc}$$

Then $$f_{Hi} = \frac{1}{2\pi R_{th1} C_i} \quad \text{(For input network)}$$

For output network $$f_{Ho} = \frac{1}{2\pi R_{th2} C_o}$$

Where $$R_{th2} = R_C \parallel R_L \parallel r_o$$

$$C_o = C_{\omega o} + C_{ce} + C_{Mo}$$

$$C_{\omega o} + C_{ce} + \left(1 - \frac{1}{A_v}\right) C_{bc}$$

$$f_\beta = \frac{1}{2\pi \beta_{mid} r_e (C_{be} + C_{bc})}$$

The frequency at which $\left| h_{fe} \right|_{dB} = 0\,\text{dB}$ is f_T

$$f_T = \beta_{mid} f_\beta$$

$$\therefore \quad f_T \cong \frac{1}{2\pi r_e (C_{be} + C_{bc})}$$

High Frequency Response FET Amplifier

The analysis is similar to that of BJT. Here, the three parasitic capacitances are C_{gd}, C_{gs}, and C_{ds}.

$$f_{Hi} = \frac{1}{2\pi R_{th1} C_i}, \quad R_{th1} = R_{sig} \parallel R_G$$

$$C_i = C_{wi} + C_{gs} + (1 - A_v) C_{gd}$$

$$f_{Ho} = \frac{1}{2\pi R_{th2} C_o}$$

$$R_{th2} = R_D \parallel R_L \parallel r_d$$

and
$$C_o = C_{wo} + C_{ds} + C_{Mo}$$

Where
$$C_{Mo} = \left(1 - \frac{1}{A_v}\right) C_{gd}$$

Problem 10.1. Determine the lower cut-off frequency for the network shown below using the following parameters
$C_S = 10\ \mu F$, $C_E = 20\ \mu F$, $C_c = 1\ \mu F$, $R_S = 1\ k\Omega$, $R_1 = 40\ k\Omega$, $R_2 = 10\ k\Omega$, $R_E = 2\ k\Omega$, $R_c = 4\ k\Omega$, $R_L = 2.2\ k\Omega$, $\beta = 100$, $V_{cc} = 20$ V, and $r_o = 40\ k\Omega$. Also find $A_{v\,mid}$.

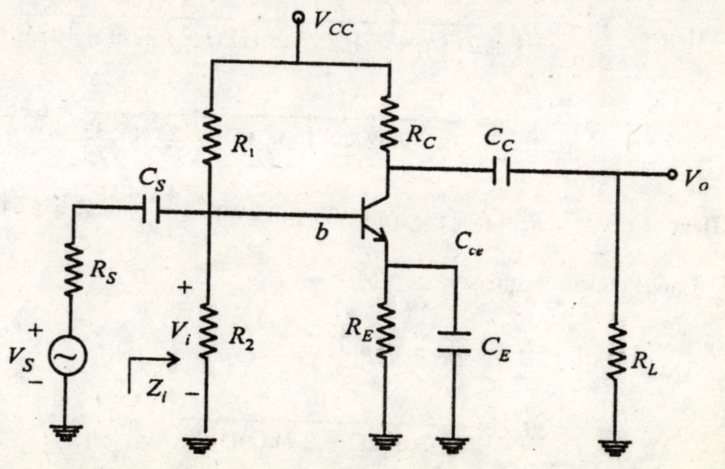

Fig. P. 10.1

Solution. From d.c. analysis, we have

$$\beta R_E = (100)\ (2\ \text{k}\Omega) = 200\ \text{k}\Omega >> 10\ R_2 = 100\ \text{k}\Omega$$

The result is $$V_{th} = V_B = V_{cc} \times \frac{R_2}{R_1 + R_2} = \frac{(20\,\text{V})10\,\text{k}\Omega}{10\,\text{k}\Omega + 40\,\text{k}\Omega}$$

or $$V_B = \frac{200}{50}\ \text{V} = 4\ \text{V}$$

Now $$V_E = V_B - V_{BE} = 4\ \text{V} - 0.7\ \text{V} = 3.3\ \text{V}$$

Hence $$I_E = \frac{V_E}{R_E} = \frac{3.3\,\text{V}}{2\,\text{k}\Omega} = 1.65\ \text{mA}$$

or $$r_e = \frac{26\ \text{mV}}{I_E} = \frac{26\ \text{mV}}{1.65\,\text{mA}} = 15.76\ \Omega$$

and $$\beta r_e = 100\ (15.76\ \Omega) = 1576\ \Omega = 1.576\ \text{k}\Omega$$

Effect of C_s: Lower cut-off frequency

$$f_{LS} = \frac{1}{2\pi(R_s + R_i)C_s}$$

Where $$R_i = R_1 \parallel R_2 \parallel \beta r_e$$

or $$R_i = (40\ \text{k}\Omega) \parallel (10\ \text{k}\Omega) \parallel 1.576\ \text{k}\Omega = 1.32\ \text{k}\Omega$$

Hence $$f_{LS} = \frac{1}{2\pi(R_s + R_i)C_s} = \frac{1}{2\pi(1\text{k}\Omega + 1.32\,\text{k}\Omega)(10\,\mu\text{F})}$$

$$= \frac{1}{2\pi \times 2.32\,\text{K} \times 10 \times 10^{-6}} = \frac{100}{2\pi \times 2.32} = 6.86\,\text{Hz.}$$

Effect of C_c $$R_o = R_c \parallel r_o = (4\ \text{kW}) \parallel (40\ \text{kW}) = \frac{4 \times 40}{44}\text{k}\Omega = 3.64\,\text{k}\Omega$$

Lower cut-off frequency

$$f_{LC} = \frac{1}{2\pi(R_o + R_L)C_c}$$

or $$f_{LC} = \frac{1}{2\pi(3.64\,\text{k}\Omega + 2.2\,\text{k}\Omega)(1\,\mu\text{F})}$$

$$= \frac{1}{6.28 \times 5.84 \times 10^3 \times 10^{-6}} = \frac{1000}{6.28 \times 5.84}$$

$\Rightarrow \qquad f_{LC} = 27.27$ Hz.

Effect of C_E $\qquad R_s' = R_s \parallel R_1 \parallel R_2 \parallel = (1 \text{ kW}) \parallel (40 \text{ kW}) \parallel (10 \text{ kW})$

$$\cong 0.889 \text{ k}\Omega$$

and
$$R_e = R_E \parallel \left(\frac{R_s'}{\beta} + r_e\right) = (2\text{k}\Omega) \parallel \left(\frac{0.889\text{k}\Omega}{100} + 15.76\Omega\right)$$

$$= (2\text{k}\Omega) \parallel (8.89\Omega + 15.76\Omega) = (2\text{k}\Omega) \parallel (24.56\Omega)$$

$$\cong 24.35\Omega$$

Hence
$$f_{LE} = \frac{1}{2\pi R_e C_E} = \frac{1}{6.28 \times 24.35 \times 20 \times 10^{-6}}$$

$$= \frac{10^5}{6.28 \times 24.35 \times 2} \cong 327 \text{Hz}$$

Midband gain

$$A_{v\text{mid}} = \frac{V_o}{V_i} = -\frac{R_c \parallel r_o \parallel R_L}{r_e}$$

or
$$A_{v\text{mid}} = -\frac{(4\text{k}\Omega) \parallel (40\text{k}\Omega) \parallel (2.2\text{k}\Omega)}{15.76\Omega} = -\frac{1370.72\Omega}{15.76\Omega} \cong -87$$

Problem 10.2. For the network shown in Fig. P. 10.2 :

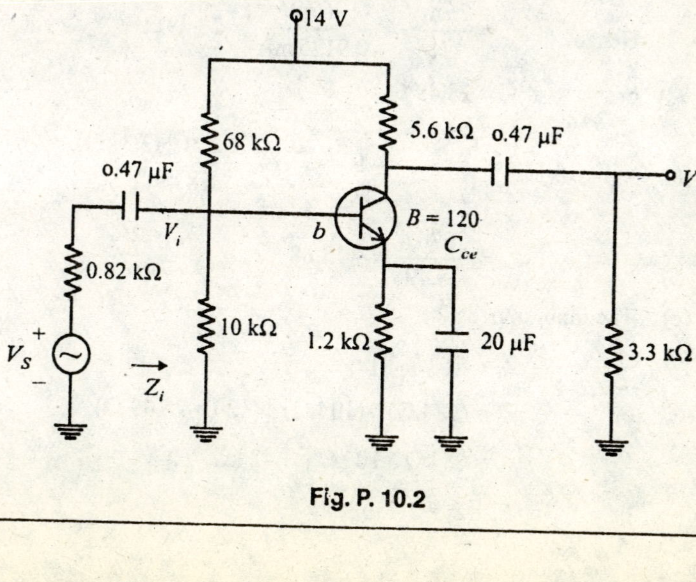

Fig. P. 10.2

(*a*) Determine r_e	(*b*) Find $A_{v\,mid} = \frac{V_o}{V_i}$
(*c*) Calculate Z_i	(*d*) Find $A_{vs\,mid} = \frac{V_o}{V_s}$
(*e*) Determine $f_{LS}, f_{LC},$ and f_{LE}	(*f*) Determine the low cut-off frequency.

Solution. From d.c. analysis

$$\beta R_E > 10\,R_2$$

$\Rightarrow$ $120 \times 1.2\text{ k}\Omega > 10 \times 10\text{ k}\Omega$

$\Rightarrow$ $144\text{ k}\Omega > 100\text{ k}\Omega$ satisfied.

So, we can apply the approximate analysis as

(*a*) $$V_{th} = \frac{R_2}{R_1 + R_2} \times V_{cc} = \frac{(10\text{k}\Omega)}{78\text{k}\Omega} \times 14\text{ V} = \frac{140}{78}\text{V}$$

or $V_{th} = 1.795\text{ V} = V_B$

and $R_{th} = R_1 \parallel R_2 = (68\text{ k}\Omega) \parallel (10\text{ k}\Omega) = 8.72\text{ k}\Omega$

Now $V_E = V_B - V_{BE} = 1.795\text{ V} - 0.7\text{ V} = 1.095\text{ V}$

and $$I_E = \frac{V_E}{R_E} = \frac{1.095\text{ V}}{1.2\text{ k}\Omega} = 0.9125\text{mA}$$

Hence $$r_e = \frac{26\text{mV}}{I_E} = \frac{26\text{mV}}{0.9125\text{mA}} = 28.49\Omega$$

or $r_e = 28.49\ \Omega$

(*b*) $$A_{v\,mid} = \frac{V_o}{V_i} = -\frac{R_C \parallel R_L}{r_e} = -\frac{(5.6\text{k}\Omega) \parallel (3.3\text{k}\Omega)}{28.49\Omega}$$

$$= \frac{2076.4}{28.49} = -72.88$$

(*c*) The input impedance

$$Z_i = R_1 \parallel R_2 \parallel \beta r_e$$

or $Z_i = (68\text{ k}\Omega) \parallel (10\text{ k}\Omega) \parallel (120 \times 28.49\ \Omega)$

$= 68\text{ k}\Omega \parallel 10\text{ k}\Omega \parallel 3.42\text{ k}\Omega$

or $R_i = Z_i = 2.456\text{ k}\Omega$

(*d*) $A_{vs\,mid} - \frac{V_o}{V_s} = \frac{V_o}{V_i} \cdot \frac{V_i}{V_s}$

Also $V_i = V_s \times \frac{R_i}{R_i + R_s}$ (voltage division rule)

or $\frac{V_i}{V_s} = \frac{R_i}{R_i + R_s}$

or $A_{vs\,mid} = A_{v\,mid} \times \frac{R_i}{R_i + R_s} = (-72.88) \times \frac{2.456}{2.456 + 0.82}$

$= (-72.88) \times 0.75 = -54.66$

(*e*) $f_{LS} = \frac{1}{2\pi(R_s + R_i)C_s}$, given $C_s = 0.47\ \mu F$

or $f_{LS} = \frac{1}{6.28(0.82\ k\Omega + 2.450\ k\Omega)0.47\mu F} \neq \pm$

$= \frac{10^8}{6.28 \times 3.276 \times 10^3 \times 47}$

$= \frac{10^5}{6.28 \times 3.276 \times 47} Hz = 103.42\,Hz$

$f_{LC} = \frac{1}{2\pi(R_o + R_L)C_c}$, $R_o = R_c = 5.6\ k\Omega$, $C_c = 0.47\ \mu F$

$f_{LC} = \frac{1}{6.28(5.6 k\Omega + 3.3 k\Omega)(0.47 \mu F)}$

$= \frac{1}{6.28 \times 8.9 \times 10^{-5} \times 47}$

$= \frac{10^5}{6.28 \times 8.9 \times 47} Hz = 38.07\,Hz$

Now $f_{LE} = \frac{1}{2\pi R_e C_E}$, $C_E = 20\ \mu F$

$R_e = R_E \,||\, \left(\frac{R'_S}{\beta} + r_e\right)$, where $R'_S = R_1 \,||\, R_2 \,||\, R_S$

Now $R'_S = (68\ k\Omega) \,||\, (10\ k\Omega) \,||\, (0.82\ k\Omega) = 0.75\ k\Omega$

$$R_e = (1.2\,k\Omega) \,||\, \left(\frac{0.75\,k\Omega}{120} + 28.49\,\Omega\right)$$

$$= (1.2\,k\Omega) \,||\, (6.246\,\Omega + 28.49\,\Omega)$$

$$= \frac{1200 \times 34.736}{1234.736} = 33.76\ \Omega$$

and $$f_{LE} = \frac{1}{2\pi \times 33.76 \times 20 \times 10^{-6}}$$

$$= \frac{10^5}{6.28 \times 33.76 \times 2}\ \text{Hz} = 235.83\ \text{Hz}$$

(*f*) Since f_{LE} is the largest of the three cut-off frequencies, it defines the low cut-off frequency

$$f_L = f_{LE} = 235.83\ \text{Hz}$$

Problem 10.3. Figure P. 10.3 for the emitter stabilized network:

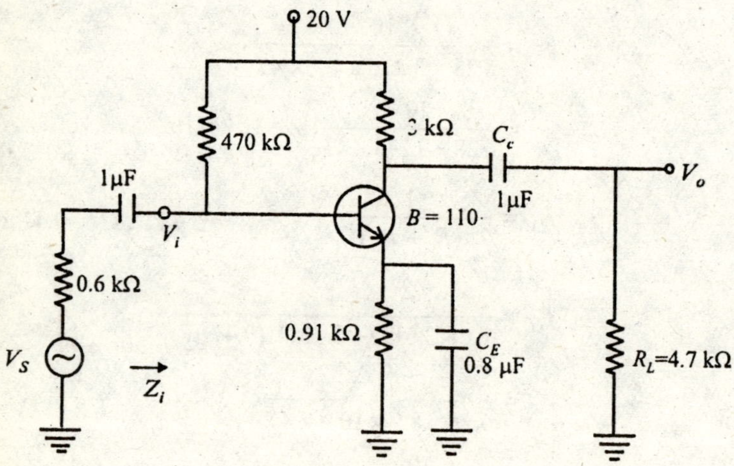

Fig. P. 10.3

(*a*) Determine r_e

(*b*) Find $A_{v\,mid} = \frac{V_o}{V_i}$

(*c*) Calculate Z_i

(*d*) Find $A_{vs\,mid} = \frac{V_o}{V_S}$

(*e*) Determine f_{Ls}, f_{LC} and f_{LE}

(*f*) Determine the low cut-off frequency.

Solution. From d.c. analysis, we have

(*a*)
$$I_B = \frac{V_{CC} - V_{BE}}{R_B + (\beta + 1)\, R_E} = \frac{20\text{ V} - 0.7\text{ V}}{470\text{ k}\Omega + 111 \times 0.91\text{ k}\Omega}$$

$$= \frac{19.3\text{ V}}{571.01\text{k}\Omega} = 33.8\mu\text{A}$$

and hence $I_E = (\beta + 1)\, I_B = 111 \times 33.8\ \mu\text{A} = 3.75\text{ mA}$

Now
$$r_e = \frac{26\text{ mV}}{I_E} = \frac{26\text{ mV}}{3.75\text{ mA}} = 6.93\ \Omega$$

(*b*)
$$A_{v\text{ mid}} = \frac{V_o}{V_i} = -\frac{R_C \,\|\, R_L}{r_e}$$

$$R_e \,\|\, R_L = (3\text{ k}\Omega) \,\|\, (4.7\text{ k}\Omega) = \frac{3 \times 4.7}{7.7}\text{ k}\Omega = 1.831\text{ k}\Omega$$

or
$$A_{v\text{ mid}} = -\frac{1.8312\text{k}\Omega}{6.93\Omega} = -264.24$$

(*c*) Input impedance

$$Z_i = R_B \,\|\, \beta_{re} = (470\text{ k}\Omega) \,\|\, (110 \times 6.93\ \Omega) = 761\ \Omega$$

(*d*)
$$A_{vs\text{ mid}} = \frac{V_o}{V_S} = \frac{V_o}{V_i} \cdot \frac{V_i}{V_S} \quad \text{or} \quad A_{vs\text{ mid}} = A_{v\text{ mid}} \times \frac{R_i}{R_i + R_S}$$

$$= (-264.26) \times \frac{761\Omega}{(0.761 + 0.6)\text{k}\Omega} = -147.76$$

(*e*)
$$f_{LS} = \frac{1}{2\pi(R_i + R_S)C_S} = \frac{1}{2\pi \times 1361 \times 10^{-6}}$$

$$= \frac{10^6}{6.28 \times 1361}\text{ Hz} \cong 117\text{ Hz}$$

$$f_{LC} = \frac{1}{2\pi(R_C + R_L)C_C} = \frac{1}{2\pi(3 + 4.7) \times 10^3 \times 10^{-6}}$$

$$= \frac{1000}{6.28 \times 7.7}\text{ Hz} = 20.68\text{ Hz}$$

$$f_{LE} = \frac{1}{2\pi R_e\, C_E}$$

Now $\quad R'_S = R_S \| R_B = (0.6\,\text{k}\Omega) \| (470\,\text{k}\Omega) = 359\,\Omega$

$$\frac{R'_S}{\beta} + r_e = \frac{359\,\Omega}{110} + 6.93\,\Omega = 10.194\,\Omega$$

Now $\quad R_e = R_E \| \left(\frac{R'_S}{\beta} + r_e\right) = (0.91\,\text{k}\Omega) \| (10.194\,\Omega)$

$$= \frac{910 \times 10.194}{920.194}\,\Omega = 10.08\,\Omega$$

Hence $\quad f_{LE} = \dfrac{1}{2\pi \times 10.08 \times 6.8 \times 10^{-6}} = \dfrac{10^6}{6.28 \times 10.08 \times 6.8}$

$$= 2322.8\text{ Hz} = 2.323\text{ kHz}$$

(*f*) Since f_{LE} is the largest out of the three frequencies, hence resultant lower cut-off frequency

$$f_L = f_{LE} = 2.323\text{ kHz}$$

Problem 10.4. For the network shown in Fig. P. 10.4:

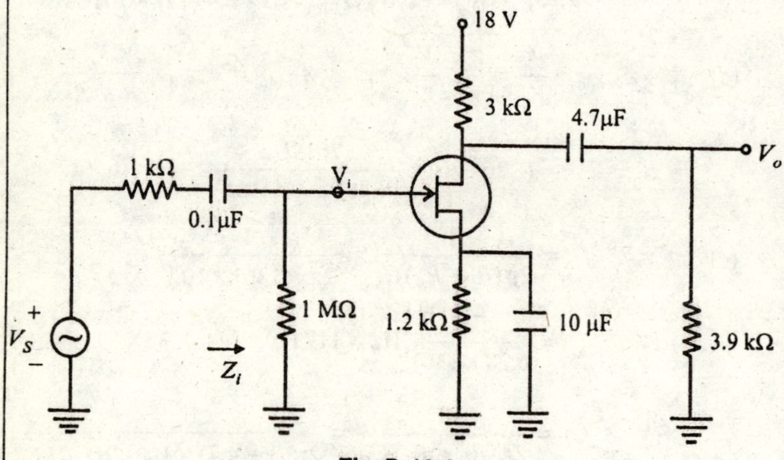

Fig. P. 10.4

Given $\quad I_{DSS} = 6$ mA and $V_P = -6$ V, $r_d = \infty\,\Omega$.

(*a*) Determine $V_{GS\text{-}Q}$ and $I_{D\text{-}Q}$.

(*b*) Find g_{mo} and g_m.

(*c*) Calculate the midband gain of $A_v = \frac{V_o}{V_i}$.

(*d*) Determine Z_i.

(*e*) Calculate $A_{vs} = \frac{V_o}{V_S}$.

(*f*) Determine f_{LG}, f_{LC} and f_{LS}.

(*g*) Determine the low cut-off frequency.

Solution. (*a*) As we know for FET that no current flows through R_G, i.e. $I_G = 0$ unlike BJT.

Hence, applying KVL to the Gate-source loop, we get

$$V_{GS} = -1.2\, I_D \quad \text{... (1)}$$

From Shockley's equation, we have

$$I_D = I_{DSS}\left(1 - \frac{V_{GS}}{V_P}\right)^2$$

or

$$I_D = 6\left(1 - \frac{V_{GS}}{-6}\right)^2 = 6\left(1 + \frac{V_{GS}}{6}\right)^2$$

$$= 6\frac{(6 + V_{GS})^2}{36} = \frac{1}{6}(6 + V_{GS})^2$$

$$\Rightarrow \quad 6I_D = (6 + V_{GS})^2 = 36 + V_{GS}^2 + 12V_{GS} \quad \text{... (2)}$$

Putting value of I_D from (1) into (2), we get

$$-\frac{6}{1.2}V_{GS} = 36 + V_{GS}^2 + 12V_{GS}$$

or $\quad V_{GS}^2 + 12V_{GS} + 36 + 5V_{GS} = 0$

$\Rightarrow \quad V_{GS}^2 + 17V_{GS} + 36 = 0$

or

$$V_{GS} = \frac{-17 \pm \sqrt{17^2 - 4 \times 36}}{2} = \frac{-17 \pm 12.04}{2}$$

$$= -2.48 \text{ V}, -14.52 \text{ V (rejected)}$$

Hence $\quad V_{GS\text{-}Q} = -2.48$ V

From (1), $I_{D-Q} = -\frac{V_{GS-Q}}{1.2} = \frac{2.48}{1.2}$ mA = 2.07 mA

(*b*) We know that $g_{mo} = \frac{2I_{DSS}}{|V_P|} = \frac{2 \times 6\,\text{mA}}{6} = 3\,\text{ms}$

and $$g_m = g_{mo}\left(1 - \frac{V_{GS-Q}}{V_P}\right)$$

$$= 2\left(1 - \left(\frac{-2.48}{-6}\right)\right) = 2\left(1 - \frac{2.48}{6}\right)\text{ms}$$

$\Rightarrow$ $$g_m = 2 \times 0.59 = 1.18 \text{ ms}$$

(*c*) Midband gain A_v for an FET is given by

$$A_v = -g_m\,(R_D \parallel R_L)$$

Given $R_D = 3$ kΩ and $R_L = 3.9$ kΩ

$\therefore$ $$R_D \parallel R_L = (3\,\text{k}\Omega) \parallel (3.9\,\text{k}\Omega) = \frac{3 \times 3.9}{6.9}\,\text{k}\Omega = 1.6956\,\text{k}\Omega$$

$\Rightarrow$ $$A_v = -1.18 \times 10^{-3} \times 1.6956 \times 10^3 = -2$$

(*d*) Input impedance for an FET is

$$Z_i = R_G = 1\,\text{M}\Omega$$

(*e*) $$A_{vs} = \frac{V_o}{V_s} = \frac{V_o}{V_i} \times \frac{V_i}{V_s} = A_v \times \frac{V_i}{V_s}$$

Now $$V_i = V_s \times \frac{R_G}{R_G + R_{sig}} = V_s \times \frac{1\text{M}\Omega}{1\text{M}\Omega + 1\text{k}\Omega}$$

or $$\frac{V_i}{V_s} = \frac{1\text{M}\Omega}{1.001\text{M}\Omega} = 0.999 \approx 1$$

and $$A_{vs} \cong A_v = -2$$

(*f*) Now $$f_{LG} = \frac{1}{2\pi(R_{sig} + R_i)\,C_G} = \frac{1}{2\pi(1\text{k}\Omega + 1\text{M}\Omega) \times 0.1\mu\text{F}}$$

$\Rightarrow$ $$f_{LG} = \frac{1}{6.28 \times 1.001 \times 10^6 \times 0.1 \times 10^{-6}} = 1.59 \text{ Hz}$$

$$f_{LC} = \frac{1}{2\pi(R_o + R_L)C_c} \qquad \text{Here, } C_c = 4.7\mu F$$

$$R_o = R_D \,\|\, r_d \qquad \because r_d = \infty$$

$$R_o = R_D$$

$$(R_D + R_L) = 3\ k\Omega + 3.9\ k\Omega = 6.9\ k\Omega$$

$$\Rightarrow \quad f_{LC} = \frac{1}{2\pi \times 6.9 k\Omega \times 4.7 \times 10^{-6}}$$

$$= \frac{1000}{6.28 \times 6.9 \times 4.7} \text{Hz} = 4.91 \text{Hz}$$

$$f_{LS} = \frac{1}{2\pi\ \text{Req.}C_S}, \quad R_{sig} = 1\ k\Omega,\ C_S = 10\ \mu F$$

$$\text{Req.} = R_{sig} \,\|\, \frac{1}{g_m} = (1\ k\Omega) \,\|\, \frac{1}{1.18\ ms} = (1\ k\Omega) \,\|\, (0.847\ k\Omega)$$

or, $\quad \text{Req.} = 458.7\ \Omega$

Hence $\quad f_{LS} = \dfrac{1}{2\pi \times (458.7\Omega) \times (10\mu F)}$

$$= \frac{1}{2\pi \times 458.7 \times 10 \times 10^{-6}}$$

$$\Rightarrow \quad f_{LS} = \frac{10^5}{6.28 \times 458.7} = 34.71\ \text{Hz}$$

(g) Since out of the three frequencies f_{LS} is highest, therefore

Lower cut-off frequency $= f_{LS} = 34.71$ Hz.

Problem 10.5. Repeat the analysis of problem (4) with $r_d = 100\ k\Omega$. Does it have any impact of any consequence on the results ? If so, which elements?

Solution. r_d comes into play wherever, we consider the output resistance.

and $\quad R_o = R_D \,\|\, r_d = (3\ k\Omega) \,\|\, (100\ k\Omega) = \dfrac{300}{103}\ k\Omega$

or $\quad R_o = 2.913\ k\Omega$

Other things remaining same, the elements which will change after including r_d are

$$A_{v\ \text{mid}},\ A_{vs\ \text{mid}},\ f_{LC} \text{ and } f_{LS}$$

$$A_{v\,\text{mid}} = \frac{V_o}{V_i} = -g_m\,(R_D \,\|\, r_d \,\|\, R_L)$$

Now $R_D \| r_d \| R_L = (3\text{ k}\Omega) \| (100\text{ k}\Omega) \| (3.9\text{ k}\Omega) = 1.667\text{ k}\Omega$

and $\quad A_{v\,\text{mid}} = -1.18 \times 10^{-3} \times 1.667 \times 10^{3} = -1.967$

and since $A_{v\,\text{mid}} \cong A_{v\,\text{mid}}$, $A_{v\,\text{mid}} \cong -1.967$

i.e. Both $A_{v\,\text{mid}}$ and $A_{vs\,\text{mid}}$ has reduced almost by 2%

Now $\quad f_{LS} = \frac{1}{2\pi\,\text{Req.}\,C_S}$, where, $\text{Req.} = \frac{R_S}{1 + R_S \frac{(1 + g_m r_d)}{(r_d + R_D \| R_L)}}$

$$R_S = \text{source resistance} = 1.2\text{ k}\Omega$$

or $\quad \text{Req.} = \frac{1.2\text{ k}\Omega}{1 + \frac{(1.2\text{ k}\Omega)(1 + 118)}{(100\text{ k}\Omega + 3113.9)}} = \frac{1.2\text{ k}\Omega}{1 + \frac{1.2 \times 119}{100 + 1.696}}$

$\Rightarrow \quad \text{Req.} = \frac{1.2}{1 + 1.4}\text{ k}\Omega = \frac{1.2}{2.4}\text{ k}\Omega \Rightarrow \text{Req.} = 0.5\text{ k}\Omega = 500\ \Omega$

$\Rightarrow \quad f_{LS} = \frac{1}{2\pi \times 500 \times 10 \times 10^{-6}} = \frac{1000}{6.28 \times 5}\text{Hz}$

$\Rightarrow \quad f_{LS} = 31.85\text{ Hz}$

Also $f_{LC} = \frac{1}{2\pi(R_o + R_L)C_C}$, where $R_o = R_D \| r_d$

or $\quad R_o = 2.913\text{ k}\Omega$ and given $C_C = 4.7\ \mu\text{F}$

$\Rightarrow \quad f_{LC} = \frac{1}{2\pi(2.913\text{k}\Omega + 3.9\text{k}\Omega)4.7\mu\text{F}}$

$\Rightarrow \quad f_{LC} = \frac{1}{2\pi \times 6.813 \times 10^{-3} \times 4.7 \times 10^{-6}}\text{Hz}$

$$= \frac{1000}{6.28 \times 6.813 \times 4.7}\text{ Hz} = 4.973\text{ Hz}$$

Thus, we observe that after including $r_d = 100\text{ k}\Omega$, f_{LC} has rises from 4.91 Hz to 4.973 Hz whereas f_{Ls} has fallen down from 34.71 Hz to 31.85 Hz, but yet the difference is minute.

Problem 10.6. For the network shown below:

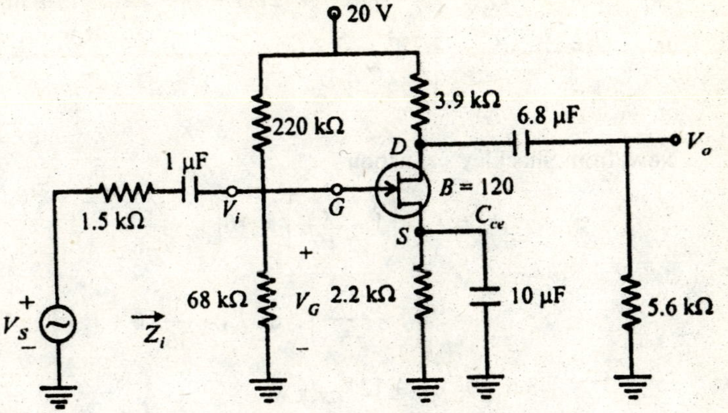

Fig. P. 10.6

(*a*) Determine V_{GSQ} and I_{DQ}.

(*b*) Find g_{mo} and g_m.

(*c*) Calculate the midband gain of $A_v = \frac{V_o}{V_i}$

(*d*) Determine Z_i.

(*e*) Calculate $A_{vs} = \frac{V_o}{V_s}$.

(*f*) Determine f_{LG}, f_{LC}, and f_{LS}.

(*g*) Determine the low cut-off frequency. Given $I_{DSS} = 10$ mA, $V_P = -6$ V.

What effect did the voltage divider configuration have on the input impedance and the gain A_{vs} compared to the biasing arrangement shown in Problem 10.4.

Solution.

(*a*)

$$V_G = V_{th} = V_{DD} \times \frac{R_2}{R_1 + R_2}$$

or

$$V_G = (20\,\text{V}) \times \frac{68\,\text{k}\Omega}{220\,\text{k}\Omega + 68\,\text{k}\Omega}$$

$$= 20 \times \frac{68}{288} = 4.722\,\text{V}$$

Now, applying Kirchhoff's voltage law in the Gate source loop, we get

$$V_G - V_{GS} - V_{RS} = 0, \quad V_{RS} = I_D R_S \quad (\because I_G = 0)$$

$$\Rightarrow \qquad V_G - V_{GS} - I_D R_S = 0$$

$$\Rightarrow \qquad V_{GS} = V_G - I_D R_S = 4.722 \text{ V} - I_D \times 2.2, \; I_D \text{ is in mA.}$$

or $$V_{GS} = 4.722 - 2.2\, I_D$$

or $$I_D = \frac{4.722 - V_{GS}}{2.2} \qquad \text{... (1)}$$

Now from Shockley's equation

$$I_D = I_{DSS}\left(1 - \frac{V_{GS}}{V_P}\right)^2 = 10\left(1 - \frac{V_{GS}}{-6}\right)^2$$

$$= 10\left(1 + \frac{V_{GS}}{6}\right)^2, \text{ or } I_D = \frac{10}{36}(6 + V_{GS})^2$$

$$\Rightarrow \qquad 3.6 I_D = V_{GS}^2 + 12 V_{GS} + 36 \qquad \text{... (2)}$$

From (1) and (2)

$$3.6\left(\frac{4.722 - V_{GS}}{2.2}\right) = V_{GS}^2 + 12 V_{GS} + 36$$

$$\Rightarrow \qquad 7.727 - 1.636\, V_{GS} = V_{GS}^2 + 12\, V_{GS} + 36$$

$$\Rightarrow \qquad V_{GS}^2 + 13.636\, V_{GS} + 28.27 = 0$$

$$\Rightarrow \qquad V_{GS} = \frac{-13.636 \pm \sqrt{(13.636)^2 - 4 \times 28.27}}{2}$$

$$\Rightarrow \qquad V_{GS} = \frac{-13.636 \pm 8.536}{2}$$

$$= -2.55 \text{ V}, \; -11.086 \text{ (rejected)}$$

Hence $$V_{GS\text{-}Q} = -2.55 \text{ V}$$

From equation (1)

$$I_{D\text{-}Q} = \frac{4.722 - V_{GS\text{-}Q}}{2.2} = \frac{4.722 + 2.55}{2.2} = 3.305 \text{ mA}$$

(*b*) $$g_{mo} = \frac{2 I_{DSS}}{|V_P|} = \frac{2 \times 10 \text{ mA}}{6} = 3.33 \text{ ms}$$

and $$g_m = g_{mo}\left(1 - \frac{V_{GS}}{V_P}\right)$$

$$= (3.33 \text{ ms})\left(1 - \frac{(-2.55)}{(-6)}\right) = 1.916 \text{ ms.}$$

(*c*) Midband voltage gain

$$A_v = -g_m (R_D \parallel R_L)$$

$$\Rightarrow \quad A_{v\,\text{mid}} = -(1.916 \text{ ms})(3.9 \text{ k}\Omega \parallel 5.6 \text{ k}\Omega)$$

$$= -(1.916 \text{ ms})(2.299 \text{ k}\Omega) = -4.4$$

(*d*) Input impedance

$$Z_i = R_G = R_1 \parallel R_2$$

$$\Rightarrow \quad Z_i = (220\,\text{k}\Omega) \parallel (68\,\text{k}\Omega) = \frac{220 \times 68}{288}\text{k}\Omega = 51.945\,\text{k}\Omega$$

(*e*)
$$A_{vs} = \frac{V_o}{V_S} = \frac{V_o}{V_i} \times \frac{V_i}{V_S} = A_{v\,\text{mid}} \times \frac{V_i}{V_S}$$

From voltage division rule

$$V_i = V_S \times \frac{R_i}{R_i + R_{sig}} = \frac{V_S \times 51.945\,\text{k}\Omega}{(51.945 + 1.5)\,\text{k}\Omega} = V_S \times 0.972$$

$$\Rightarrow \quad \frac{V_i}{V_S} = 0.972$$

and $\quad A_{vs} = -4.4 \times 0.972 = -4.276$

we see that the effect of this configuration is to reduce Z_i from 1 MΩ (from problem 10.4) to 51.945 kΩ and to raise the magnitude of A_{vs} from 2 to 4.276 as is compared from the results of problem 10.4.

(*f*)
$$f_{LG} = \frac{1}{2\pi(R_{sig} + R_i)C_G}$$

$$= \frac{1}{2\pi(1.5\,\text{k}\Omega + 51.945\,\text{k}\Omega) \times 10^{-6}}$$

$$= \frac{1}{6.28 \times 53.445 \times 10^3 \times 10^{-6}}$$

$$= \frac{1000}{6.28 \times 53.445} = 2.98\,\text{Hz}$$

$$f_{LC} = \frac{1}{2\pi\,(R_o + R_L)\,C_C}$$

$$= \frac{1}{2\pi(3.9\,\text{k}\Omega + 5.6\,\text{k}\Omega)6.8\,\mu\text{F}} \qquad (\because R_o = R_D)$$

$$= \frac{1}{6.28 \times 9.5 \times 10^3 \times 6.8 \times 10^{-6}} = \frac{1000}{6.28 \times 9.5 \times 6.8}$$

$\Rightarrow \quad f_{LC} = 2.465$ Hz.

and $\quad f_{LS} = \dfrac{1}{2\pi \text{ Req. } C_S}$, $\quad$ where Req. $= R_{sig} \parallel \dfrac{1}{g_m}$

$$\text{Req.} = (1.5\,\text{k}\Omega) \parallel \frac{1}{1.916\text{ mS}} = (1.5\text{ k}\Omega) \parallel (0.522\text{ k}\Omega) = 387.2\ \Omega$$

Hence $\quad f_{LS} = \dfrac{1}{2\pi \times 387.2 \times 10\mu\text{F}}$

$$\Rightarrow \quad f_{LS} = \frac{1}{6.28 \times 387.2 \times 10^{-5}} = \frac{10^5}{6.28 \times 387.2}\text{Hz}$$

$\Rightarrow \quad f_{LS} = 41.125$ Hz

(*g*) Out of the three frequencies, f_{LS} is the largest one, hence

Lower cut-off frequency $\equiv f_{LS}$

$\Rightarrow \quad f_L = 41.125$ Hz

Problem 10.7. For the network of Fig. P. 10.7 shown in Problem 10.2 the high frequency parameters are given below:

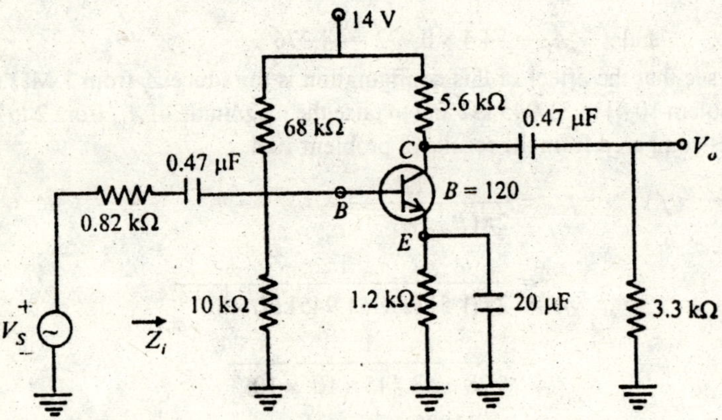

Fig. P. 10.7

$C_{\omega_i} = 5\,\text{PF}$, $C_{\omega o} = 8\,\text{PF}$, $C_{bc} = 12\,\text{PF}$, $C_{be} = 40$ PF, $C_{ce} = 8\,\text{PF}$

(*a*) Determine f_{Hi} and f_{Ho}

(*b*) Assuming that $C_{b'e} = C_{be}$ and $C_{b'c} = C_{bc}$, find f_β and f_T.

Solution. (*a*) $$f_{Hi} = \frac{1}{2\pi R_{th1} C_i}$$

Where $R_{th1} = R_S \,||\, R_1 \,||\, R_2 \,||\, R_i$

$R_i = \beta r_e$, from solution of Problem 10.2

$R_i = 120 \times 28.49\ \Omega = 3.42\ k\Omega$, and $A_{v\,mid} = -72.88$

$$\Rightarrow \quad R_{th1} = (0.82\,k\Omega) \,||\, (68\,k\Omega) \,||\, (10\,k\Omega) \,||\, (3.42\,k\Omega)$$

$$= (0.82\ k\Omega) \,||\, (2.456\ k\Omega)$$

$$= 0.6147\ k\Omega = 614.75\ \Omega$$

and $C_i = C_{\omega i} + C_{be} + C_{Mi}$ $\qquad C_{Mi}$ = Miller input capacitance

$= C_{\omega i} + C_{be} + (1 - A_v)C_{bc}$ $\qquad C_{Mo}$ = Miller output capacitance

$= 5 + 40 + (1 - (-72.88)) \times 12$

$= 45\ PF + 12 \times 73.88\ PF = 931.56\ PF$

Hence $$f_{Hi} = \frac{1}{2\pi R_{th1} C_i}$$

$$\Rightarrow \quad f_{Hi} = \frac{1}{2\pi \times 614.75 \times 931.56 \times 10^{-12}}\ Hz$$

$$= \frac{10^{12}}{6.28 \times 614.75 \times 931.56}\ Hz = 278.055\ kHz$$

$$f_{Ho} = \frac{1}{2\pi R_{th2} C_o}$$

$$R_{th2} = R_C \,||\, R_L = (5.6\,k\Omega) \,||\, (3.3\,k\Omega)$$

$$= \frac{5.6 \times 3.3}{8.9}\ k\Omega = 2076.4\ \Omega$$

and $$C_o = C_{\omega o} + C_{Ce} + C_{Mo} = C_{\omega o} + C_{Ce} + \left(1 - \frac{1}{A_v}\right) C_{bc}$$

$$= 8\,PF + 8\,PF + \left(1 - \frac{1}{-72.88}\right) 12\,PF$$

$$= 16\text{ PF} + 12.165\text{ PF} = 28.165\text{ PF}$$

Hence $$f_{Ho} = \frac{1}{2\pi \times 2076.4 \times 28.165\text{PF}}$$

$$\Rightarrow \quad f_{Ho} = \frac{1}{2\pi \times 2076.4 \times 28.165 \times 10^{-12}}$$

$$= \frac{10^{12}}{6.28 \times 2076.4 \times 28.165} = 2.723\text{ MHz}$$

(*b*) $$f_\beta = \frac{1}{2\pi\beta_{\text{mid}}\, r_e\,(C_{be} + C_{bc})}$$

$$= \frac{1}{2\pi \times 120 \times 28.49\,(40\text{PF} + 12\text{PF})}$$

$$= \frac{10^{12}}{2\pi \times 120 \times 28.49 \times 52}$$

$$\Rightarrow \quad f_\beta = \frac{10^{12}}{6.28 \times 120 \times 28.49 \times 52} = 0.896\text{ MHz}$$

$$f_T = \beta_{\text{mid}}\ f_\beta = (120)\,(0.896)\text{ MHz}$$

$$\Rightarrow \quad f_T = 107.484\text{ MHz}$$

Problem 10.8. For the network shown in Fig. P. 10.3:

(*a*) Determine f_{Hi} and f_{Ho}

(*b*) Find f_β and f_T

Given $C_{\omega i} = 7\text{ PF}$, $C_{\omega o} = 11\text{PF}$, $C_{bc} = 6\text{ PF}$, $C_{be} = 20\text{ PF}$, and $C_{Ce} = 10\text{ PF}$

Solution. As calculated in Problem 10.3

$$r_e = 6.93\ \Omega \text{ and } \beta\, r_e = (110)\,(6.93\ \Omega) = 762.3\ \Omega$$

and $$A_{v\,\text{mid}} = -264.24$$

(*a*) $$f_{Hi} = \frac{1}{2\pi R_{\text{th}1}\, C_i},\quad R_{\text{th}1} = R_S \,\|\, R_B \,\|\, \beta r_e$$

$$= (0.6\text{ k}\Omega)\,\|\,(470\text{ k}\Omega)\,\|\,(0.7623\text{ k}\Omega)$$

$$= 0.3355 \text{ k}\Omega = 335.5 \,\Omega$$

$$C_i = C_{\omega i} + C_{be} + C_{Mi} = C_{\omega i} + C_{be} + (1 - A_v)\, C_{bc}$$

or $$C_i = 7\,\text{PF} + 20\,\text{PF} + (1 - (-264.24))\,(6\,\text{PF})$$

$$= 27 \text{ PF} + 1591.44 \text{ PF} = 1618.44 \text{ PF}$$

Therefore $$f_{Hi} = \frac{1}{2\pi \times 335.5 \times 1618.44 \times 10^{-12}}$$

$\Rightarrow$ $$f_{Hi} = \frac{10^{12}}{6.28 \times 335.5 \times 1618.44} \text{ Hz} = 293.26\,\text{kHz}$$

and $$f_{Ho} = \frac{1}{2\pi R_{\text{th2}}\, C_o}, \text{ where}$$

$$R_{\text{th2}} = R_C \,\|\, R_L = (3\,\text{k}\Omega) \,\|\, (4.7\,\text{k}\Omega)$$

$$= \frac{3 \times 4.7}{7.7} \text{ k}\Omega = 1831.17 \,\Omega$$

and $$C_o = C_{\omega o} + C_{Ce} + C_{Mo}$$

$$= C_{\omega o} + C_{Ce} + \left(1 - \frac{1}{A_v}\right) C_{bc}$$

or $$C_o = 11\,\text{PF} + 10\,\text{PF} + \left(1 - \frac{1}{-264.24}\right) 6\,\text{PF}$$

$$= 21 \text{ PF} + 6.023 \text{ PF} = 27.023 \text{ PF}$$

$\therefore$ $$f_{Ho} = \frac{1}{2\pi R_{\text{th2}}\, C_0} = \frac{1}{2\pi \times 1831.17 \times 27.023\,\text{PF}}$$

$$= \frac{1}{6.28 \times 1831.17 \times 27.023 \times 10^{-12}}$$

$$= \frac{10^{12}}{6.28 \times 1831.17 \times 27.023} \text{Hz} = 3.218 \text{ MHz.}$$

(*b*) $$f_\beta = \frac{1}{2\pi \beta_{\text{mid}}\, r_e\, (C_{be} + C_{bc})}$$

$$\Rightarrow \qquad f_\beta = \frac{1}{2\pi \times 110 \times 6.93 \times (20\,\text{PF} + 6\,\text{PF})}$$

$$= \frac{1}{2\pi \times 110 \times 6.93 \times 26\,\text{PF}}$$

$$= \frac{10^{12}}{6.28 \times 110 \times 6.93 \times 26}\ \text{Hz} = 8.034\,\text{MHz}$$

$$f_T = \beta_{\text{mid}}\ f_\beta = (110)\,(8.034)\ \text{MHz}$$

$$\Rightarrow \qquad f_T = 883.7\ \text{MHz}$$

Problem 10.9. For the network shown in Fig. P. 10.9, determine f_{Hi} and f_{Ho}.

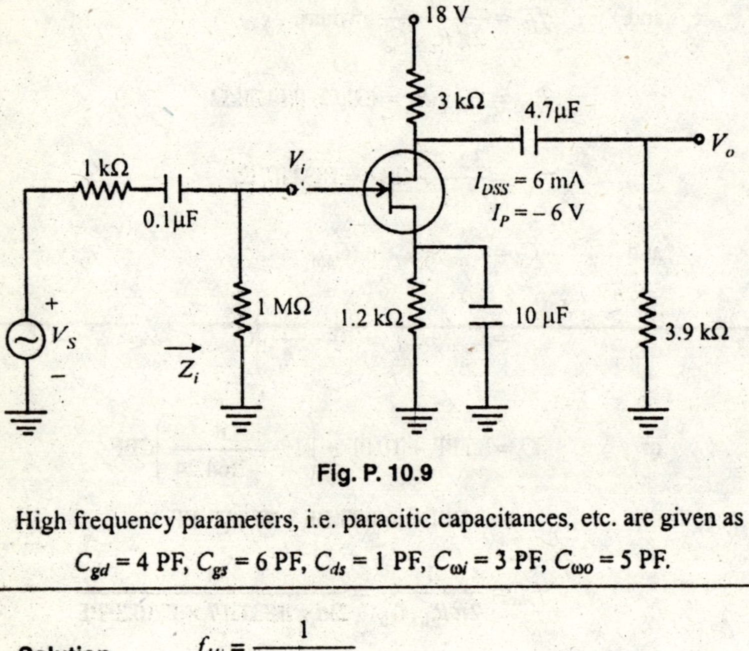

Fig. P. 10.9

High frequency parameters, i.e. paracitic capacitances, etc. are given as

$C_{gd} = 4$ PF, $C_{gs} = 6$ PF, $C_{ds} = 1$ PF, $C_{\omega i} = 3$ PF, $C_{\omega o} = 5$ PF.

Solution. $$f_{Hi} = \frac{1}{2\pi R_{\text{th1}} C_i}$$

Where $$R_{\text{th1}} = R_{sig} \,\|\, R_G$$

$$\Rightarrow \qquad R_{\text{th1}} = (1\text{ k}\Omega) \,\|\, (1\text{ M}\Omega) = 0.999\,\text{k}\Omega \approx 1\,\text{k}\Omega = R_{sig}$$

$$C_i = C_{\omega i} + C_{gs} + C_{Mi} = C_{\omega i} + C_{gs} + (1 - A_v)\, C_{gd}$$

As calculated in Problem 10.4

$$A_v = -2$$

Hence $\quad C_i = 3\text{PF} + 6\text{PF} + (1-(-2))\,4\text{ PF}$

$$= 9\text{ PF} + 12\text{ PF} = 21\text{ PF}$$

Hence $\quad f_{Hi} = \dfrac{1}{2\pi \times 999 \times 21 \times 10^{-12}} = \dfrac{10^{12}}{6.28 \times 999 \times 21}\text{Hz}$

$\Rightarrow \quad f_{Hi} = 7.59 \times 10^6\text{ Hz} = 7.59\text{ MHz}$

f_{Ho} is given by $f_{Ho} = \dfrac{1}{2\pi R_{\text{th}2} C_o}$

$$R_{\text{th}2} = R_D \,\|\, R_L \,\|\, r_d, \text{ as } r_d = \infty$$

$$R_{\text{th}2} = R_D \,\|\, R_L = (3\text{k}\Omega) \,\|\, (3.9\text{k}\Omega) = \frac{3 \times 3.9}{6.9}\text{k}\Omega$$

or, $\quad R_{\text{th}2} = 1.6956\text{ k}\Omega = 1695.65\,\Omega$

and $\quad C_o = C_{\omega o} + C_{ds} + C_{Mo} = C_{\omega o} + C_{ds} + \left(1 - \dfrac{1}{A_v}\right) C_{gd}$

$\Rightarrow \quad C_o = 5\text{ PF} + 1\text{PF} + \left(1 - \dfrac{1}{(-2)}\right) 4\text{PF} = 6\text{PF} + 6\text{ PF} = 12\text{ PF}.$

Hence $\quad f_{Ho} = \dfrac{1}{2\pi \times 1695.65 \times 12 \times 10^{-12}} = \dfrac{10^{12}}{6.28 \times 1695.65 \times 12}\text{ Hz}$

$\Rightarrow \quad f_{Ho} = 7.826\text{ MHz}$

Problem 10.10. Find the cut-off frequencies f_{H_o} for the network shown in Problem 10.6 with the high frequency parameters given below:

$$C_{\omega i} = 4\text{PF},\ C_{\omega o} = 6\text{PF},\ C_{gd} = 8\text{PF},\ C_{gs} = 12\text{PF},$$

$C_{ds} = 3\text{PF}$. Other parameters remaining the same.

Solution.

Here $\quad R_{\text{th}_1} = R_{sig} \,\|\, R_G = (1.5\text{k}\Omega) \,\|\, (220\text{k}\Omega \,\|\, 68\text{k}\Omega)$

$$R_G = 220\text{ k}\Omega \,\|\, 68\text{ k}\Omega = 51.94\text{ k}\Omega$$

or $\quad R_{\text{th}_1} = 1.5\text{k}\Omega \,\|\, 51.94\text{ k}\Omega = 1.4579\text{ k}\Omega$

$$\Rightarrow \qquad R_{th_1} = 1457.9\ \Omega$$

$$C_i = C_{\omega i} + C_{gs} + C_{Mi} = 4\,\text{PF} + 12\,\text{PF} + (1 - A_v)8\,\text{PF}$$

As calculated in Problem 10.6,

$$A_v = -4.4$$

$$\therefore \qquad C_i = 16\ \text{PF} + (1 - (-4.4))\ 8\ \text{PF} = 16\ \text{PF} + 43.2\ \text{PF} = 59.2\ \text{PF}$$

Now $$f_{Hi} = \frac{1}{2\pi R_{th_1} C_i} = \frac{1}{2\pi \times 1457.9 \times 59.2 \times 10^{-12}}$$

$$\Rightarrow \qquad f_{Hi} = \frac{10^{12}}{6.28 \times 1457.9 \times 59.2}\ \text{Hz} = 1.845\ \text{MHz}$$

and $$R_{th_2} = R_D \,\|\, R_L = (3.9\,\text{k}\Omega) \,\|\, (5.6\,\text{k}\Omega) = 2.2989\ \text{k}\Omega$$

or, $$R_{th_2} = 2298.95\ \Omega$$

$$C_o = C_{\omega o} + C_{ds} + C_{Mo} = C_{\omega o} + C_{ds} + \left(1 - \frac{1}{A_v}\right) C_{gd}$$

$$\Rightarrow \qquad C_o = 6\,\text{PF} + 3\,\text{PF} + \left(1 - \frac{1}{(-4.4)}\right) 8\,\text{PF}$$

$$= 9\ \text{PF} + 9.818\ \text{PF} = 18.818\ \text{PF}$$

Hence $$f_{Ho} = \frac{1}{2\pi \times R_{th_2} C_o} = \frac{1}{2\pi \times 2298.95 \times 18.818 \times 10^{-12}}$$

$$= \frac{10^{12}}{6.28 \times 2298.95 \times 18.818} = 3.68 \times 10^6\ \text{Hz}$$

or, $$f_{Ho} = 3.68\ \text{MHz}$$

Problem 10.11. Derive a formula for the output voltage of a inverting amplifier circuit outing to voltage offset effects. Estimate the maximum output voltage due to an input voltage offset, if a 741 op-amp is used and the circuit is designed to have a gain of 10. The offset voltage is 5 mV. Assume the only op-amp non-ideality present is offset voltage V_{os}. Is the output saturated on account of offset voltage of the saturation level is ± 10V?

Solution. The circuit of an inverting op-amp with offset voltage V_o is shown in Fig. 10.11.

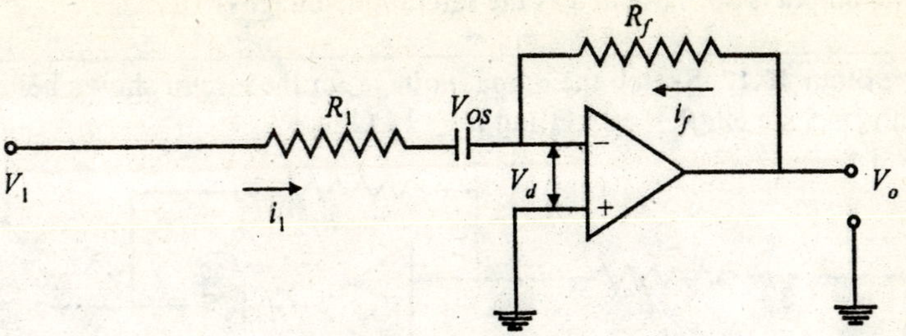

Fig. 10.11

We have $\quad i_1 = \dfrac{V_1 + V_{os}}{R_1}$ and $i_f = \dfrac{V_o + V_{os}}{R_f}$

Now, from KCL at the junction, $i_1 + i_f = 0$

$$\Rightarrow \quad \frac{V_1 + V_{os}}{R_1} + \frac{V_o + V_{os}}{R_f} = 0$$

$$\Rightarrow \quad \frac{V_o}{R_f} = -\frac{V_1}{R_1} - \frac{V_{os}}{R_f} - \frac{V_{os}}{R_1} = -\frac{V_1}{R_f} - V_{os}\left(\frac{1}{R_1} + \frac{1}{R_f}\right)$$

$$\Rightarrow \quad V_o = -V_1\left(\frac{R_f}{R_1}\right) - V_{os}\left(1 + \frac{R_f}{R_1}\right)$$

Therefore, output voltage on account of offset voltage V_{os} is

$$V_o' = -V_{os}\left(1 + \frac{R_f}{R_1}\right)$$

The closed loop gain of an inverting amplifier is

$$\frac{R_f}{R_1} = 10$$

$\therefore$ output voltage due to offset voltage is

$$V_o' = -5 \times 10^{-3}(1 + 10) = 55 \times 10^{-3}\text{ V}$$

or $\quad V_o' = -55\,\text{mV}$

For a real circuit the actual output voltage owing to offset lies in the range −55 mV to +55 mV.

The output is not saturated as the saturation voltage is 10V.

Problem 10.12. Sketch the output voltage for the circuit shown below with $v_1 = 5 \sin \omega t$, $R_f = 10\ k\Omega$ and $R_1 = 1\ k\Omega$.

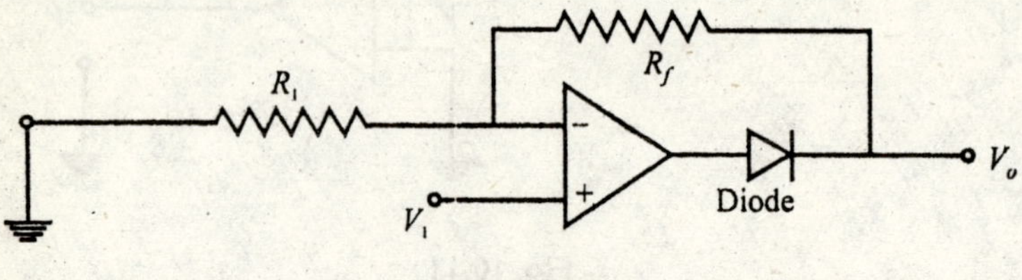

Fig. P. 10.12

Solution. The circuit shown above is an active rectifier circuit and here it is a half-wave circuit.

The sine wave is amplified by a factor

$$\left(\frac{R_f}{R_1}+1\right) = \left(\frac{10}{1}+1\right) = 11$$

The input and output voltages are shown below.

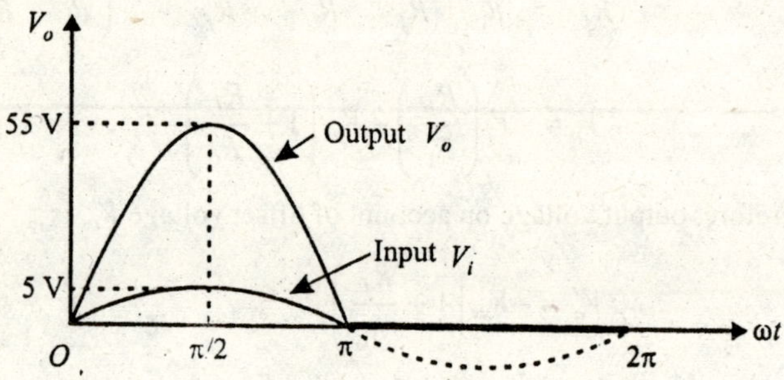

Fig. 10.12 (a)

Maximum value of input voltage = 5 V,

Hence maximum value of output voltage

$$= 5 \times 11 = 55\ V$$

Problem 10.13. An operational amplifier is used as an integrator to produce a ramp signal of –10 V/ms. Derive the expression for the output voltage and corresponding design the circuit.

Solution. The circuit diagram for the op-amp acting as an integrator is shown in Fig. 10.13 (a):

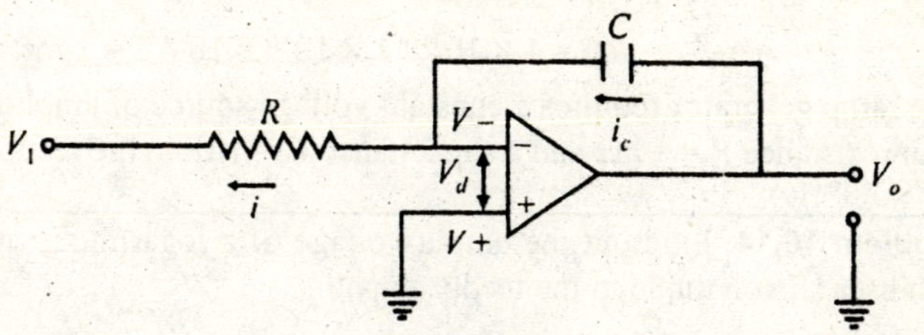

Fig. 10.13 (a)

Applying KCL at node v_-,

$$\frac{v_- - v_1}{R} = C\frac{d}{dt}(v_o - v_-)$$

For infinite differential gains, $v_- = 0$

$$\Rightarrow \qquad -\frac{v_1}{R} = c\frac{d}{dt}v_o$$

Integrating the both sides, we get

$$v_o = -\frac{1}{RC}\int v_1\, dt$$

The expression for output voltage v_o is proportional to the integral of the input voltage.

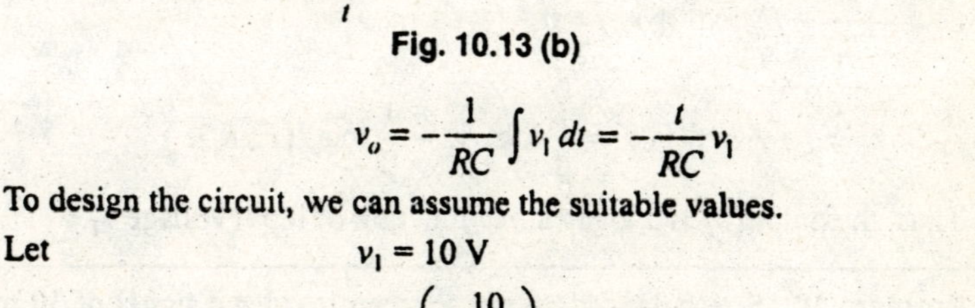

Fig. 10.13 (b)

$$v_o = -\frac{1}{RC}\int v_1\, dt = -\frac{t}{RC}v_1$$

To design the circuit, we can assume the suitable values.

Let $\qquad v_1 = 10\text{ V}$

$$\therefore \qquad v_o = \left(-\frac{10}{RC}\right)t$$

Thus, for production of a ramp of

$$\frac{v_o}{t} = -\frac{10}{RC} = -10\text{ V/ms}$$

$$\Rightarrow \qquad RC = 1 \text{ ms}$$

with $\qquad C = 1\ \mu\text{F and } R = 1\ \text{k}\Omega$

$$RC = 1 \times 10^3 \times 1 \times 10^{-6} = 10^{-3}\ \text{S} = 1\ \text{ms}.$$

The ramp generator requires a constant voltage source of amplitude 10 V, an input resistance $R = 1\ \text{k}\Omega$ and a capacitance $C = 1\mu\text{F}$ in the feedback path.

Problem 10.14. Find out the output voltage of a logarithmic amplifier which uses silicon diode in the feedback path.

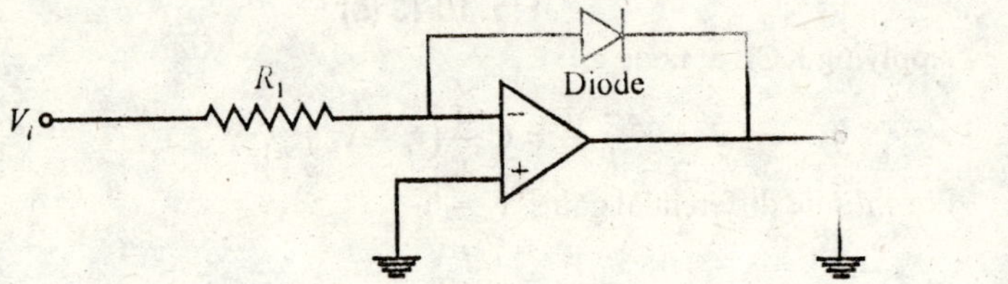

Fig. P. 10.14

Solution. The current of the diode depends upon the output voltage. From the circuit diagram, we get

$$\frac{v_i}{R} + f(v_o) = 0$$

where $f(v_o)$ = variation of the current with voltage and is exponential in the case of a diode

$$f(v_o) = F_o \exp[\beta v_o]$$

F_o = a constant and β = an exponent constant

i.e. $\quad -\frac{v_1}{R_1} + F_o \exp(\beta v_o) = 0$

or $$v_o = \frac{1}{\beta}\log_e v_1 - \frac{1}{\beta}\log_e (F_o R_1).$$

Thus, the output v_o is a logarithmic function of input voltage v_1.

Problem 10.15. A balanced output source provides a signal of 30 mV from each terminal to ground, this provides a difference signal of 60 mV for a difference amplifier. The noise signal common to both terminals is 600 mV. The differential gain of the amplifier is 150, while the common mode gain is 0.04. What is the ratio of signal to noise voltage at the output? Comment upon the result.

Solution. The balanced source provides an output of 30 mV from each terminal to ground.

Thus $$v_2 = 30 \text{ mV}, v_1 = -30 \text{ mV}$$

∴ Difference mode input

$$V_d = 30 - (-30) = 60 \text{ mV}$$

Signal output voltage $V_{OS} = A_d V_d = 150 \times 60 \times 10^{-3} = 9$ V

The noise at the input terminals has a magnitude of 600 mV.

∴ Noise output voltage

$$V_{ON} = A_C V_C = 0.04 \times 600 \times 10^{-3} \text{ V}$$

$$\Rightarrow \quad V_{ON} = 24 \text{ mV}$$

∴ Signal to noise ratio $= \dfrac{V_{OS}}{V_{ON}} = \dfrac{9}{24 \times 10^{-3}} = 375$

Common mode rejection ratio

$$\text{CMRR} = \frac{A_d}{A_C}$$

$$\Rightarrow \quad \text{CMRR} = \frac{150}{0.04} = 3750$$

Comment: We find that the noise signal is 10 times as great as the desirable signal at the input but at the output, the noise signal is $\dfrac{1}{375}$ of the desired signal.

Therefore, the noise has been reduced by a factor $10 \times 375 = 3750$, which is equal to CMRR by the use of a difference amplifier.

Therefore, difference amplifier are very useful where elimination of common mode undesirable signal is concerned.

Reader's Notes

CHAPTER

11

Power Amplifiers

Power amplifiers are categorised on the basis of class. Basically, amplifier classes represent the amount the output signal varies over one cycle of operation for full cycle of input signal.

There are five such classes, viz. Class A, Class B, Class AB, Class C and Class D.

1. Class A: The output signal varies for a full 360° of the cycle. This type of amplifiers requires a d.c. bias level, so that at least half the signal swing of the output may vary up and down without going to a high-enough voltage to be limited by the supply voltage level, as too low to approach the lower supply level. Figure 11.1 shows the concept.

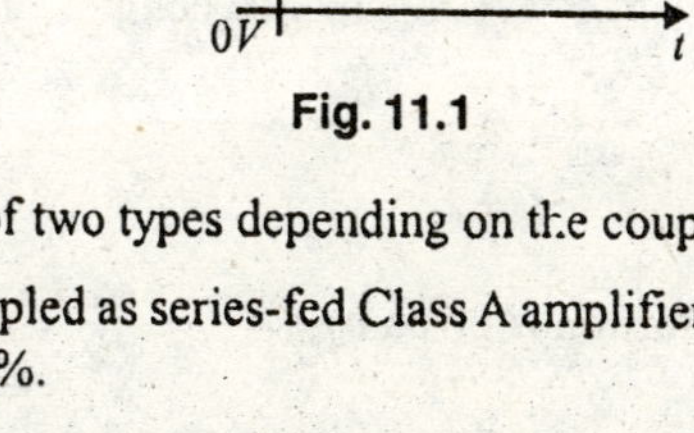

Fig. 11.1

This is further of two types depending on the coupling with the load:

(*i*) Direct coupled as series-fed Class A amplifier: This has efficiency of around 25%.

(*ii*) Transformes coupled: This has efficiency of 60 %.

Thus, Class A amplifiers have poor efficiency but have least distortion.

2. Class B: The output signal varies over one-half the input signal cycle or for 180° of signal. The d.c. bias point for Class B is therefore at 0 V. Two Class B operations—one to provide output on the positive output half-cycle and another to provide operation on the negative output half-cycle are necessary. This type of connection is referred to as push-pull operation. Class B operation, by itself, creates a very distorted output signal, since reproduction of the input takes place for only 180° of the output signal swing.

This type of amplifier has efficiency of about 78.5%.

3. Class AB: An amplifier whose output signal swing occurs between the two extremes defined for Class A and Class B, i.e. between 180° and 360° is Class AB amplifier. This may be biased at a d.c. level above the zero base current level of Class B and above one-half the supply voltage level of Class A. This too requires a push-pull connection to achieve a full output cycle. The cross-over distortions produced by Class B is eliminated by Class AB.

Its efficiency is between that of Class A and B.

4. Class C: The output of this amplifier is biased for operation at less than 180° of the cycle and will operate only with a tuned (resonant) circuit, therefore, these are used in ratio or communications.

It has efficiency as high as 80–85% but contains the maximum distortion in the output signal.

5. Class D: This amplifier provides the most efficient operation of all the Class. This has efficiency over 90%. This amplifier uses pulses for its operation, which is on for a short interval and off for a longer interval.

Distortion of second harmonic:

$$D_2 = \frac{\text{Second harmonic component}}{\text{Fundamental component}} = \frac{|A_2|}{|A_1|}$$

Similarly $$D_3 = \frac{|A_3|}{|A_1|}, \ldots\ldots$$

Total harmonic distortion is given by

$$\text{THD} = \sqrt{D_2^2 + D_3^2 + D_4^2 + \ldots\ldots}$$

Problem 11.1. Calculate the input and output power for the circuit of Fig. P. 11.1.

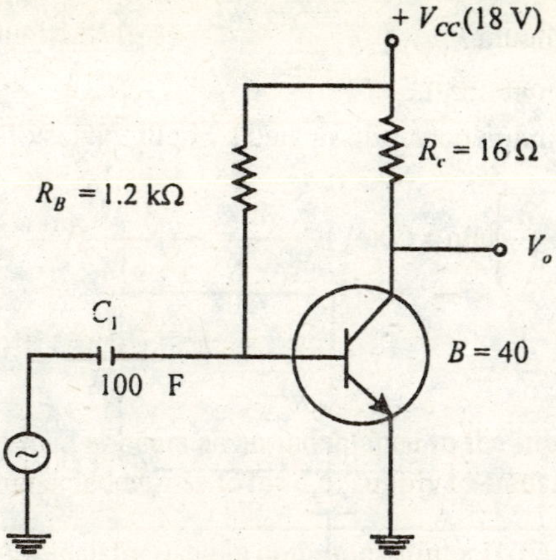

Fig. P. 11.1

Solution. This is a problem of Series-fed Class A amplifier.

$$I_B = \frac{V_{CC} - 0.7\text{ V}}{R_B} = \frac{18\text{ V} - 0.7\text{ V}}{1.2\Omega} = 14.42\text{ mA}$$

Thus, Q-point can be obtained as below:

$$I_{C\text{-}Q} = \beta I_B = 40 \times 14.42\text{ mA} = 576.8\text{ mA} = 0.5768\text{ A}$$

$$V_{CE\text{-}Q} = V_{CC} - I_C R_C$$

$$\Rightarrow \quad V_{CE\text{-}Q} = 18\text{ V} - (576.8\text{ mA})\ 16\ \Omega = 8.7712\text{ V}$$

Now, given $\quad I_B = 5$ mA rms

$$\therefore \quad I_{C\text{-}rms} = \beta I_B = 40 \times 5 = 0.2\text{ A}$$

$$P_{o\,(\text{a.c.})} = \left(I_{C\,(rms)}\right)^2 \times R_C = \frac{V_{C\,rms}^2}{R_C}$$

$$\Rightarrow \quad P_{o\,(\text{a.c.})} = (0.2)^2 \times 16\text{ W} = 0.04 \times 16\text{ W} = 0.64\text{ W}$$

$$P_{1\,(\text{d.c.})} = V_{CC}\, I_{C\text{-}Q} = (18\text{ V}) \times (0.5768\text{ A}) = 10.38\text{ W}$$

Also

Power efficiency $\quad \%\eta = \dfrac{P_{o\,(\text{ac})}}{P_{i\,(\text{d.c.})}} \times 100\% = 6.16\%$

Problem 11.2. For the Class B power amplifier of Fig. P. 11.2, calculate.

(*a*) Maximum $P_{o\,(a.c.)}$. (*b*) Maximum $P_{i\,(d.c.)}$.

(*c*) Maximum %η

(*d*) Maximum power dissipated by both transistors.

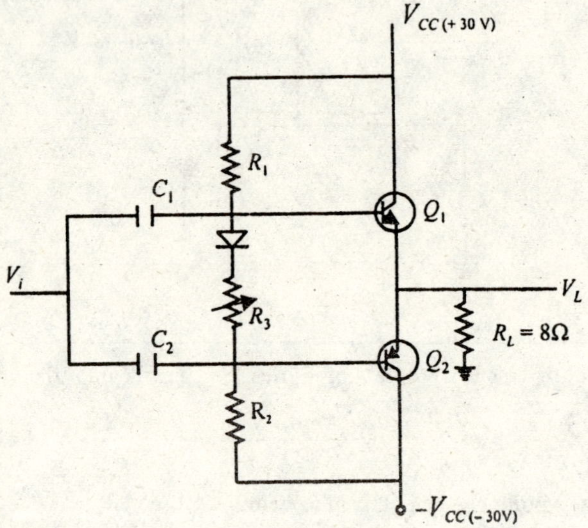

Fig. P. 11.2

Solution. For Class B operation, the maximum output power is delivered to the load, when

$$V_{L\,(peak)} = V_{CC}$$

(*a*) and hence, maximum

$$P_{o\,(a.c.)} = \frac{V_{CC}^2}{2\,R_L}$$

$$\Rightarrow \qquad P_{o-max} = \frac{30^2}{8 \times 2} = 56.25\text{ W}$$

(*b*) The corresponding peak a.c. current I_{peak} is given by

$$I_{peak} = \frac{V_{CC}}{R_L}$$

maximum value of d.c. current from power supply is

$$I_{max.\,(d.c.)} = \frac{2I_{peak}}{\pi} = \frac{2V_{CC}}{\pi R_L}$$

maximum value of input power $P_{i\text{-}max\,(d.c.)} = V_{CC}\left(I_{max\,(d.c.)}\right)$

$$\text{maximum} \quad P_{i\,(d.c.)} = V_{CC}\left(\frac{2V_{CC}}{\pi R_L}\right) = \frac{2V_{CC}^2}{\pi R_L}$$

$$\Rightarrow \quad P_{i\,(d.c.)-max} = \frac{2\times 30^2}{\pi\times 8} = 71.62\ \text{W}$$

(*c*) maximum %

$$\eta = \frac{P_{o(a.c.)-max}}{P_{i(d.c.)-max}}\times 100$$

$$= \frac{56.25}{71.62}\times 100 = 78.54\%$$

(*d*) The maximum power dissipated by the two output transistors occurs, when the output voltage across the load is

$$V_{L\,(peak)} = \frac{2\,V_{CC}}{\pi}$$

$$\text{maximum} \quad P_{2Q} = \frac{2\,V_{V_{CC}}^2}{\pi^2 R_L} = \frac{2\times 30^2}{\pi^2\times 8} \quad \Rightarrow \text{max. } P_{2Q} = 22.8\ \text{W}$$

Problem 11.3. For the Class B amplifier shown in Fig. P. 11.2, if the input voltage is 8 V_{rms}, calculate:

(*a*) $P_{i\,(d.c.)}$ (*b*) $P_{o\,(a.c.)}$

(*c*) % η

(*d*) Power dissipated by both power output transistors.

Solution. From the given figure

$$V_{CC} = 30\ \text{V},\ R_L = 8\ \Omega$$

(*a*) Given $V_{rms} = 8$ V $\quad \therefore V_{peak} = 8\times\sqrt{2} = 11.314$ V

$$\text{and} \quad I_{L\,(peak)} = \frac{V_{L(peak)}}{R_L} = \frac{11.314\ \text{V}}{8\Omega} = \sqrt{2} = 1.414\ \text{amp}$$

The d.c. value of the current drawn $I_{d.c.} = \dfrac{2\,I_{L(peak)}}{\pi}$

or $$I_{\text{d.c.}} = \frac{2}{\pi}(\sqrt{2}) = 0.9\,\text{A}$$

Hence $$P_{i\,(\text{d.c.})} = V_{CC}\, I_{\text{d.c.}} = (30\ \text{V})\,(0.9) = 27\ \text{W}$$

(*b*) $$P_{o(\text{a.c.})} = \frac{V_{L(\text{rms})}^2}{R_L} = \frac{8^2}{8} = 8\,\text{W}$$

(*c*) $$\%\,\eta = \frac{P_{0(\text{a.c.})}}{P_{i(\text{d.c.})}} \times 100 = \frac{8}{27} \times 100 = 29.63\%$$

(*d*) Power dissipated by the two output power transistors

$$P_{2Q} = P_{i\,(\text{d.c.})} - P_{o\,(\text{a.c.})}$$
$$= 27\ \text{W} - 8\ \text{W} = 19\ \text{W}$$

Problem 11.4. Calculate the harmonic distortion components for an output signal having fundamental amplitude of 2.1 V, second harmonic amplitude of 0.3 V, third harmonic component of 0.1 V, and fourth harmonic component of 0.05 V. Also, calculate the total harmonic distortion THD.

Solution. $$\%\,D_2 = \frac{|A_2|}{|A_1|} \times 100\% = \frac{0.3\,\text{V}}{2.1\,\text{V}} \times 100\% = 14.3\%$$

$$\%\,D_3 = \frac{|A_3|}{|A_1|} \times 100\% = \frac{0.1\,\text{V}}{2.1\,\text{V}} \times 100\% = 4.8\%$$

$$\%\,D_4 = \frac{|A_4|}{|A_1|} \times 100\% = \frac{0.05\,\text{V}}{2.1\,\text{V}} \times 100\% = 2.4\%$$

Total harmonic distortion

$$\%\,\text{THD} = \sqrt{D_2^2 + D_3^2 + D_4^2} \times 100\%$$

Where $D_2 = 0.143$, $D_3 = 0.048$, and $D_4 = 0.024$

∴ $$\%\,\text{THD} = \sqrt{(0.143)^2 + (0.048)^2 + (0.024)^2} \times 100\%$$

$$= 15.274\,\% \approx 15.3\,\%$$

Reader's Notes

Reader's Notes